Dept
De Cambridge.

John N. Maina

The Lung-Air Sac System of Birds –
Development, Structure, and Function

John N. Maina

The Lung-Air Sac System of Birds
Development, Structure, and Function

With 94 Figures, 5 in Color, and 6 Tables

 Springer

Professor JOHN N. MAINA
School of Anatomical Sciences
Faculty of Health Sciences
University of the Witwatersrand
7 York Road, Parktown 2193
Johannesburg
South Africa

ISBN-10 3-540-25595-8 Springer Berlin Heidelberg New York
ISBN-13 978-3-540-25595-6 Springer Berlin Heidelberg New York

Library of Congress Control Number: 2005927146

This work is subject to copyright. All rights are reserved, whether the whole or part of the material is concerned, specifically the rights of translation, reprinting, reuse of illustrations, recitation, broadcasting, reproduction on microfilm or in any other way, and storage in data banks. Duplication of this publication or parts thereof is permitted only under the provisions of the German Copyright Law of September 9, 1965, in its current version, and permissions for use must always be obtained from Springer-Verlag. Violations are liable for prosecution under the German Copyright Law.

Springer-Verlag is a part of Springer Science+Business Media
springeronline.com

© Springer Berlin Heidelberg 2005
Printed in Germany

The use of general descriptive names, registered names, trademarks, etc. in this publication does not imply, even in the absence of a specific statement, that such names are exempt from the relevant protective laws and regulations and therefore free for general use.

Editor: Dr. Dieter Czeschlik, Heidelberg, Germany
Desk-Editor: Anette Lindqvist, Heidelberg, Germany
Production and typesetting: Friedmut Kröner, Heidelberg, Germany
Cover design: *design & production* GmbH, Heidelberg, Germany

Printed on acid free paper   31/3152 YK   5 4 3 2 1 0

Dedication

To my wife and children for their unstinting support

Acknowledgements

My work and interest on the functional morphology of the gas exchangers and that of the avian lung in particular started under the mentorship of Professor (Emeritus) A.S. King at the University of Liverpool. I owe him immense gratitude for his guidance and continued friendship over the years. I have had the good fortune of collaborating with colleagues who gave me their time and shared ideas unreservedly. The ones that particularly stand out are: M.A. Abdalla, King Saud University; G.M.O. Maloiy, University of Nairobi; S.P. Thomas, Duquesne University; D.M. Hyde, University of California (Davis) and California Primate Institute; J.B. West, University of California, San Diego; and A.N. Makanya and S.K. Kiama, University of Nairobi. I should, however, hasten to add that I bear full responsibility for the final form of this account. Any infelicities of judgment that may have occurred are entirely mine. I beg the reader's indulgence and would be most grateful if such aspects are pointed out to me.

Preface

> *The respiratory system of any animal species has its own characteristics, and there are more than a million species. The physiological diversity parallels the phenotypic diversity.* Dejours (1989)

The field of avian respiratory biology is vast and research in the area is progressing rapidly. Although plentiful data, especially on its structure and function, now exist, they are scattered in many publications, some of which, especially the older ones, are unfortunately difficult to find. With the last comprehensive review prepared about 15 years ago (*A.S. King and J. McLelland, eds: Form and function in birds, volume IV. Academic Press, London, 1989, pp 591*), consideration of the progress since made is necessary to help identify and fill the gaps in the knowledge. Timely reviews of literature help minimize isolation between investigators. By pointing out the unresolved issues, focus is maintained and sustained on unresolved aspects, avoiding possible costly duplication of effort. While summarizing and building on earlier observations and ideas, this book provides cutting-edge details on the development, structure, function, and to a significant extent the evolutionary design of the avian respiratory system, the lung–air sac system. It is hoped that while outlining the mechanisms and principles through which biological complexity and functional novelty have been crafted in a unique gas exchanger, this account will provoke further inquiries on the many still uncertain issues. The specific goal here is to highlight the distinctiveness of the design of the avian respiratory system and the factors that obligated it. In order to allow readers without prior knowledge on the subject to follow the arguments easily, the book is well illustrated. It was written with a broad readership in mind: students (novices) and specialists, be they biologists, anatomists, zoologists, physiologists, evolutionary biologists, ecologists, or persons in other disciplines who may work on or contemplate studying respiratory aspects, especially from comparative perspectives, will find the book useful. Due to limitation of space, it was not possible to give an encyclopedic review of all aspects of the biology of the avian respiratory system. Only the most important characteristics are included here. There are excellent accounts, monographs, treatises, and perspectives on different areas of its biology. Many such works are given in the list of references. They should be consulted where specific details are required.

J.N. Maina   University of Witwatersrand
30 March 2005, Johannesburg

Contents

Chapter 1	**Flight**	1
1.1	Energetic Cost of Flight	1
1.2	Flight Speed and Endurance	9
1.3	Flight at High Altitude	10
Chapter 2	**Development**	13
2.1	General Considerations	13
2.2	Bronchial (Airway) System	15
2.3	Air Sacs	34
2.4	Pulmonary Vasculature	35
2.4.1	Hematogenesis	36
2.4.2	Vasculogenesis and Angiogenesis	45
2.5	Blood–Gas Barrier (BGB)	48
2.6	Molecular and Genetic Aspects in Lung Development	53
2.6.1	General Observations	53
2.6.2	Fibroblast Growth Factors (FGFs)	55
2.6.3	Vascular Endothelial Growth Factor (VEGF)	59
2.6.4	Wnt Genes and Signaling	61
Chapter 3	**Qualitative Morphology**	65
3.1	General Observations	65
3.2	Lung	66
3.3	Airway (Bronchial) System	73
3.3.1	Primary Bronchus	73
3.3.2	Secondary Bronchi	75
3.3.3	Parabronchi (Tertiary Bronchi)	78
3.3.4	Atria, Infundubulae, and Air Capillaries	82
3.4	Blood–Gas Barrier	95
3.5	Surfactant	95
3.6	Air Sacs	96

3.6.1	Topographical and Structural Morphology	96
3.6.2	Ostia	98
3.6.3	Cytoarchitecture of the Wall of Air Sacs	99
3.7	Paleopulmo and Neopulmo	100
3.8	Pulmonary Vasculature	101
3.9	Arrangement of the Structural Components for Gas Exchange	104
3.10	Cellular Defenses of the Lung	107
3.11	Control of Air Flow	116

Chapter 4	**Quantitative Morphology (Morphometry)**	**125**
4.1	General Observations	125
4.2	Volume of the Lung	128
4.3	Respiratory Surface Area	145
4.4	Thickness of the Blood–Gas Barrier	149
4.5	Pulmonary Capillary Blood Volume	151
4.6	Modeling a Gas Exchanger: Integrative Morphometry	153
4.6.1	General Principles	153
4.6.2	Modeling the Avian Lung	154
4.6.3	Pros and Cons of Pulmonary Modeling	157

Chapter 5	**Comparative Respiratory Morphology**	**159**
5.1	General Observations	159
5.2	Comparison of the Structure of the Avian Respiratory System with Those of Some Other Animals	161
5.2.1	Dipnoan Lung	161
5.2.2	Amphibian Lung	164
5.2.3	Reptilian Lung	165
5.2.4	Insectan Tracheal System	168
5.3	Conclusions	173

References 175

Subject Index 203

Abbreviations

AAS(s)	abdominal air sac(s)
ABPA	accessory branch of the pulmonary artery
AC(s)	air capillary/air capillaries
AHP	air-haemoglobin pathway
AM(s)	alveolar macrophage(s)
AS(s)	air sac(s)
AVH	avian venous hematocrit
BALu(s)	bronchioalveolar lung(s)
BC(s)	blood capillary/blood capillaries
BGB	blood–gas barrier
BL	basement lamina
BV(s)	blood vessel(s)
CaTAS(s)	caudothoracic air sac(s)
CBPA	cranial branch of the pulmonary artery
CeAS(s)	cervical air sac(s)
CL	capillary loading
ClAS(s)	clavicular air sac(s)
CLBPA	caudolateral branch of the pulmonary artery
CMBPA	caudomedial branch of the pulmonary artery
CO_2	carbon dioxide
CrTAS(s)	craniothoracic air sac(s)
D_eO_2	morphometric diffusing capacity of the erythrocyte
D_LO_{2m}	total morphometric diffusing capacity of the lung
D_LO_{2p}	physiological diffusing capacity of the lung
D_mO_2	membrane diffusing capacity of the lung
DO_2	diffusing capacity of oxygen
D_pO_2	morphometric diffusing capacity of the plasma layer of the lung
D_tO_2	morphometric diffusing capacity of the tissue barrier (BGB) of the lung
EC	erythrocyte cytoplasm
EnC(s)	endothelial cell(s)
EPPB	extrapulmonary primary bronchus
ET(s)	exchange tissue(s)
FGF(s)	fibroblast growth factor(s)

FGF-2	fibroblast growth factor-2
GF(s)	growth factor(s)
HGC(s)	hematogenetic cell(s)
IAV	inspiratory aerodynamic valving
IPPB	intrapulmonary primary bronchus
IPRS	interparapronchial septum/septa
IPRBAo(s)	interparapronchial arteriole(s)
IPRBAr(s)	interparapronchial artery/arteries
IPRBV(s)	interparabronchial blood vessel(s)
IPRV(s)	interparabronchial vein(s)
K_pO_2	Krogh's oxygen permeation constant through the plasma layer (PL)
K_tO_2	Krogh's oxygen permeation constant through the tissue barrier (BGB)
LDSB	laterodorsal secondary bronchus/laterodorsal secondary bronchi
LOBs	laminated osmiophilic bodies
LVSB	lateroventral secondary bronchus/lateroventral secondary bronchi
MCSAS	multicapillary serial arterialization system
MDSB	mediodorsal secondary bronchus/mediodorsal secondary bronchi
MR	metabolic rate
MSRSA	mass specific respiratory surface area
MVSB	medioventral secondary bronchus/medioventral secondary bronchi
Mya	million years ago
NPP	neopulmonary part
NPPR	neopulmonary parabronchi
O_2	oxygen
PA	pulmonary artery
$PaCO_2$	partial pressure of carbon dioxide in the arterial blood
PaO2	partial pressure of oxygen in the arterial blood
PB	primary bronchus
PCB	pulmonary capillary blood
PCBV	pulmonary capillary blood volume
$PECO_2$	partial pressure of carbon dioxide in the end expired air
PEO_2	partial pressure of oxygen in the end expired air
PIVM(s)	pulmonary intravascular macrophage(s)
PL	plasma layer
PO_2	partial pressure of oxygen
PPPR	paleopulmonary parabronchi
PR	parabronchus/parabronchi
PRL	parabronchial lumina/lumen

Abbreviations

PRLu(s)	parabronchial lung(s)
PSEM(s)	pulmonary subepithelial (interstitial) macrophage(s)
PV	pulmonary vein
RBC	red blood cell
RSA	respiratory surface area
SA	segmentum accerelans
S_a	surface area
SAPM(s)	surface avian pulmonary macrophage(s)
SB	secondary bronchus/secondary bronchi
S_c	surface area of the capillary endothelium
S_e	surface area of the erythrocytes
S_p	surface area of the plasma layer (PL)
S_t	surface area of the blood–gas barrier (BGB)
$S_V BGB$	surface density of the blood–gas barrier
Shh	sonic hedgehog
TLS	trilaminar substance
VEC(s)	vasculoendothelial cell/vasculoendothelial cells
VEGF	vascular endothelial growth factor
VO_{2c}	oxygen consumption
VO_{2f}	oxygen flow rate
$VO_{2\,max}$	maximum oxygen consumption
VL	volume of the lung
PCBV	volume of the pulmonary capillary blood
ΔP	partial pressure gradient
ΔPCO_2	partial pressure gradient of carbon dioxide
ΔPO_2	partial pressure gradient of oxygen
τ_{hp}	harmonic mean thickness of the plasma layer (PL)
τ_{ht}	harmonic mean thickness of the blood-gas barrier (BGB)
θO_2	oxygen uptake coefficient of whole blood

1
Flight

> *Muscle powered flight requires a high metabolic rate and a very efficient respiratory system.* Constable (1990)

1.1
Energetic Cost of Flight

Defined as capacity to produce lift, accelerate, and maneuver at various speeds, powered flight is an elite form of locomotion. It has compelled incomparable structural specializations and refinements and striking functional integration of practically all the organs and organ systems, especially the gastrointestinal, cardiovascular, respiratory, muscular, and nervous systems. Exerting substantial metabolic, mechanical, and aerodynamic demands, flight adaptively evolved in response to particular selective pressures in order to meet specific survival needs. To morphologists and physiologists, in many ways volant animals offer exceptional an opportunity in determination and understanding of the upper limits of biological design and performance of vertebrates.

A novel mode of locomotion, flight sets birds well apart from other vertebrates. Historically, humans have coveted ability to fly since the first moment that the cave person looked up into the sky and observed birds soaring effortlessly. In the Greek mythology, some 3500 years ago, using fake wings of which the feathers were affixed to their arms with bee's wax, Daedalus and his son Icarus escaped from King Mino's island prison dungeons (Fig. 1). Unfortunately, as the story goes, with youthful daring, Icarus disregarding his father's wise counsel flew too close to the sun, the wax melted, the 'wings' disintegrated, and he fell to his death. Without the relentless provocation and attestation presented by birds that gravity could be overcome, it would have taken much longer before human beings thought of flight as a feasible form of locomotion. In his 1505 treatise entitled *'On bird flight'*, Leonardo da Vinci, who strongly believed that human beings would never learn to fly until they thoroughly understood the secrets of bird flight, wrote: *"I remember that in my earliest childhood, I once dreamed that a vulture flew towards me, opened my mouth and stroked it a number of times with its feathers, so that I could talk about wings for the rest of my life"*. For understandable, though not defensible reasons, the physiology and to some extent the morphology of the human being is the conventional reference point against which other animals are compared and measured. From this anthropocentric position, factual errors have resulted from direct extrapolation of the mammalian functional morphology to that of birds and other animals. Failing to appreciate nature's remarkable resourcefulness of conceiving different solutions to various selective pressures, a great deal of time and immense resources have been wasted pursuing costly, unproductive inquiries. For example, on strong belief that all that was needed to fly (just like in the case of the mythological angels, dragons, and the 'flying' horse Pegasus) was to strap on 'wings' and flap, fatalities after jumping from heights using various contraptions (including a raincoat!) are well documented in the recent past. It took time and great frustration before the fact that humans were never built to fly at long last sank in. Literally going back to the drawing boards, it was ultimately realized that active flight would never be achieved by directly emulating birds but rather by alternatively engineering flying machines. Such designs had to be formulated on the universal laws of nature that apply to all moving objects, including birds.

Fig. 1. A fourteenth century illustration showing the mythical flight of Daedalus and Icarus from King Mino's prison island. From an anonymous nineteenth century wood carving

1.1 Energetic Cost of Flight

Paying back a well-deserved tribute to their 'feathered instructors', Orville Wright (of the Wright brothers) wrote: *"Learning the secret of flight from a bird was a good deal like learning the secret of magic from a magician. After you once know the trick, you see things that you didn't notice when you didn't know exactly what to look for"*. Despite the great advances that have now been achieved in the field of avionics, materials and technology needed to construct machines that can fly efficiently using flapping deformable wings are still in the offing.

An underlying principle in biology is that manifestation of exclusivity, e.g. in structure and/or function, offers important insight into the opportunistic paths that evolution has taken. Obliging many constraints on structure, flight is the single most important factor that has decreed the form and function of birds. As they got bigger, it necessitated more energy for birds to fly. At odds with what would be expected, doubling-up body mass does not result in a doubling-up of the energy needed to fly at minimum speeds but rather by an increase by a higher factor of 2.25. Realization of certain intricate adaptive traits like volancy entails certain trade-offs and compromises. For flight in birds, these changes included the following: (a) aerodynamic streamlining of the body, (b) total commitment of the forelimbs (wings) to flight, (c) development of a long flexible neck to perform the activities previously executed by the wings (forelimbs), and (d) inevitably with development of a long neck a long trachea. A long trachea precipitated respiratory limitations regarding air flow resistance and large dead-space volume. To offset the restrictions, birds developed a wider trachea (about three times larger than that of a mammal of equivalent body mass) and a slower respiratory rate.

The exacting demands that flight compelled on birds (the entire class of Aves) obligated greater uniformity in their external morphology than has been possible in single orders of fishes, amphibians, and reptiles (Marshall 1962, p 555). Yapp (1970, p. 40) observes that, in the entire population of birds, there is less variation in structure than in the 90 or so species of primates and 290 species of *Carnivora*. Even those birds that have lost capacity for flight, e.g. ostrich, cassowaries, rheas, emu, and kiwi, cannot be mistaken for any other vertebrate group! Paradoxically, while a harmonious external form exists, flight has imposed exceptional diversity in the internal, anatomical structural details of birds (King and King 1979, p 2). The features pertaining to the respiratory system are mentioned in this account. In some cases, they are of uncertain or no functional consequence. The benefits reaped by birds from capacity of flight were enormous. Capable of overcoming geographical obstacles, they occupied diverse ecological niches and habitats, consequently undergoing remarkable adaptive radiation that has culminated in about 9000 species (e.g. Morony et al. 1975; Gruson 1976). In contrast, the contemporary reptiles total about 6000 species (Bellairs and Attridge 1975, p. 17) whilst mammals (Yapp 1970, p. 40) and amphibians (Bellairs and Attridge 1975, p. 17) have fewer species. Bats, the only volant mammals, comprise some 800 species – in specific numerical diversity they are only exceeded by the order

Rodentia with 1660 species. Of all known mammalian species, one in five species is a bat! After the human being, *Myotis* (family: Vespertilionidae) is said to be the most widely spread naturally occurring mammalian genus on earth (Yalden and Morris 1975). Constituting about 75% of the envisaged 5 to 50 million animal species (e.g. May 1992), both numerically and specifically, the insects (another volant taxon) are the most abundant (e.g. Wigglesworth 1972). The fact that birds and bats are the only existing vertebrates capable of powered flight clearly attests to the extreme selective pressure that the mode of locomotion imposes.

After evolving independently from reptiles much later than mammals (e.g. de Beer 1954; Ostrom 1975) and achieving endothermic homiothermy, birds reached metabolic scopes between resting and maximal rates of exercise or cold-induced thermogenesis that are 4 to 15 times higher than those of their progenitors at same body temperature (e.g. Lasiewski 1962). Moreover, among the endotherms, birds, especially the passerines, the most successive taxon with 5739 (over 60% of the total number of extant avian species; e.g. Sibley and Alhquist 1990; Barker et al. 2004), operate at a relatively higher body temperature of 40–42 °C compared with the lower one (38 °C) of mammals (e.g. Lasiewski and Dawson 1967; Aschoff and Pohl 1970). A significant metabolic barrier separates ectothermic from endothermic animals and volant from nonvolant ones. The evolution of flight in birds is thus unmistakably associated with development of an exceptionally efficient respiratory system. Whether by default or by design, the lung-air sac system appears to have been the optimal solution to the metabolic needs of birds. Since bats, a taxon with a characteristic mammalian lung (e.g. Maina 1985, 1986) but a highly refined one (e.g. Maina et al. 1982; Maina and King 1984; Maina et al. 1991), fly as well as birds (e.g. Norberg 1990), it is patently clear that the design of the avian respiratory system is not a prerequisite for flight.

The grace and majesty with which birds in particular fly are extremely deceptive of the severe constraints that had to be surmounted to achieve and maintain flight. In practically all active vertebrate groups, locomotion exerts the highest demands on the respiratory system (e.g. Banzett et al. 1992). Skeletal muscle accounts for 96% of a flying animal's total oxygen consumption (VO_{2c}) during flight (Thomas and Suthers 1972; Thomas et al. 1984). The need to lower the cost of transport may have been the foremost selective pressure that compelled the evolution of flight (Scholey 1986). During migration or while foraging, compared with other vertebrates, birds and bats travel over longer distances at faster speeds (Fenton et al. 1985).

Active flight is innately a highly energetically demanding form of locomotion (Ellington 1999; Nudds and Bryant 2000; Tobalske et al. 2003). The mass-specific aerobic capacities of flying birds and bats are 2.5 to 3 times greater than those of mammals of the same body mass running fast on the ground (e.g. Thomas 1987). During continuous flight, e.g., bats increase their VO_{2c} by a factor of 20–30 (e.g. Thomas and Suthers 1972) and, in turbulent air or when ascending, a bird can increase VO_{2c} for brief periods by the same magnitude (Tucker 1970). At an ambient temperature of 20 °C, a 12-g bat, *Myotis velifer*, amazingly increases VO_{2c} 130 times that at rest (Riedesel and Williams 1976). A budgerigar, *Mellopsitacus undulatus*, in level flight, i.e. at its most economical speed, increases its VO_{2c} 13 times its standard metabolic rate (MR), a value that is about 1.5 times that of a similar sized mouse, *Mus musculus*, running hard on a tread mill (e.g. Tucker 1968). In the pigeon, *Columba livia*, when running on the ground, VO_{2c} is 27.4 ml.min^{-1} and, during flight at a speed of 10 m s^{-1}, VO_{2c} is 77.8 ml min^{-1}, a factorial difference of 3 (e.g. Grubb 1982). In the herring gull, *Larus argentatus*, during flight, the MR is twice the resting value while, in the grey-headed albatross, *Diomedia chrysostoma*, the rate is about three times the predictable basal one (e.g. Costa and Prince 1987).

Flight has only been realized by two phyla, namely the Chordata and the Arthropoda. Chronologically, volancy was achieved by the insects about 350 million years ago (mya; e.g. Wigglesworth 1972), the now extinct pterodactyls some 220 mya (e.g. Bramwell 1971), in birds around 150 mya (e.g. de Beer 1954), and bats 50 mya (e.g. Yalden and Morris 1975). *Archeopteryx lithographica* of the Upper Jurassic and *Icaronycteris index* of the Eocene are respectively the oldest bird and bat. The variety of animals wrongly said to 'fly', e.g. the freshwater butterfly-fish, *Pantodon buchholzii*, of the West African rivers, the parachuting frog of Borneo, *Rhacophorus dulitensis*, the flying snakes of the jungles of Borneo, *Chrysopelea* sp., the flying squirrel, *Glaucomys volans*, of North America, the flying lemur, *Cyanocephalus volans*, and the East Indian gliding lizard, *Draco volans*, are strictly acrobatic passive gliders or parachutists that use modifications of certain parts of their body to delay a fall by using drag and lift. Such animals have not had to grapple with the daunting aerodynamic and metabolic imperatives for active flight.

While costly in terms of absolute demands for energy, powered flight is an exceptionally efficient form of locomotion. At high speed, the distance covered per unit energy expended is much less than that incurred during most other kinds of locomotion (e.g. Schmidt-Nielsen 1972; Rayner 1981). In the bat species *Phyllostomus hastatus* and *Pteropus gouldii*, e.g., the energy needed to cover a given distance is respectively between one-sixth and one-quarter of that required by similarly sized, nonflying mammals (Thomas 1975). At their optimal speeds, the minimum cost of flying for a 380-g bird is about 30 % of the energetic cost of a 380 g nonvolant mammalian runner (e.g. Hainsworth 1981). On the whole, birds have a larger daily physical activity, i.e. field MR, than mammals (King 1974). A 380 g bird spends about 74 % more energy daily than a 380-g mammal (Powell 1983). Further to achieving a more economical mode of foraging, animals that accomplished flight reaped extra advantages from it. They were able to: (1) inhabit a less crowded and almost limitless ecological niche, (2) escape from ground predation, and (3) experience unimpeded geographical dispersal, allowing extensive adaptive radiation that culminated in remarkable speciation. By the end of Eocene (about 35 mya), e.g., 26 of the modern 27 avian orders were well established, a process that occurred in an amazingly short evolutionary time of only 100 million years.

All the modern birds, including groups like the penguins, the rhea, the kiwi, the ostrich, the cassowaries, and the emu that are now flightless, arose from flying progenitors (e.g. Welty 1979). The upper limit of body mass for flight is set by birds such as the Californian condor, *Gymnogyps californianus*, the kori bustard, *Ardeotis kori*, the white pelican, *Pelecanus anocrotalus*, and the mute swan, *Cygnus olor*, birds that weigh between 12 and 18 kg. The great bustard, *Otis tarda*, has, however, been reported to reach a body mass of 21 kg (Martin 1987). Unlike birds, to varying extents, virtually all bats fly. Signifying definite constraints of design, the heaviest bat, the flying fox, *Pteropus edulis*, weighs a maximum of 1.5 kg (e.g. Yalden and Morris 1975; Carpenter 1985): the heaviest volant bird weighs about an order of magnitude more than the heaviest bat. Interestingly, in the past, outstandingly large animals flew. The super-condor, *Teratornis incredibilis*, of the North American Pleistocene, that is estimated to have weighed in excess of 20 kg, and the giant *Argentavis magnificens* at 120 kg (e.g. Martin 1987), are such examples. While a possible lower gravity on earth, as suggested by, e.g., Carey (1976), may have permitted it, theoretical calculations of maximal power generation by the flight muscles (e.g. Marden 1994; Ellington 1999) suggest that, even under the present measure of gravity, at least take off and probably anaerobically powered short burst flights may have been possible by some of these birds and probably for the heaviest (250 kg) known dinosaur, *Quetzalcoatlus northropi* (e.g. Paul 1991).

A range of refinements of the respiratory systems of the extant birds, features that correspond with their various metabolic capacities, occur. Body mass, an important factor that sets the MR, ranges from a mere 2 g of the Cuban bee hummingbird, *Calypte helenae*, to the approximately 150-kg African ostrich, *Struthio camelus* (e.g. Clark 1979), a colossal factorial difference of 7.5×10^4. For bats, body mass ranges from about 1.6 g of the Thai bat, *Craeonycteris thonglongyai*, to the approximately 1.5-kg flying fox, *Pteropus edulis*, a factorial difference of only 10^3. Except perhaps for the most extreme habitats, birds permanently or fleetingly occupy practically all parts of earth. Means and ways of enhancing flight efficiency have evolved (e.g. Pennycuick 1975; Norberg 1990). The V-shaped flight formation commonly employed by migrating birds like geese is, e.g., an energy-saving strategy that allows individual birds, except for the leader, to rest their wing tips on the rising vortex of air displaced by the wings of the bird in front (e.g. Cutts and Speakman 1994). Long wing spans, complex pectoral girdle (Fig. 2A,B), bounding flight (Rayner 1985), and in species that soar and glide over long periods of time a humeral locking mechanism (e.g. Fisher 1946), are but some of the specializations that allow birds to conserve energy. In the highly energetic species of birds, e.g. hummingbirds, flight muscles may form more than 25% of the body mass and are highly aerobic (Suarez et al. 1991; Suarez 1992). Comprising about two-thirds of the flight muscle mass, the pectoralis muscle is the

1.1 Energetic Cost of Flight

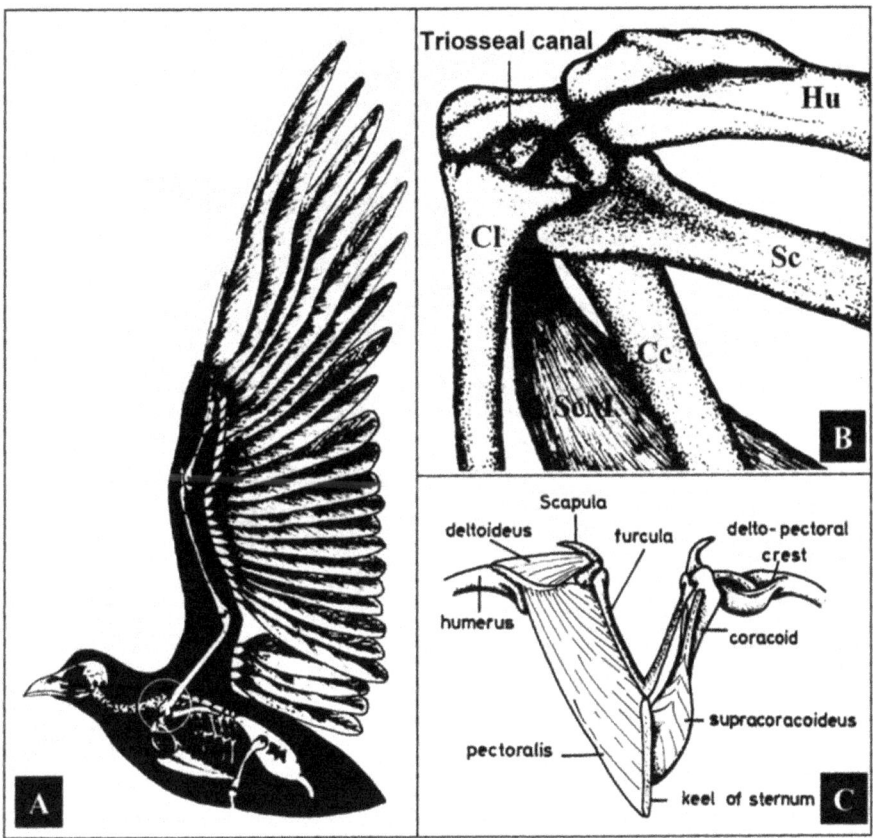

Fig. 2. A The complex pectoral girdle of birds (*circle*). **B** The pectoral girdle consists of four bones namely the humerus (*Hu*), the scapula (*Sc*), the coracoid (*Cc*), and the clavicle (*Cl*). **B, C** Although both the pectoralis and the supracoracoideus muscles arise from the sternum (the keel), the supracoracoideus muscle (*ScM*) passes through the triosseal canal to insert into the dorsal part of the humerus, where it exerts an upward pull on the wing. **A**, modified from Fisher and R.T. Peterson (1988); **B**, sketched in a modified form from Wallace (1955); **C**, sketched in a modified form from Pettingill (1969)

depressor of the wing while the supracoracoideus is the elevator (Fig. 2C). Interestingly, while both muscles attach essentially at the same point, the keel (sternum), the supracoracoideus finds its way to the dorsal (upper) surface of the wing, where it effects an upward pulley-like pull, after passing through the triosseal canal (Fig. 2B,C). Hovering, i.e. stationary flight relative to the surrounding air, where the downward air displacements (movements) that support the weight of an animal are generated by the wing beats alone, is energetically the most demanding mode of flight (e.g. Wells 1993; Fig. 3): it is utilized by only a few species of birds for foraging. A hovering bird consumes O_2 at a rate 2.5 times that during forward flapping flight (Wells 1993). From their

Fig. 3. A hovering bird supports its weight (against gravity) entirely by the power generated by the beating of its wings. Remaining stationary, the wing rotates at pectoral girdle in a figure of eight producing lift

miniature size and fast wing-beat frequency, the VO_{2c} of a hovering and forward flying hummingbird ranges from 40–85 ml O_2 g^{-1} h^{-1} (e.g. Bartholomew and Lighton 1986; Wells 1993). These values are remarkably higher than those of the maximum oxygen consumption ($VO_{2\,max}$) in a running pygmy mouse (7 g), kangaroo rat (1.1 kg), and dog (21 kg), that are respectively 15.7, 10.6, and 9.5 ml O_2 g^{-1} h^{-1} (Seeherman et al. 1981) but appreciably lower than those of flying insects (e.g. the euglossine bees) of 66–154 $mlO_2 \cdot g^{-1} \cdot h^{-1}$ (e.g. Casey et al. 1985). Hummingbirds typically hover more than 100 times per day (e.g. Krebs and Harvey 1986) and each bout normally lasts for less than a minute. Owing to their exceptionally high MR, 20% of the daylight hours are spent foraging (Diamond et al. 1986). If not utilized to acquire energy-rich foods such as nectar, hovering is largely avoided in preference to the less energetically demanding forward flight.

The aerobic capacities of the small hummingbirds in flight appear to constitute the upper limit of vertebrate mass-specific MR that is attuned with efficient design of the endothermic homeotherms. The hummingbird flight muscle is the most metabolically active vertebrate skeletal muscle (Suarez et al. 1991; Mathieu-Costello et al. 1992; Suarez 1992) and the maximum enzyme activities during flight are by average avian standards exceptional (Suarez et al. 1986). To support the enormously large energetic demands, especially during flight, a hummingbird weighing 4–5 g ingests about 2 g of sucrose per day (Powers and Naggy 1988) and, before migration, the fat content in the body increases enormously (Carpenter et al. 1983). The ruby-throated hummingbird, *Archilochus colubris*, accumulates about 0.15 g of triacyglycerols per day per gram body weight, a value that in a human being would be equivalent to a weight gain of 10 kg/day (Hochachka 1973)! To enhance the absorptive rate, the gastrointestinal system of the hummingbird has a higher sucrose activity per cm^2 of surface area and higher densities of intestinal glucose transporters than any other vertebrate species known (e.g. Karasov et al. 1986; Martinez del Rio 1990).

1.2
Flight Speed and Endurance

While most birds generally fly at moderate speeds, with the smaller and more agile ones (e.g. passerines) achieving speeds of 15–40 km h^{-1}, the swifts (Apodidae), the loons (Gaviidae), and the pigeons (Columbidae) reach and sustain speeds of between 90–150 km h^{-1} (Norberg 1990). Given enough time and distance, it is estimated that a diving peregrine falcon, *Falco peregrinus*, can reach a speed of 403 km h^{-1} (112 m s^{-1}; Tucker 1998). On close scrutiny, even the avian flight speeds that may appear mediocre by average standards are quite fast. Normalized with the body lengths covered per unit time, a passeriform bird flying at a moderate speed of 40 km h^{-1} covers about 100 body lengths per second compared with only 5 in a highly athletic human being. Among bats, a speed of 16 km h^{-1} in *Pipistrellus pipistrellus* has been reported by Jones and Rayner (1989). In its annual migration, the Arctic tern, *Sterna paradisea*, flies from pole to pole, a distance of 35,000 km, between breeding seasons (e.g. Berger 1961; Salomonsen 1967) while the American golden plover, *Pluvialis dominica*, flies 3300 km nonstop from the Aleutian Islands to Hawaii in only 35 h (Johnston and McFarlane 1967). The minute 3-g ruby-throated hummingbird, *Archilochus colubris*, incredibly flies nonstop for more than 1000 km across the Gulf of Mexico from the eastern United States, a distance that takes about 20 h to cover. Many passerine species fly nonstop for 50–60 h on the trans-Saharan migratory route (Berger 1961) and swifts (Apodidae) and the wandering albatross of the southern oceans are reported to fly continuously (day and night) sleeping, eating, drinking, and mating on the wing (Jameson 1958; Lockley 1970) and only landing when nesting and inevitably falling down when they die!

1.3
Flight at High Altitude

The process by which birds solved the challenges posed by the high O_2 demand for flight, mostly at high altitude, elicits interest. In the rarefied high-altitude atmosphere, while the aerodynamic drag is reduced, the bird has to do more work to generate the thrust required to maintain forward level flight (Tucker 1974). Birds that fly or live at high altitude not only face extreme hypoxia, but also low ambient temperatures. A Ruppell's griffon vulture, *Gyps rueppellii*, was sucked into the engine of a jet-craft at an altitude of 11.3 km (Laybourne 1974): at that altitude the barometric pressure is about 24 kPa (180 mmHg; i.e. about 24% of that at sea level), the PO_2 in the expired air is less than 5.3 kPa (39.8 mmHg) – closer to 2.7 kPa (20 mmHg) if hyperventilation brings the PCO_2 to about 0.67 kPa (5 mmHg) – and the ambient temperature is about –60 °C. The ability to survive, let alone exercise under such conditions is beyond the reach of many animals. A flock of swans (probably whooper, *Cygnus cygnus*) were observed by radar flying at an altitude of 8.5 km (e.g. Elkins 1983). The alpine chough, *Pyrrhocorax granulus*, permanently resides and nests above an altitude of 6.5 km (Swan 1961) where it encounters an average temperature of –27 °C while breathing air with a PO_2 of 9 kPa (69 mmHg). Perhaps the most striking high-altitude well-documented flight behavior in birds is that of the bar-headed goose, *Anser indicus*. The bird makes semiannual migratory flights from its wintering grounds in the Indian subcontinent to the breeding ones around the large lakes in the south-central regions of Asia (elevations about 5.5 km). With great admiration and astonishment, climbers near collapse have sighted birds flying high above the summits of Mt. Everest and Annapurna 1 (alt. 8.8 km). At the altitude that they cross the highest of the Himalayan peaks, the barometric pressure is about 31.1 kPa (233 mmHg), i.e. about one-third of that at sea level, and the (P) of oxygen (PO_2) in dry air is 6.5 kPa (48.8 mmHg; West 1983). The bar-headed geese spectacularly cover the journey in one nonstop flight: they take off from virtually sea level and almost directly reach an altitude of about 10 km without acclimatization (e.g. Swan 1961). If during these excursions the geese maintain a constant body core temperature of about 41 °C and the inhaled air is warmed to that of the body and fully saturated with water vapor, the PO_2 in the air reaching the respiratory surface should not exceed 4.9 kPa (36.6 mmHg). In a hypobaric chamber, the goose experimentally tolerates hypoxia at a simulated altitude of 11 km (Black and Tenney 1980): up to an altitude of 6.1 km, the bird maintains normal Vo_{2c} without hyperventilating and, up to 11 km, where the concentration of O_2 is only 1.4 mmol l^{-1}, the bird extracts adequate amounts of O_2 to necessitate only minimal increase in ventilation.

The greylag goose, *Anser anser*, a close relative that subsists at lower altitudes, at 39 °C and pH 7.4, has an O_2 affinity (P_{50}) lower (39.5) than that of the blood of the bar-headed goose (29.7; Petschow et al. 1977; Black et al. 1978).

Fedde et al. (1989) observed that, in the bar-headed goose, muscle blood supply and O_2 loading from the muscle blood capillaries (BCs) rather than ventilation or pulmonary gas transfer are the limiting steps in the supply of O_2 to the contracting flight muscles under hypoxia. The O_2 extraction efficiency of the lung of the bar-headed goose is very high: at a simulated altitude of 11.6 km, the PO_2 in the arterial blood is only 0.13 kPa (1 mmHg) less than that in the inhaled air (Black and Tenney 1980). Though bats are not known to fly at very high altitude, for among other reasons they would lose heat from their bare bodies excessively, especially across the wings, they have been shown to tolerate experimental hypoxia very well (Thomas et al. 1985).

In this account, the evolutionary, developmental, structural, and functional morphologies of the lung-air sac system of birds are examined. Comparison with other gas exchangers are made to emphasize how the avian respiratory system differs from or fits into the general pattern of design of vertebrate gas exchangers.

2
Development

> *The morphology of the avian lung can be made clear only by observations of its development and it is through this channel alone that one becomes acquainted with the nature of the modifications of the bird's lung that place it in a class by itself.* Locy and Larsell (1916a)

2.1
General Considerations

The embryonic vertebrate lung inaugurates as a ventral out pocketing of the foregut endoderm into the splanchnic mesoderm. The primordium is thought to arise from a select group of endodermal cells set aside early in the development of the gut (e.g. Hackett et al. 1996). A variety of epithelial organs begin to form when primitive epithelium interacts with specific mesenchymal tissues at different stages of development, producing organ-specific structures and functions. Mediated by growth factors (GFs) and cytokines which act as paracrine signals that direct specific gene responses that regulate cell division and differentiation, the epithelial–mesenchymal interactions may be permissive or phenotypically instructive. Lack of contact with the pulmonary epithelial cells causes mesenchymally derived cells to fail to produce cells expressing endothelial-specific genes and ultimately necrose (Gebb and Shannon 2000). The embryogenesis of the lung calls for coordinated development of the airway and vascular systems, a process essential for matched ventilation and perfusion in the fully functional, mature organ. In the developing lung bud, epithelial cells proliferate quickly and undergo reiterative branching, producing an array of tubes that decrease in diameter as they project outwards (e.g. Alsberg et al. 2004; Schittny and Burri 2004). Described as 'growth and branching of epithelial buds' by Saxen and Sariola (1987), branching morphogenesis is a fundamental embryonic process that occurs in many developing complex organs like the lung, the mammary gland, the kidney, the tooth, the pancreas, the tracheal system of insects, and the salivary gland whereby characteristic three-dimensional architecture of the functional parts is generated. Such organs provide excellent models for studying and understanding morphological patterning and cell-type differentiation (Shannon et al. 1998; Hogan 1999; Warburton et al. 2000; Schittny and Burri 2004).

Although epithelial development and branching morphogenesis have undergone intense investigation (e.g. Alsberg et al. 2004; Schittny and Burri 2004; Tuyl et al. 2004), relatively little is known about the molecular mechanisms that regulate pulmonary vascular formation (e.g. Perl and Whitsett 1999; Warburton et al. 2000; Taichman et al. 2002). Spatially and temporally, lung development is a highly dynamic process. Cross-talk, i.e. cell-to-cell signaling, between the mesenchymal and the epithelial cells determine cell-specific developmental pathways that are critical for the normal morphogenesis of the trachea, bronchi, and respiratory components of the lung as well as differentiation of distinct pulmonary epithelial cell lines (e.g. Wessels 1970; Minoo and King 1994). Distal lung mesenchyme is able to reprogram tracheal epithelium to branch in a lung-like manner and express distal lung phenotype while tracheal mesenchyme can prevent distal lung morphogenesis (Shannon 1994; Shannon et al. 1998; Yuan et al. 2000). Branching morphogenesis epitomizes an evolutionary conserved mechanism that is exploited to amplify functional surface area in the lung and other organs. The proposition by Weibel (1997) that the formation of proper dimensions during branching morphogenesis of the airway and the vascular elements may be programmed in the genetic instructions for design and that by West (1987) that an ordered mechanism of bifurcation is easier to genetically program have been well substantiated, especially by molecular biology techniques. Regarding the development of the avian (parabronchial) lung, however, only scanty and in some cases irreconcilable data exist (e.g. Romanoff 1960; Duncker 1978; Maina 2003a,b, 2004a,b; Maina and Madan 2003; Maina et al. 2003). The means and manner by which the structural components form and assemble to grant an exceptionally complex morphology (Chap. 3) are unsubstantiated. Thorough grasp of the development of the avian lung is vital to understanding its functional design.

In addition to air sacs (ASs) that among the vertebrate gas exchangers are exclusive to the avian respiratory system, in various other ways, morphologically, the parabronchial lung (PRLu) is profoundly different from the bronchioalveolar lung (BALu; e.g. Maina 1994, 1996, 1998, 2002a, 2004 c). The foremost morphogenetic differences are: (1) in the BALu the airway (bronchial) system forms by successive budding, elongation, and iterating dichotomous bifurcation while, in the PRLu, a three-tier contiguous system of air conduits that comprise primary, secondary, and tertiary bronchi (PR, parabronchi) and complex vascular architecture form; (2) while in the BALu the airway system ends blindly, fashioning what is commonly termed 'respiratory tree' or 'bronchioalveolar tree', in the avian lung the airways form a continuous loop: strictly, while the alveoli are the terminal gas-exchange components of the mammalian lung, the air capillaries (ACs) of the avian lung are not culs de sac; and (3) in complete departure from the perceptibly phased (quantum) developmental process in which the mammalian lung passes through glandular/pseudoglandular, canalicular, and ultimately alveolar stages, distinct developmental phases are lacking in the avian lung: the developmental changes are diffuse and progressive. Like all the other kinds of gas exchangers, the PRLu and the BALu are designed to allow the respiratory media, inspired air and venous blood to be exposed to each other over a large surface area across which respiratory gases, O_2 and CO_2, are exchanged by passive diffusion along prevailing Ps. Extensive respiratory surface area (RSA) and a thin blood–gas barrier (BGB) are prototypical structural features of gas exchangers: they are generated through a sequence of well-programmed developmental processes.

Regrettably, of the large number of extant avian species (9000; e.g. Gruson 1976; Morony et al. 1975), only a handful of species have been well studied. These include the domestic fowl, *Gallus gallus* variant *domesticus*, the pigeon, *Columba livia*, the guinea fowl, *Numida meleagris*, and the domestic muscovy duck, *Cairina moschata*. These species are frequently chosen rather for convenience (they are more accessible and are sufficiently large for ease of experimental manipulation) than for their actual taxonomic representivity. An important matter-of-fact caveat is that interpretations based on data and observations based or made on these species should be circumspectly extrapolated to other species of bird and to the avian taxon in general.

2.2
Bronchial (Airway) System

While the molecular and genetic controls of the development of the tracheal system of insects (e.g. Sutherland et al. 1996; Jarecki et al. 1999; Sato and Kornberg 2002) and the BALu have now been well documented (e.g. Cardoso 2000, 2001; Hislop 2002; Xiao et al. 2003), little is known about the processes that occur in the avian respiratory system. In the domestic fowl, the embryonic lungs first become perceptible from day 3.5 (about stage 26; Hamburger and Hamilton 1951) of embryogenesis (e.g. Duncker 1978; Maina 2003a,b). They appear like small, ridge-like outgrowths (Fig. 4A). Between days 3.5 and 5.5, the swellings approximate (Fig. 4B) and fuse on the ventral midline. Thereafter, they divide into left and right primordial lung buds (Fig. 4C,D). The lung buds form by multiplication and projection of the epithelial (endodermal) cells (that initially lined the primitive foregut) into the surrounding mesenchyme (Fig. 5A,B). The buds are covered by a somatopleural mesothelium (Fig. 5B). As they progressively advance into the coelomic cavity, the developing lungs diverge towards the respective lateral aspects of the body wall (Fig. 4D), reaching their definitive topographical locations on the dorsolateral aspect of the coelomic cavity at about day 8 (Fig. 6A). At that stage, the lung buds have enlarged and transformed from a saccular to a rather wedge shape

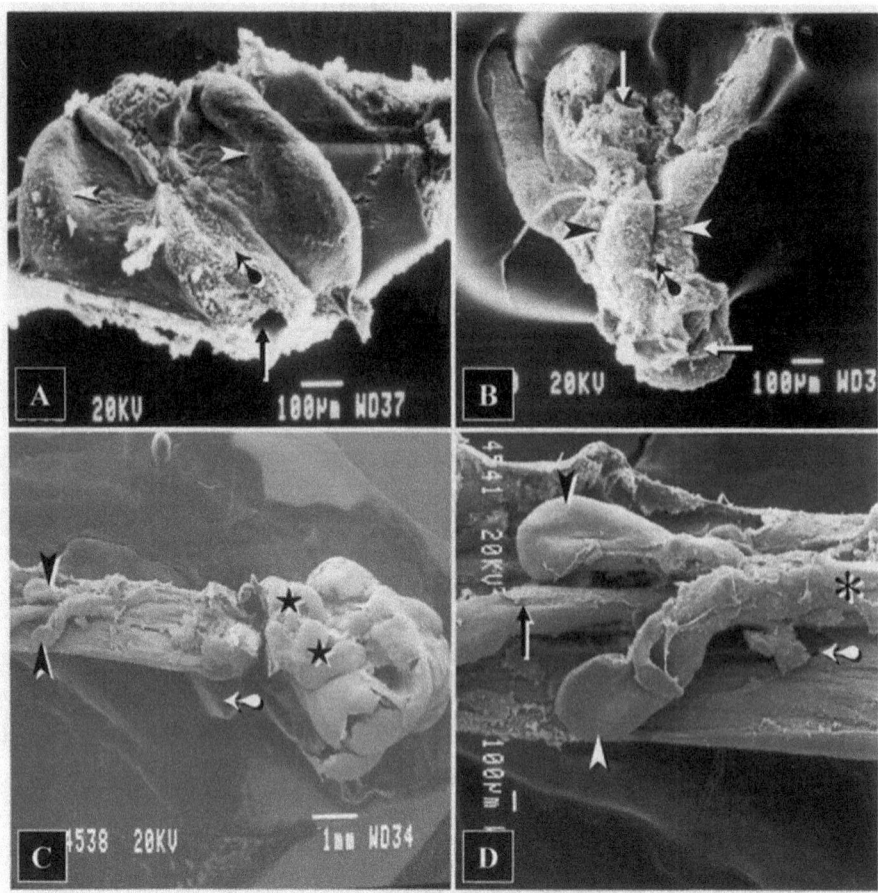

Fig. 4. A Lung buds (*arrowheads*) developing on the ventral aspect of the floor of the primitive pharynx. *Arrow* Gastrointestinal system. Day 3. **B** Lung buds approximating on the ventral midline (*arrowheads*). *Arrows* Gastrointestinal system. Day 4. **C** Saccular lungs (*arrowheads*) diverging and advancing towards the dorsolateral aspects of the coelomic cavity. *Stars* Branchial arches. Day 5. **D** Close-up of the lungs (*arrowheads*). *Arrow* Gastrointestinal system; *asterisk* trachea. Day 5. The *knobbed arrow* shows the cranial-caudal orientation of the specimen. Figures 4–20 and 22–36 are preparations from the domestic fowl, *Gallus gallus* variant *domesticus*. (Maina 2003b)

(Fig. 6A). Next, the lungs rotate along longitudinal and transverse axes and start to engage and insert into the underlying ribs. From the end of day 9, they are firmly affixed to the ribs and the hilus lies on the craniomedial aspect. Costal impressions (sulci) are conspicuous on the dorsolateral aspect of the lung (Fig. 6B–D).

Forming as an extension of the trachea and the extrapulmonary primary bronchus (EPPB), the intrapulmonary primary bronchus (IPPB) is the first airway to form (Fig. 7A). It runs from the hilus to the caudal edge of the devel-

2.2 Bronchial (Airway) System

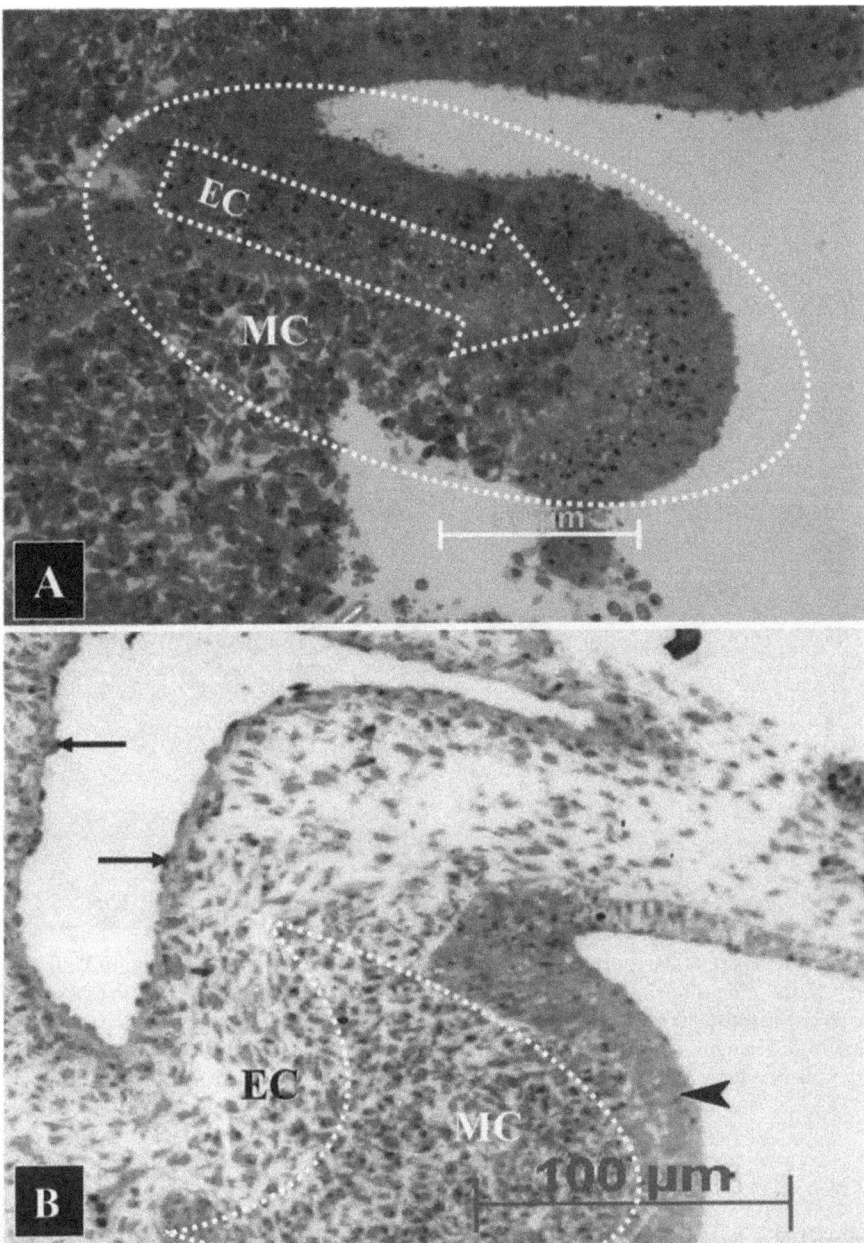

Fig. 5. A Lung bud (*dashed oval enclosure*) developing on the ventrolateral aspect of the foregut. *Dashed arrow* Endothelial cells extending into the underlying mesenchymal cells (*MC*). Day 3. **B** Close-up of the developing lung (after the fusion of the right and left lung buds) showing the trachea lined by an epithelium with conspicuously large epithelial cells (*arrows*). *EC* Endothelial cells; *dashed area* mesenchymal cells (*MC*); *arrowhead* mesothelial cells covering the formative lung. Day 5. (Maina 2003a)

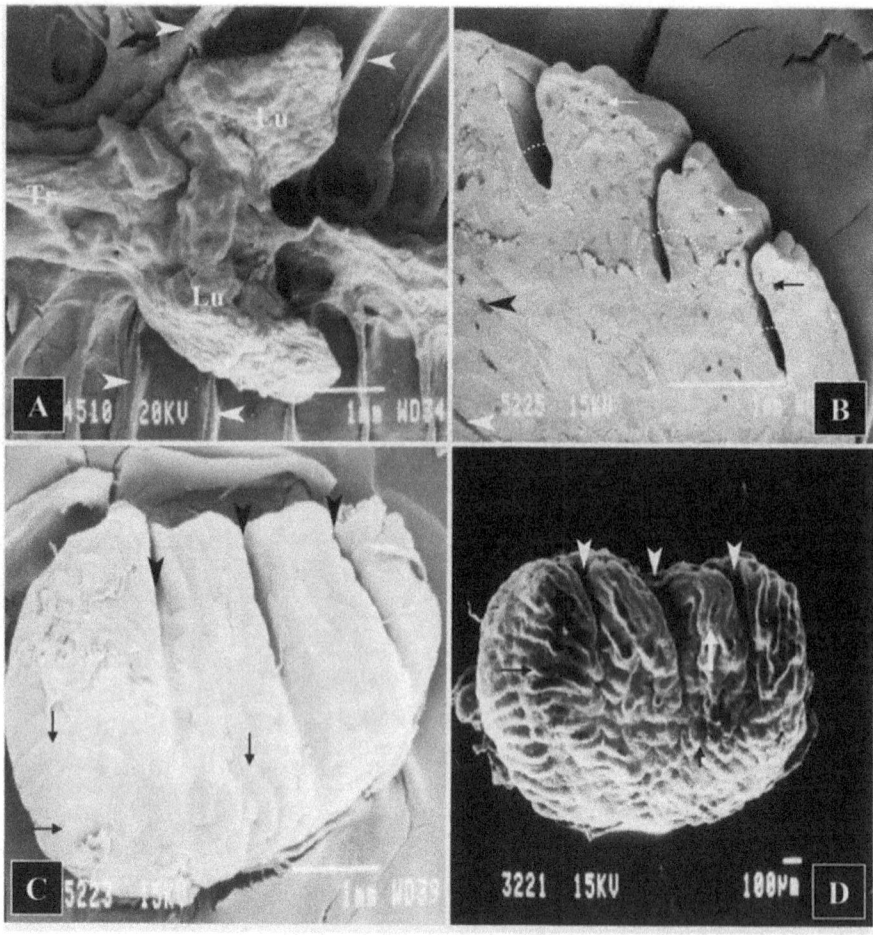

Fig. 6. A Lungs (*Lu*) starting to insert into the ribs (*arrowheads*). Day 6. **B** Longitudinal section of the lung showing costal sulci (*dashed circles*). *Arrows* Developing parabronchi; *arrowheads* secondary bronchi. Day 9. **C** Lateral surface of the lung showing anastomosing parabronchi (*arrows*). *Arrowheads* Costal sulci. Day 10. **D** Lateral surface of the lung showing well-developed parabronchi (*arrows*) and costal sulci (*arrowheads*). Day 13. (Maina 2003b)

oping lung. From day 9, secondary bronchi (SB) progressively sprout from the various aspects of the luminal (circumferential) surface of the IPPB and radiate outwards (Figs. 7B–D and 8A–F). The IPPB terminates in the formative abdominal air sac (AAS; Fig. 8D,F–H). Inaugurating as solid cords of epithelial cells (Fig. 9A), parabronchi (PR; tertiary bronchi) form. As the PR lengthen, they develop a lumen that is encircled by epithelial cells (Fig. 9B,C). External to the developing PR are cytoarchitecturally heterogeneous mesenchymal cells (Mc, Fig. 9D). Successively, the PR grow and anastomose pro-

2.2 Bronchial (Airway) System

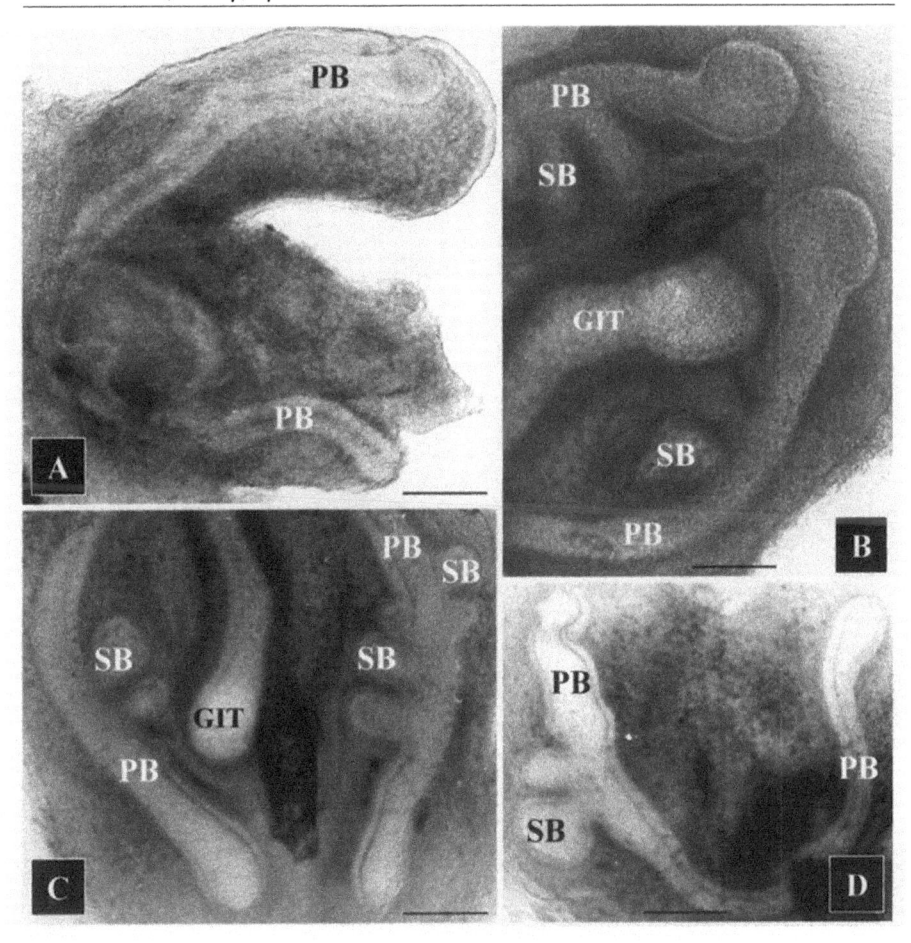

Fig. 7A–D. Embryonic avian lungs grown on matrigel rafts at day 7 of development (stage 28). *PB* Primary bronchus; *SB* secondary bronchus; *GIT* gastrointestinal system. *Scale bars* 0.1 mm. J.N. (Maina and B.K. Kramer, unpubl. observ.)

fusely as they interconnect the SB (Figs. 10 and 11), ultimately forming the bulk of the lung. Initially, columnar epithelial cells line the parabronchial lumina (PRL, Fig. 12A–C). The cells attach onto a basement lamina (BL, Fig. 12B,C). Mesenchymal cells affix onto the outer surface of the BL (Fig. 12B,C). In due course, epithelial cells transform and organize into cords that delineate the atria (Fig. 11D), out pocketings from the luminal surface of the PR (Figs. 11D and 12D). Initially, the epithelial cells that line the PRL join across laterally located bridge-like extensions (Fig. 13A), large intercellular spaces exist between the epithelial cells that project prominently into the PRL (13A,B). As the atria form, they give rise to a variable number of infundibulae (Figs. 14 and 15). The atria are conspicuous on day 15, the infundibulae on day

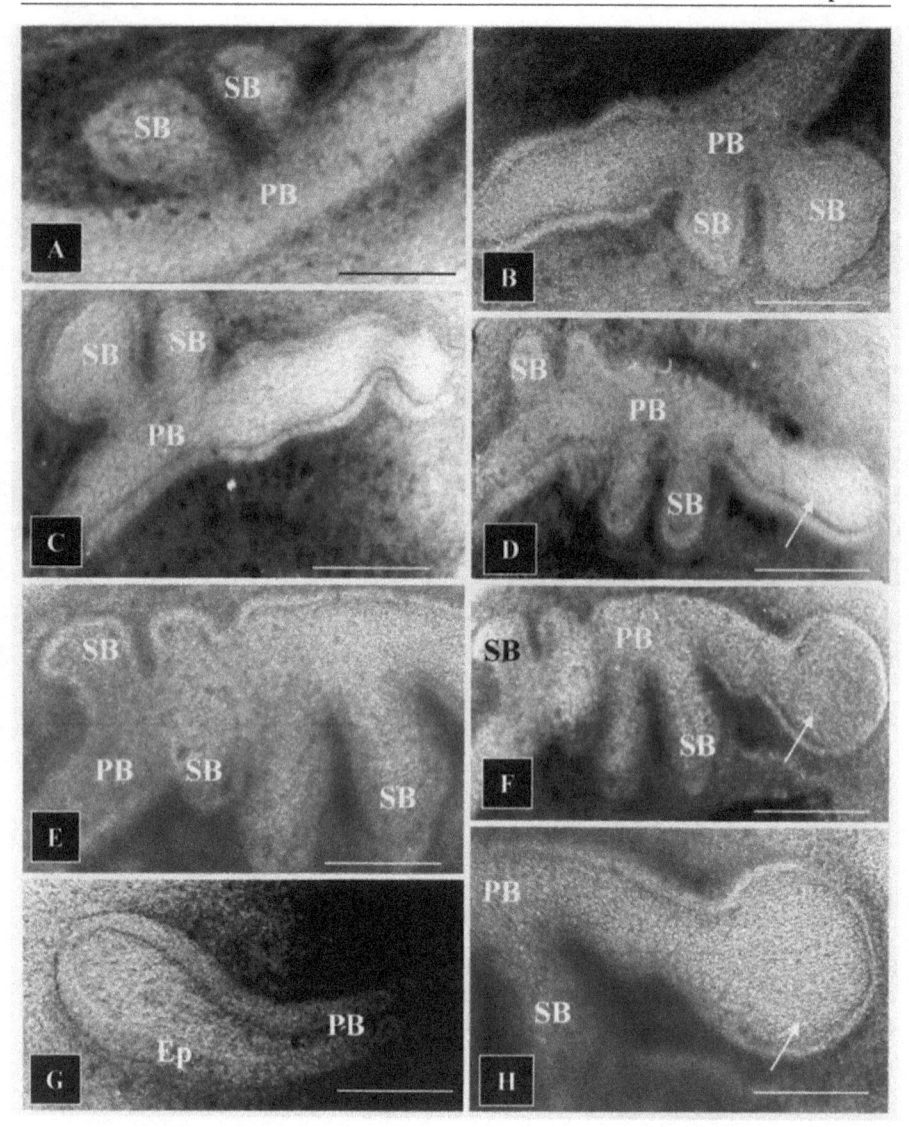

Fig. 8A-H. Embryonic avian lungs grown on matrigel rafts at day 10 showing a primary bronchus (*PB*) giving rise to secondary bronchi (*SB*). The expansions at the ends of the primary bronchus (*arrows*) may give rise to air sacs. *Ep* Epithelium. *Scale bars* 0.5 mm. J.N. (Maina and B.K. Kramer, unpubl. observ.)

2.2 Bronchial (Airway) System

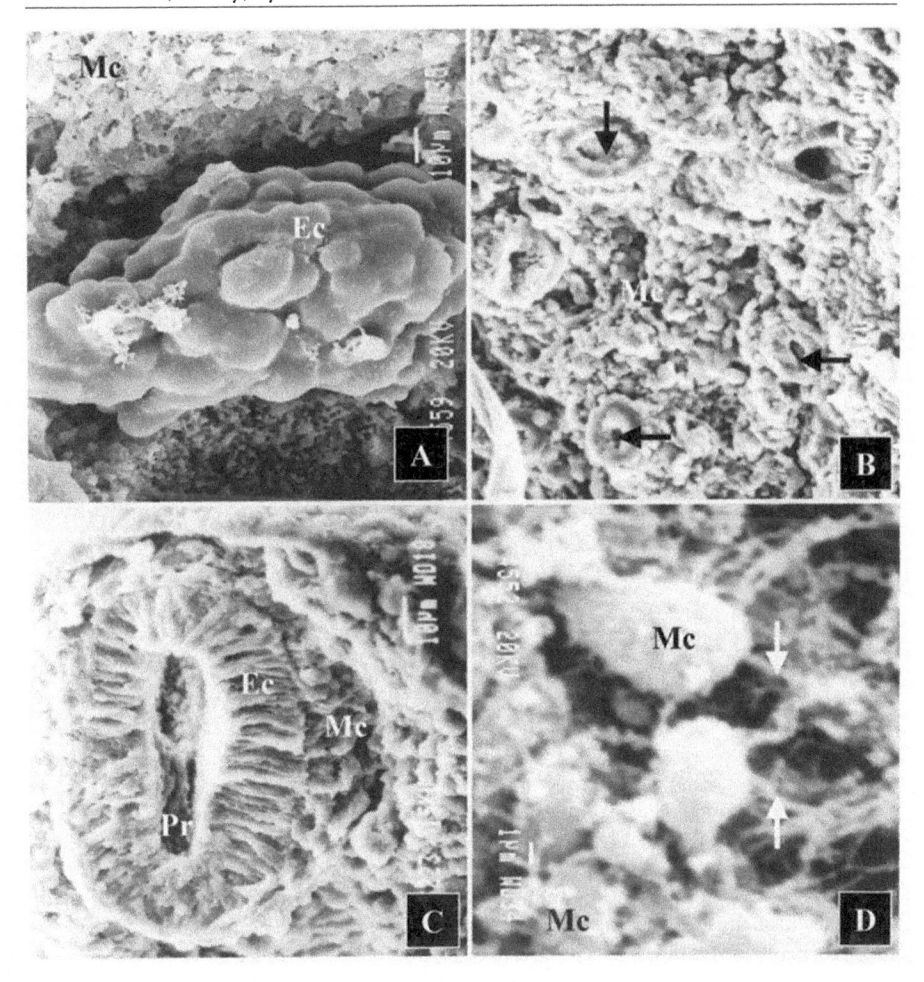

Fig. 9. A Cluster of epithelial cells (*Ec*) surrounded by mesenchymal cells (*Mc*). Day 9. **B** Parabronchi (*arrows*) with a formed lumen. *Mc* Mesenchymal cells. Day 11. **C** Close-up of a developing parabronchus (*Pr*) surrounded by a columnar epithelium (*Ec*). *Mc* Mesenchymal cells. Day 15. **D** Mesenchymal cells of the stroma of the developing avian lung showing interconnections by thin cytoplasmic extensions (*arrows*). Day 16. (Maina 2003b)

16, and the AC and the blood capillaries (BC) on day 18 (Figs. 16–18). The atria, infundibulae, and AC form by massive remodeling of the mesenchymal cell stroma of the 'undifferentiated' exchange tissue (ET, parenchyma) of the developing lung (Figs. 14–18). In the rat lung, the canaliculi enlarge with condensation of the mesenchyme, a process brought about by programmed cell death (apoptosis; Stiles et al. 2001). Progressively, in the lung of the domestic fowl, the AC approximate the BC, as the thickness of the BGB attenuates. By time of hatching (day 21), the AC and BC have lavishly intertwined with each other and the BGB is remarkably thin (Fig. 18C,D).

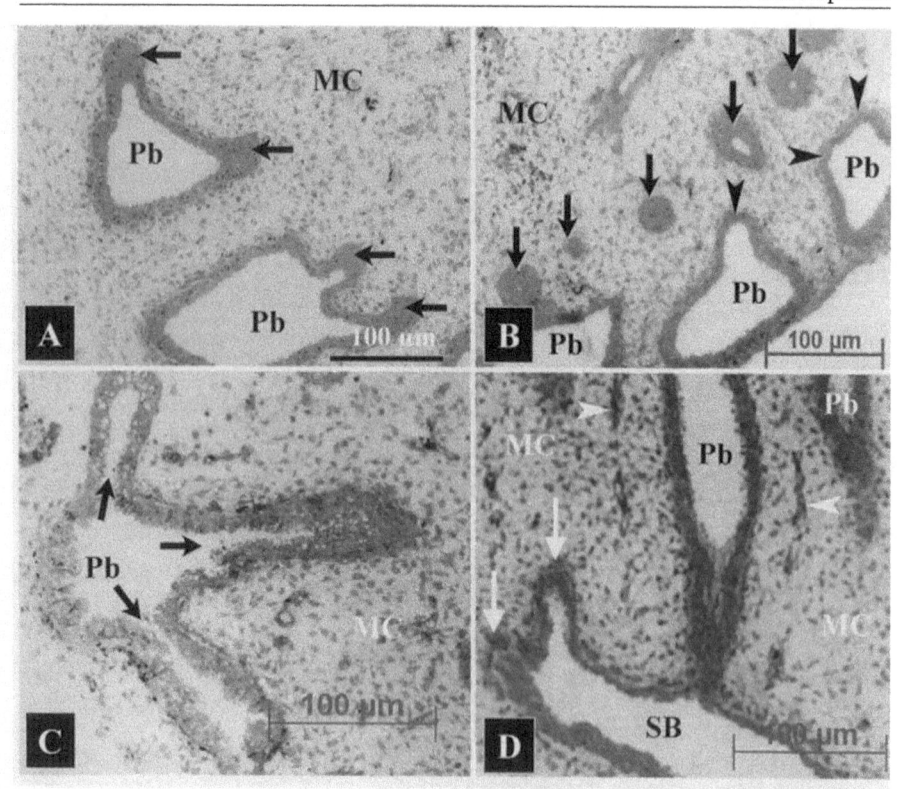

Fig. 10. A Parabronchi (*Pb*) extending (*arrows*) into the surrounding mesenchymal cells (*MC*). Day 15. **B** Parabronchi (*Pb*) at the periphery of the lung budding (*arrowheads*) and projecting inwards (*arrows*). *MC* Mesenchymal cells. Day 16. **C** A parabronchus (*Pb*) branching (*arrows*) and extending into the mesenchymal cells (*MC*). Day 16. **D** Close-up of secondary bronchus (*SB*) from which parabronchi (*arrows*) are sprouting. *MC* Mesenchymal cells; *arrow heads* developing blood vessels. Day 10. (Maina 2003a)

Starting with a narrow lumina and proportionately thick ET (Fig. 19A,C), gradually, the PRL enlarge at the expense of the gas ET (Fig. 19D). In birds like the domestic fowl (galliforms), interparabronchial septa (IPRS), bands of connective tissue that separate PR, form (Fig. 19C). The paleopulmonary parabronchi (PPPR) and the neopulmonary parabronchi (NPPR; Chap. 3.6) are well formed and have intensely anastomosed by day 14 (Fig. 20). The PPPR are hoop-like in orientation and are located dorsal to the primary bronchus (PB), while the NPPR are generally located ventral to the PB and anastomose more profusely (Fig. 20C). Where they exist, the IPRS bestow a rather hexagonal (geodesic) shape to the PR (Fig. 19D). The shape may optimize the packaging of the PR in the lung. In the avian lung, a large number of PR gives a higher volume of ET that in turn should confer more extensive RSA. In human engineering (e.g. French 1988), geodesic design is known to

2.2 Bronchial (Airway) System

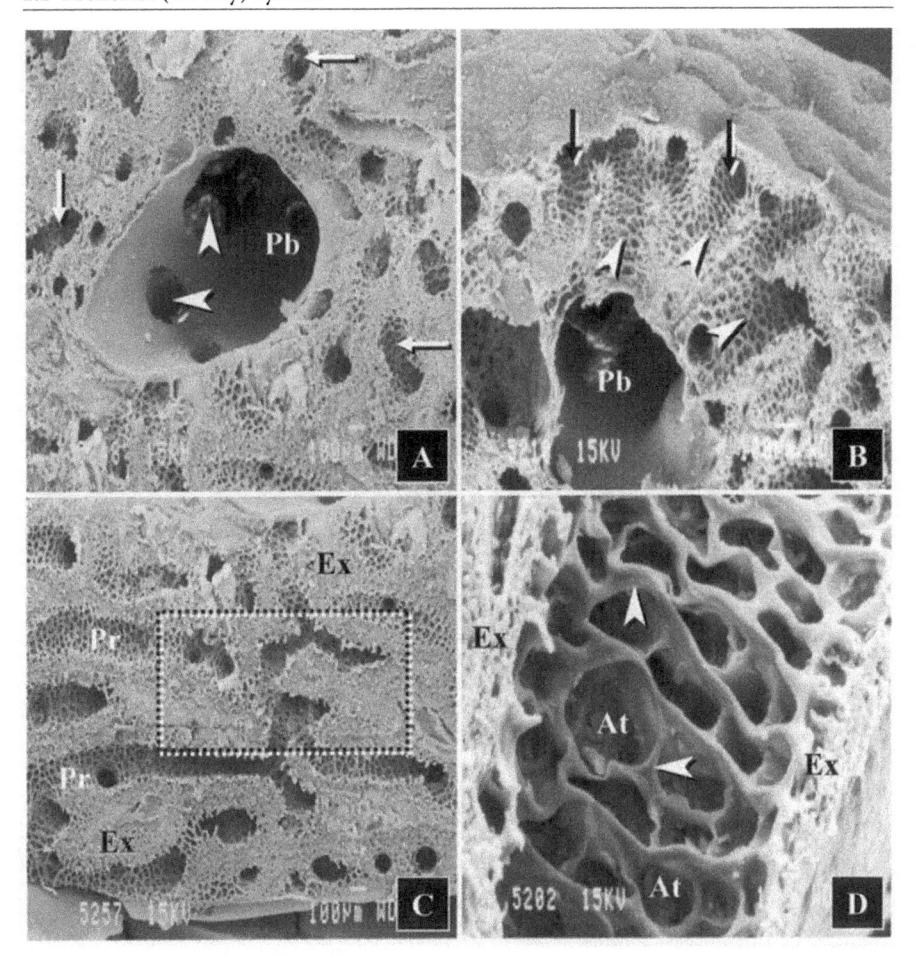

Fig. 11. A Cross section of the primary bronchus (*Pb*) giving rise to secondary bronchi (*arrowheads*) that in turn give rise to parabronchi (*arrows*). Day 15. **B** A primary bronchus (*Pb*) giving rise to secondary bronchi (*arrowheads*) that in turn give rise to parabronchi (*arrows*). Day 16. **C** Parabronchi (*Pr*). *Dashed area* Anastomosing parabronchi; *Ex* exchange tissue. Day 16. **D** Longitudinal view of a parabronchus showing atria (*At*) that are separated by atrial muscles (*arrowheads*). *Ex* Exchange tissue. Day 18. (Maina 2003b)

impart stronger construction. The cells of the bee's wax are perhaps the best known example of exploitation of geodesic design in nature. The avian lung is reported to be very strong: compression of the lung does not cause significant collapse of the AC (Macklem et al. 1979). In species of birds where IPRS are well formed, the AC and BC of adjacent PR do not communicate. The granular pneumocytes (type II cells) are conspicuous from day 17 of development (Figs. 15B–D and 16B). Thereafter, the surfactant should appear on the respiratory surface.

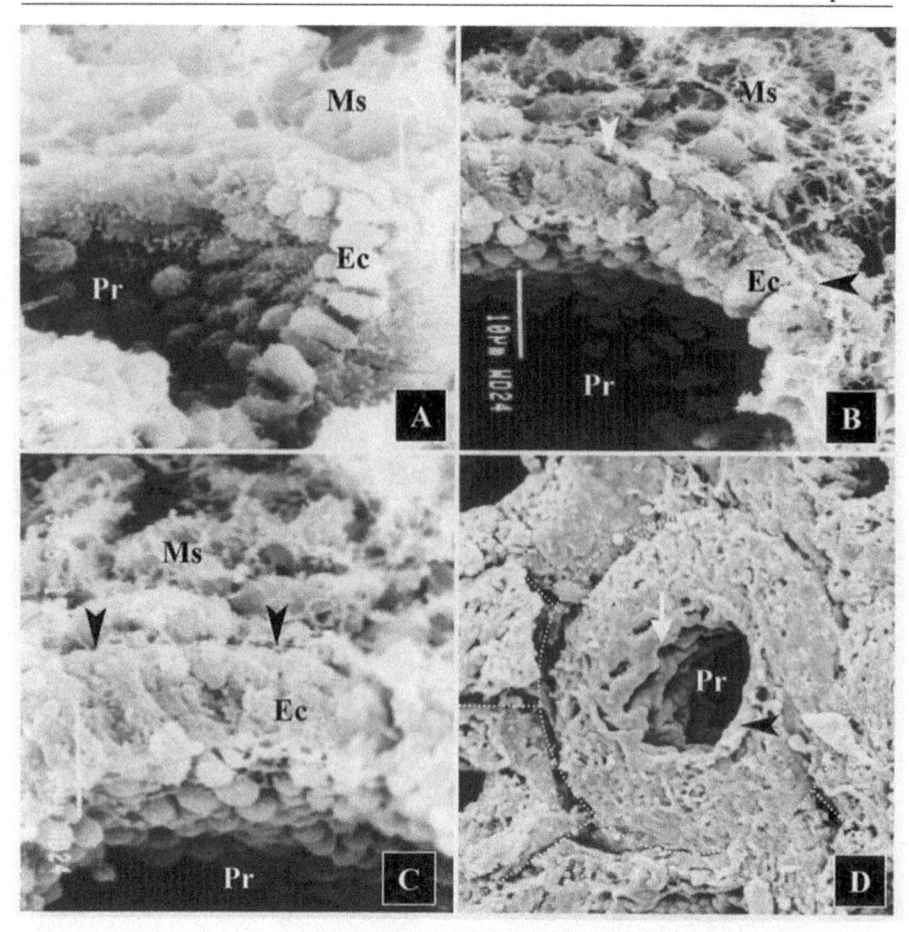

Fig. 12. A Parabronchus (*Pr*) lined by an epithelium (*Ec*) that comprises of columnar cells. *Ms* Mesenchymal cells. Day 12. **B, C** Parabronchi (*Pr*) surrounded by an epithelium (*Ec*) that comprises simple columnar cells. *Arrowheads* Basement lamina; *Ms* mesenchymal cells. Day 13. **D** Cross section of a parabronchus (*dashed circle*) showing developing atria between atrial muscles (*arrows*). *Arrowhead* The parabronchial epithelial lining from which the atrial muscles develop; *dashed lines radiating from the dashed circle* positions of the inter-parabronchial septa that separate adjacent parabronchi. Day 14. (Maina 2003b)

In the BALu substantial development, growth, and remodeling of the terminal airways and alveoli occur postnatally, with the BGB thinning and the RSA extending at the end of the embryonic life and for a certain time postnatally (e.g. Burri and Weibel 1977; Schittny and Burri 2004). Moreover, the degree of lung development at birth varies greatly among mammalian species. The marsupial quokka wallaby, *Setonix brachyurus*, e.g., is born with the lung at the canalicular stage of development (Makanya et al. 2001) and the highly neotenic naked mole rat, *Heterocephalus glaber*, carries a rela-

2.2 Bronchial (Airway) System

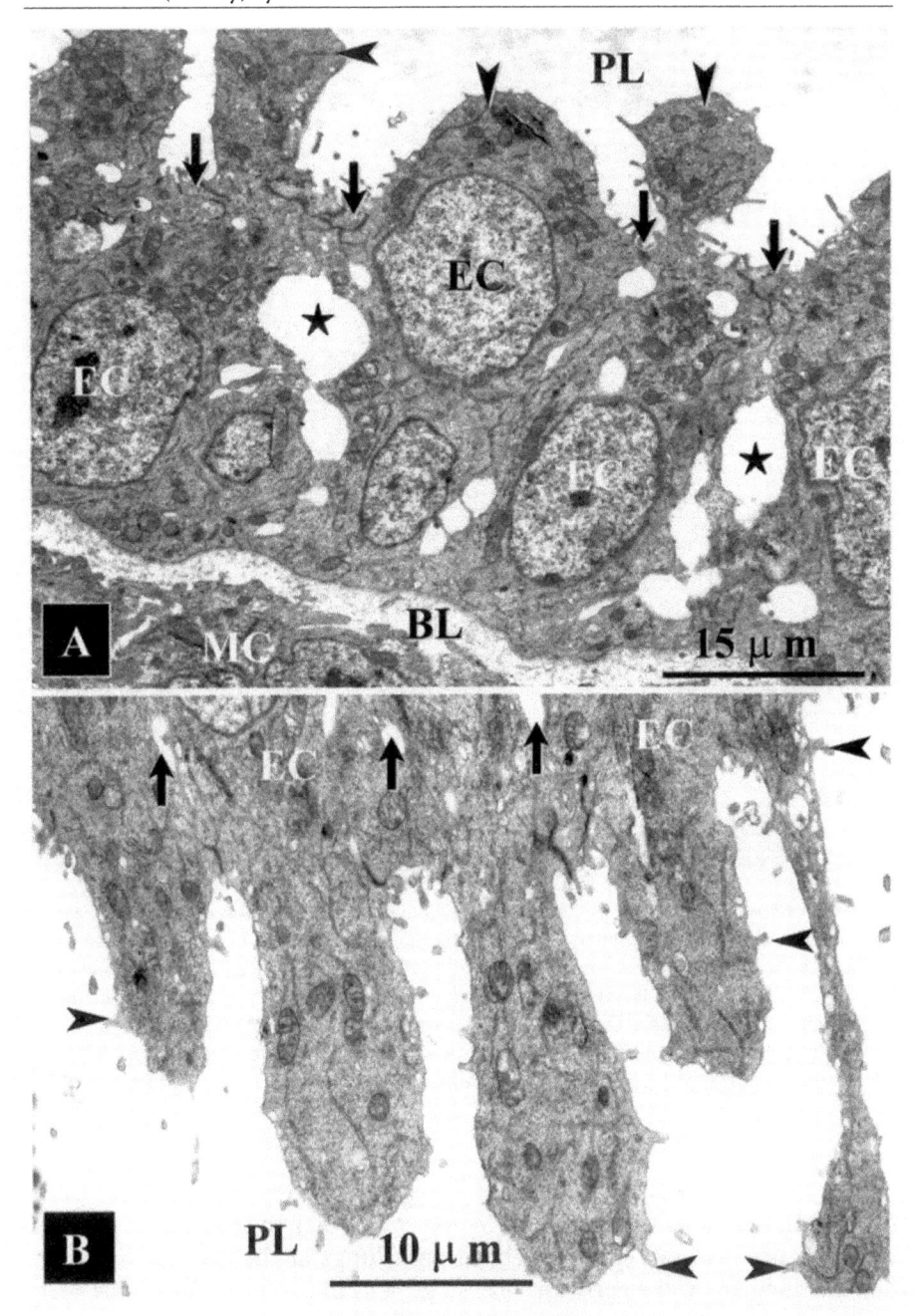

Fig. 13. A Epithelial cells (*EC*) lining the lumen of a developing parabronchus. The apical aspects (*arrowheads*) project deeply into the parabronchial lumen (*PL*). The cells are interconnected through laterally located intercellular junctions (*arrows*). *Stars* Intercellular spaces; *BL* basement lamina; *MC* mesenchymal cells. Day 15. **B** Close-up of the apical aspects of the epithelial cells (*EC*) extending into the parabronchial lumen (*PL*). *Arrowheads* Microvilli; *arrows* intercellular spaces. Day 16. (Maina 2003a)

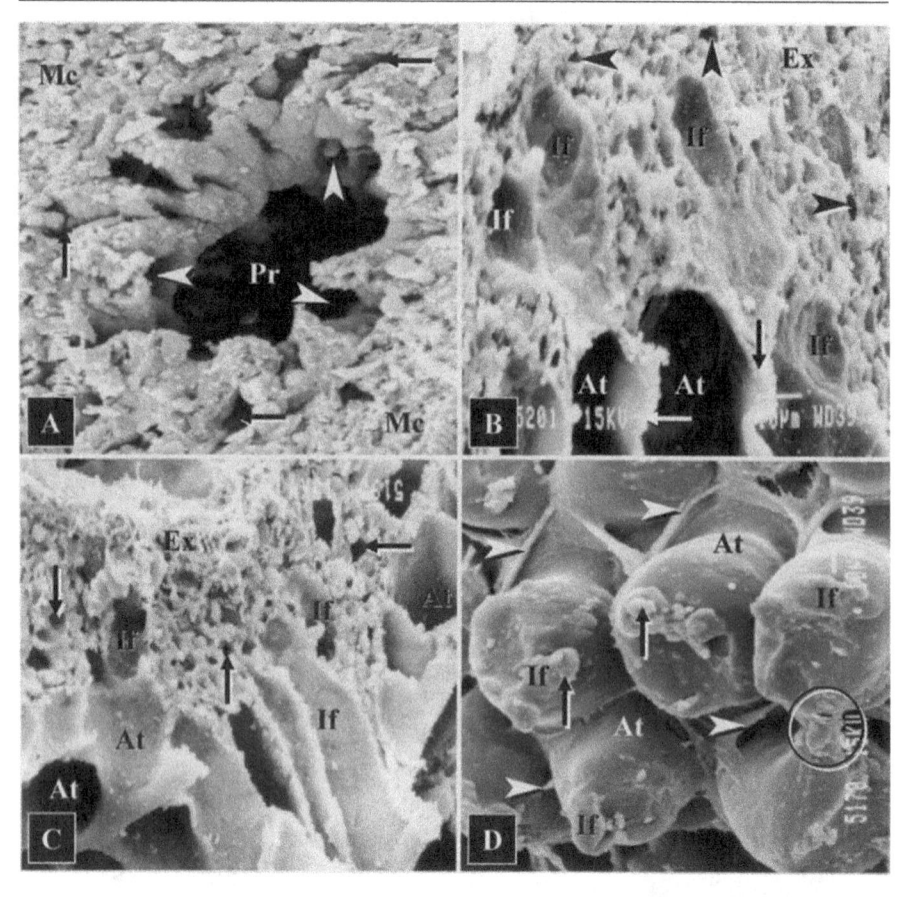

Fig. 14. A Cross section of a developing parabronchus (*Pr*) in which atria (*arrowheads*) have formed. *Arrows* Infundibulae; *Mc* mesenchymal cells. Day 15. **B** Atria (*At*) separated by atrial muscles (*arrows*) giving rise to infundibulae (*If*). *Ex* Exchange tissue; *arrowheads* developing air capillaries. Day 16. **C** Infundibulae (*If*) giving rise to air capillaries (*arrows*). *Ex* Exchange tissue; *At*, atria. Day 17. **D** Atria (*At*) forming infundibulae (*If*) that in turn give rise to air capillaries (*arrows*). *Arrowheads* Strut-like muscular structures that may support the developing atria; *circle*, a site where adjacent atria anastomose. Day 19. (Maina 2003b)

tively immature lung to adulthood (Maina et al. 1992). In sensorial mammals such as mice and rats, pups are born with lungs at the saccular stage, in humans they are at the early alveolar stage, and in precocial mammals such as sheep they are at the late alveolar stage (Schittny and Burri 2004). On the whole, the PRLu is well developed and functionally competent at the end of the incubation period (Duncker 1978; Maina 2003a,b). The structural components, especially the AC and the BC, are well formed and reach their greatest number and size, while the BGB is noticeably thin. Thereafter, only streamlining and consolidation of the constitutive components occur: no

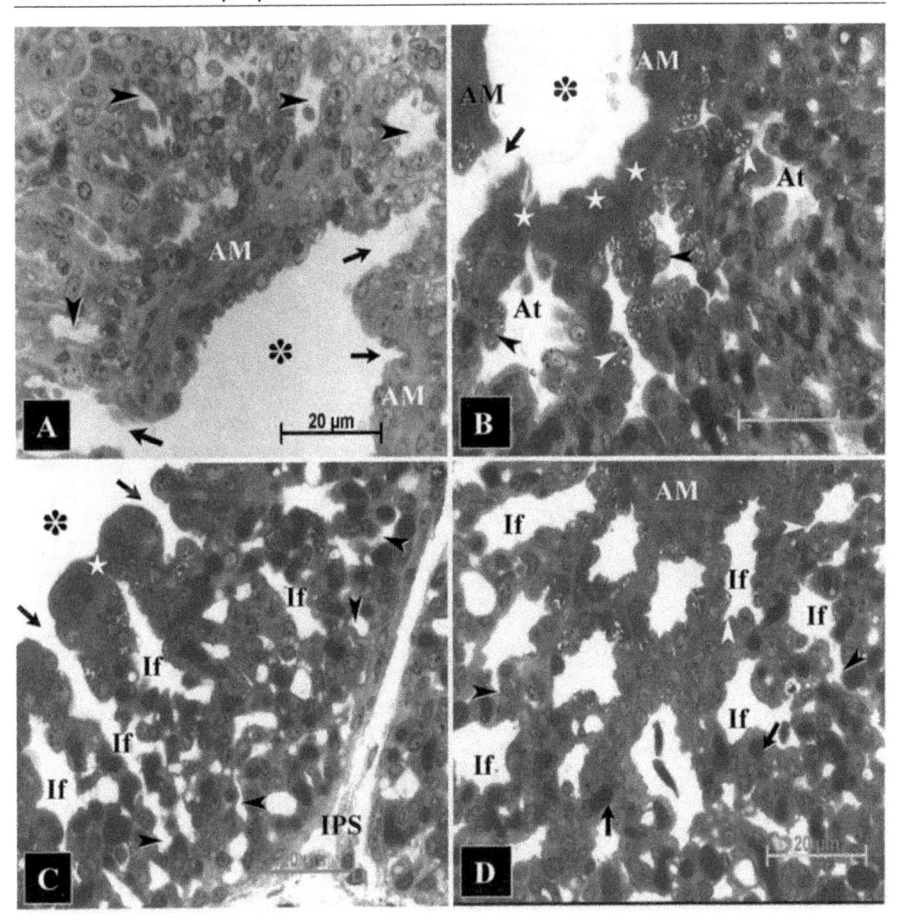

Fig. 15. A A developing parabronchus showing a lumen (*asterisk*) surrounded by atrial muscles (*AM*). *Arrows* Communication between the atria and infundibulae with the parabronchial lumen; *arrowheads*, atria. Day 15. **B** Parabronchial lumen (*asterisk*) surrounded by atrial muscles (*AM*). *Arrow* Connection between an atrium and parabronchial lumen; *stars* sites where atrial muscles interconnect. *At* Atria; *arrowheads* putative type II (granular) pneumocytes. Day 17. **C** Exchange tissue of a parabronchus showing the parabronchial lumen (*asterisk*) and forming atria (*arrows*) leading into the infundibulae (*If*). *Star* Site atrial muscles connect; *arrowheads* developing air capillaries; *IPS* interparabronchial septum. Day 17. **D** Close-up of the exchange tissue showing infundibulae (*If*) giving rise to air capillaries (*arrowheads*). *AM* Atrial muscles; *arrows* red blood cells contained in blood capillaries. Day 17. (Maina 2003a)

important new structures appear after hatching. The final remodeling of the parenchyma occurs through extensive apoptosis, especially during the last 3 days of development. In the mammalian (rat) lung, apoptosis occurs towards the third postnatal week of development (Schittny et al. 1998; Bruce et al. 1999). Duncker (1978) attributed the accelerated embryonic development of

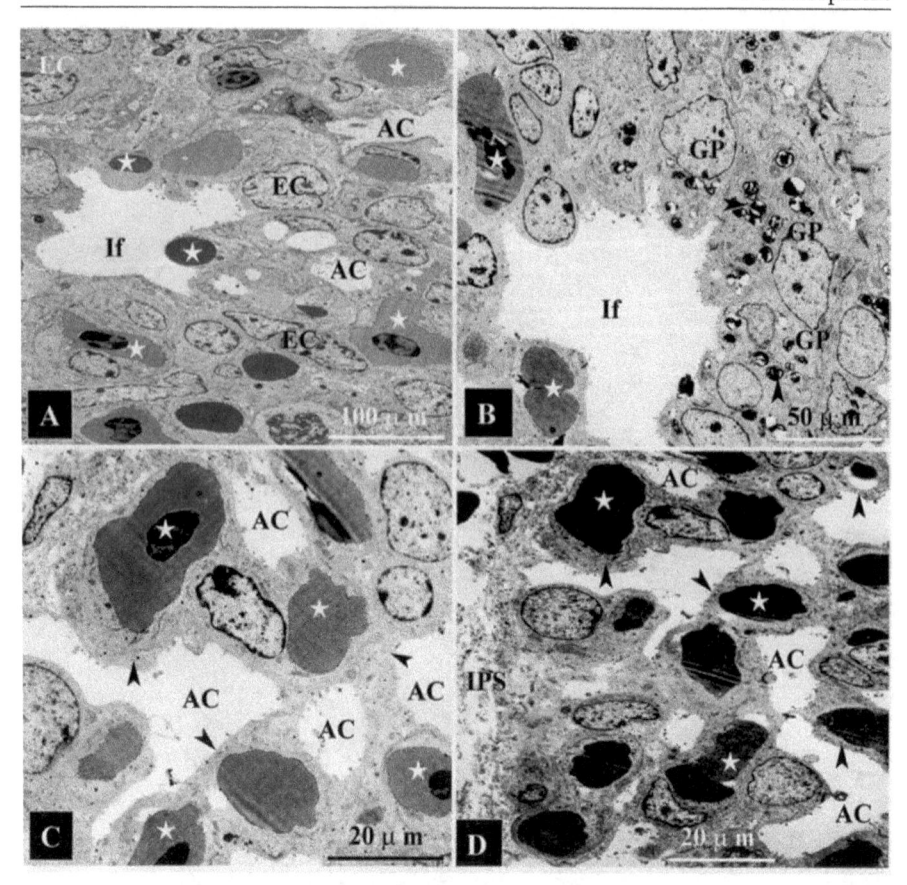

Fig. 16. A A formative infundibulum (*If*) giving rise to air capillaries (*AC*). *EC* Epithelial cells; *stars* red blood cells. Day 16. **B** Close-up of an infundibulum (*If*) with type II (granular) pneumocytes (*GP*) in the surrounding epithelium. *Arrowheads* Osmiophilic lamellated bodies; *stars* red blood cells. Day 17. **C** Exchange tissue showing blood capillaries containing red blood cells (*stars*) surrounded by developing air capillaries (*AC*). *Arrowheads* Conspicuously thick blood-gas barrier (BGB). Day 18. **D** Exchange tissue at the periphery of a parabronchus showing blood capillaries containing red blood cells (*stars*) surrounded by air capillaries (*AC*). The blood-gas barrier (*arrowheads*) is much thinner. *IPS* Interparabronchial septum. Day 20. (Maina 2003a)

the PRLu, particularly during the final days, to mechanical ventilation of the lung, a process that is reported to occur before hatching (e.g. Visschedijk 1968).

Among birds, evident differences exist in the degrees of lung development between altricial and precocial birds (Duncker 1978). In the latter, e.g., the domestic fowl and the herring gull, *Larus argentatus*, of which the chicks receive little or no parental care and the chicks are able to walk and feed soon after hatching, the atria and the gas ET are well developed at hatching. In altri-

2.2 Bronchial (Airway) System

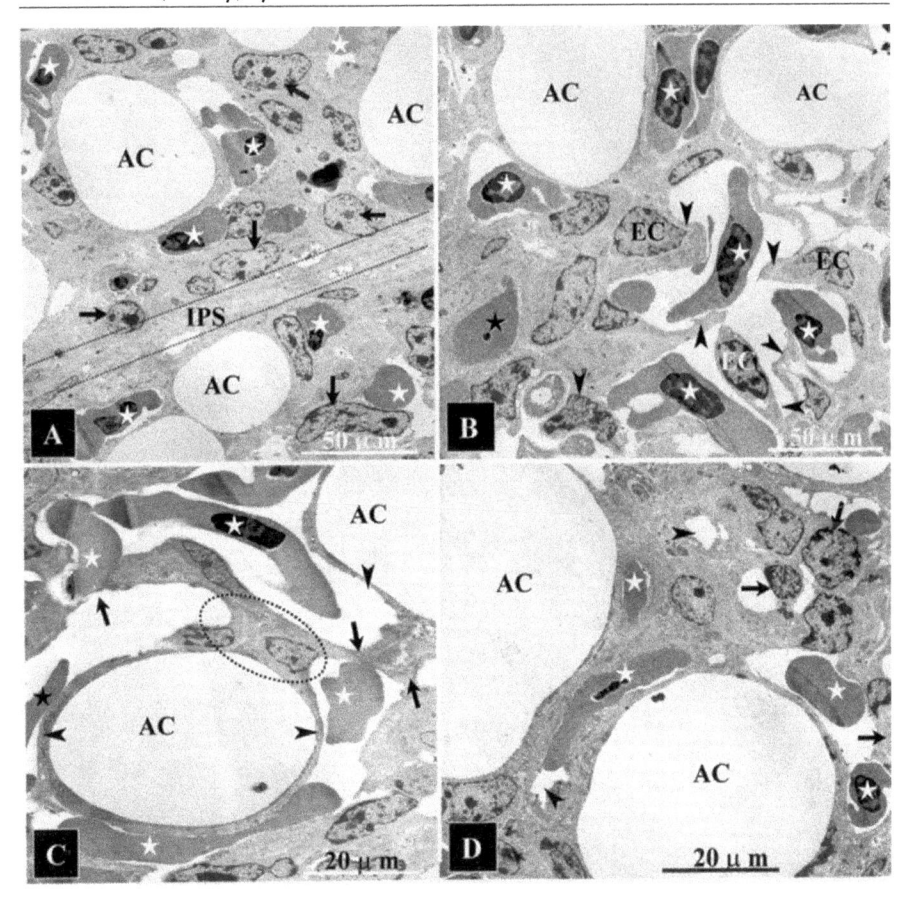

Fig. 17. A Air capillaries (*AC*) surrounded by developing blood capillaries that contain red blood cells (*stars*). *Arrows* Nuclei of forming endothelial cells; *IPS* interparabronchial septum. Day 17. **B** Developing blood capillaries that contain red blood cells (*stars*) extending and surrounding the air capillaries (*AC*). *Arrowheads* Cells and connective tissue elements breaking down and leading to interconnection of the blood capillaries; *EC* endothelial cells. Day 17. **C** Close-up of developing air capillaries (*AC*) and blood capillaries containing red blood cells (*stars*). *Arrows* Sites where connective tissues and cells have broken down to allow connection between the developing blood capillaries; *dashed area* a site where cells are breaking down to lead to an air capillary being totally surrounded by a blood capillary; *arrowhead* formative BGB. Day 18. **D** Air capillaries (*AC*) and developing blood capillaries containing red blood cells (*stars*). *Arrows* Disintegrating mesenchymal cells; *arrowheads*, extension of and formation of vascular spaces. Day 19. (Maina 2003a)

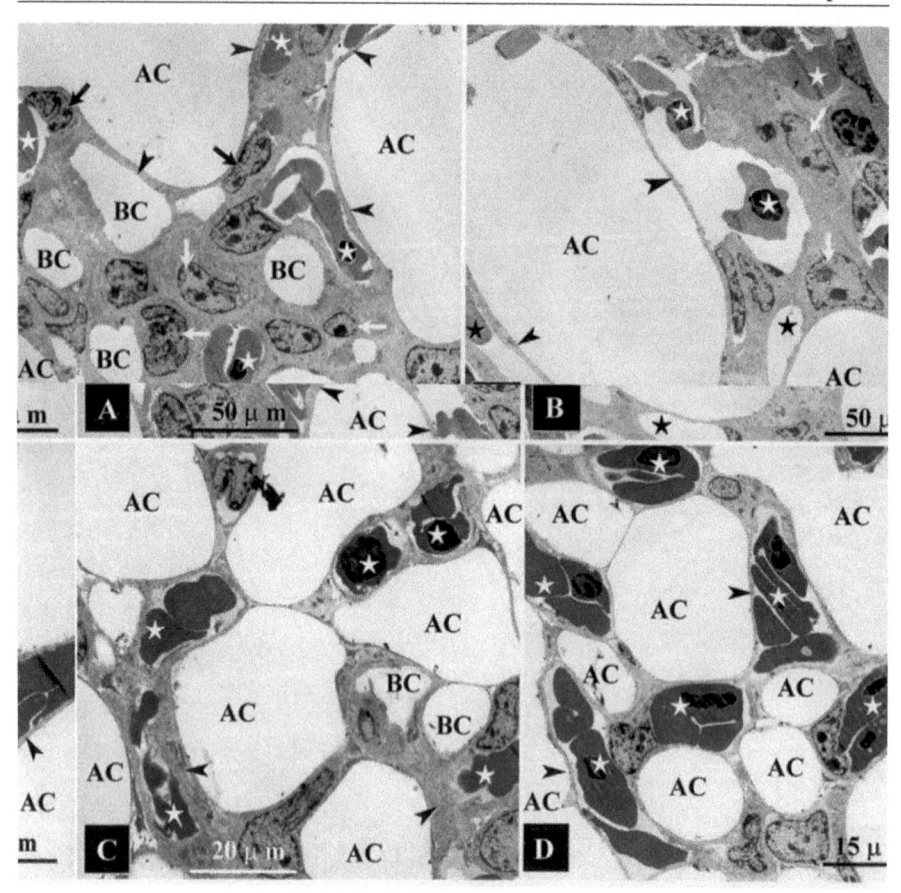

Fig. 18. A Developing blood capillaries (*BC*) with some of them containing red blood cells (*stars*). *AC* Air capillaries; *arrows* formative endothelial cells; *arrowheads* BGB. Day 18. **B** Thin BGB (*arrowheads*) between air capillaries (*AC*) and developing blood capillaries some of which contain red blood cells (*stars*). *Arrows* Formative endothelial cells or disintegrating mesenchymal cells. Day 19. **C** Air capillaries (*AC*) and blood capillaries containing red blood cells (*stars*). *Arrowheads* Thick parts of the BGB. Day 21. **D** Air capillaries (*AC*) and blood capillaries containing red blood cells (*stars*). *Arrowheads* Thin BGB. Day 26. (Maina 2003a)

cial birds, where the chicks are helpless and unable to feed themselves, e.g., the pigeon, *Columba livia*, the PR are small, the atria shallow, and the gas-exchange mantle is thin. While some growth of the PR continues after hatching in both the precocial and altricial birds, the process is more marked in later taxon where the development of the PR is completed just before the birds start to fly. With minimal or no parental care, adequately efficient lungs are essential for the survival of chicks of precocial birds to enable them to run and/or fly in order to escape from predators and procure food.

2.2 Bronchial (Airway) System

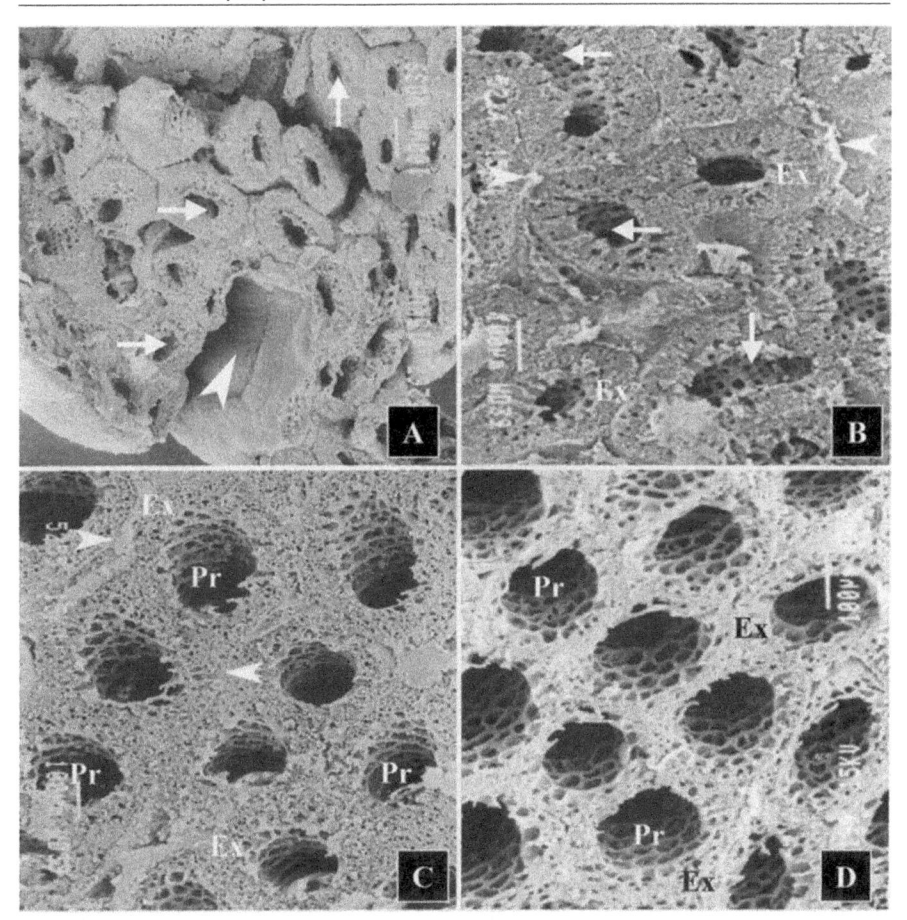

Fig. 19. A Stacks of parabronchi (*arrows*). *Arrowhead* Costal sulcus. Day 15. **B** Developing parabronchi (*arrows*) separated by interparabronchial septa (*arrowheads*). *Ex* Exchange tissue. Day 16. **C** Parabrochi (*Pr*) with parabronchial lumina surrounded by relatively thick exchange tissue mantle (*Ex*). *Arrowheads* Interparabronchial septa. Day 18. **D** Parabronchi (*Pr*) with wide parabronchial lumina and thin gas exchange tissue (*Ex*). Day 26. (Maina 2003a)

On the whole, in the avian lung, a 'tree-like' arrangement where arteries and veins strictly track the airways, as occurs in the mammalian lung (Fig. 21), does not form (e.g. Abdalla 1989). The PR anastomose and interconnect the SB, establishing continuity of the air conduits: there are no cul-de-sacs (blind-ends) in the PRLu. While in the BALu the airway system forms by regular dichotomous bifurcation, with branches following predictable dorsoventral, mediolateral, and proximal-distal trajectories (e.g. Adamson 1997; Cardoso 2001; Schittny and Burri 2004; Fig. 21A,C), in the PRLu the

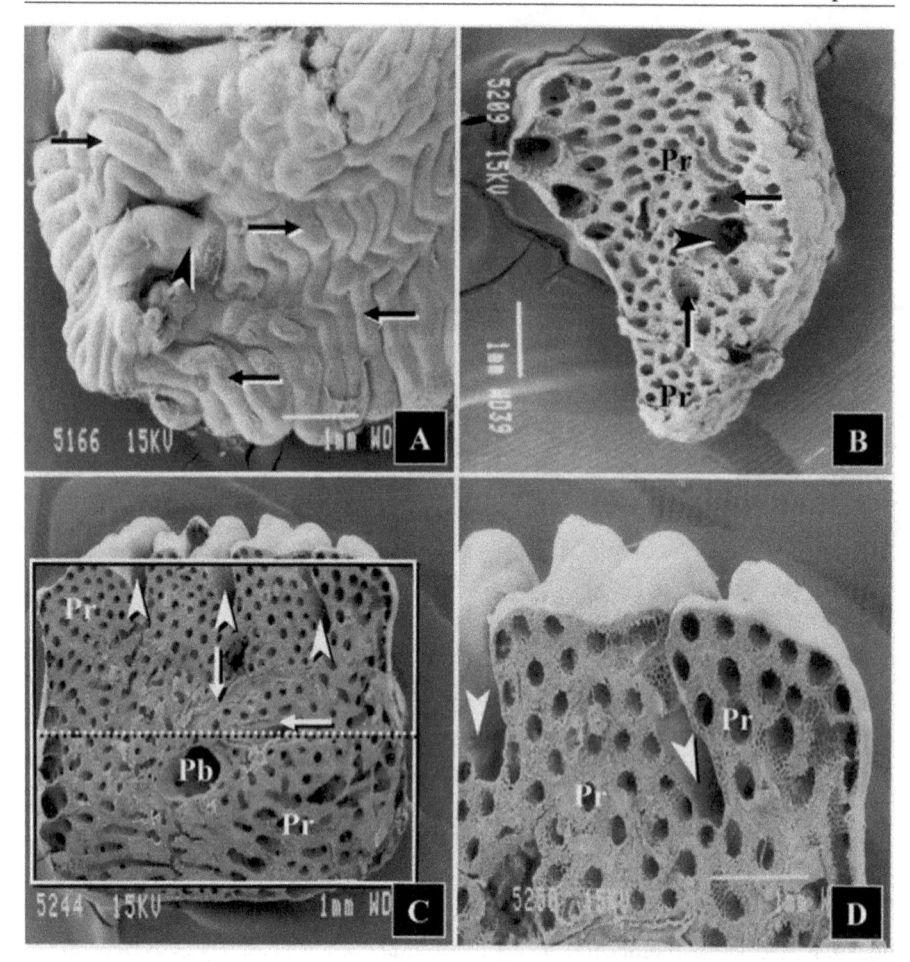

Fig. 20. A Parabronchi (*arrows*) on the craniomedial aspect of the lung. *Arrowhead* First medioventral secondary bronchus. Day 13. **B** Transverse section of the lung showing a primary bronchus (*arrowhead*) giving rise to secondary bronchi (*arrows*). *Pr* Parabronchi. Day 15. **C** Longitudinal section of the lung showing a primary bronchus (*Pb*) giving rise to secondary bronchi (*arrows*). The *dashed line* separates the paleopulmonic region of the lung (*above*) and the neopulmonic one (*below*). *Pr* Parabronchi; *arrowheads* costal sulci. Day 15. **D** Craniodorsal aspect of the lung showing deep costal impressions (*arrowheads*) and paleopulmonic parabronchi (*Pr*) separated by the costal sulci. Day 16. (Maina 2003a)

IPPB runs through the lung (Figs. 7 and 8), giving rise to SB and PR that interconnect the SB (Figs. 9 and 20): a continuous loop of the airway system exists. This design is fundamental to the back-to-front through-flow of air (continuous and unidirectional ventilation) in the avian lung that is effected by the concerted action of the air sacs (e.g. Fedde 1980) and an inherent aerodynamic valving mechanism (Chaps. 3.9 and 3.11).

2.2 Bronchial (Airway) System

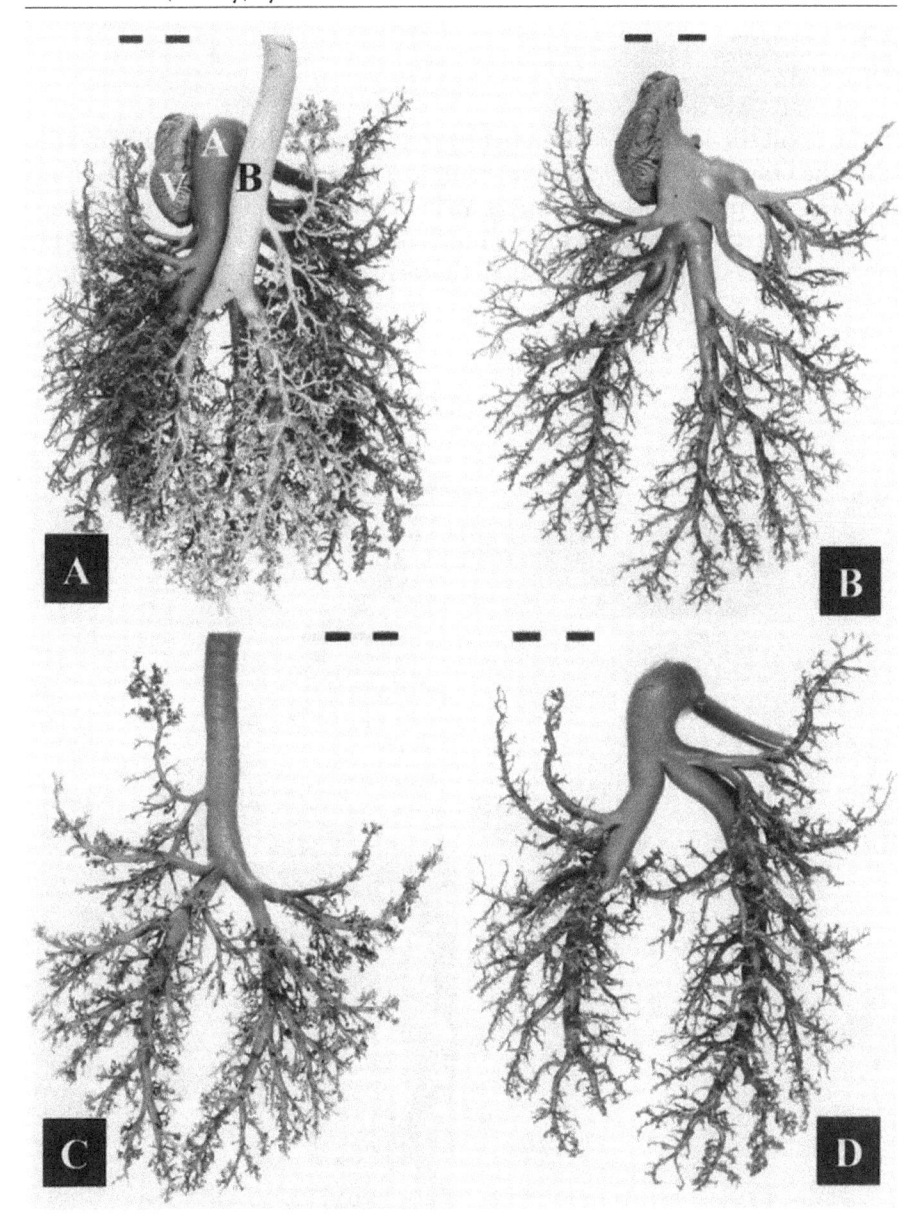

Fig. 21A-C. Triple latex cast preparations of the lung of the pig, *Sus sucrofa*, showing that the bronchial system (*B*), the arterial system (*A*), and the venous systems (*V*) pattern each other. Respectively, **B**, **C**, and **D** show the venous, bronchial, and arterial systems. *Scale bars* 1 cm. (Maina and van Gils 2001)

2.3
Air Sacs

In the domestic fowl, the air sacs (ASs) first form as blister-like outgrowths from the cranial, ventral, and caudal surfaces of a rather wedge-shaped lung (Fig. 22). The promordia of the AASs form between days 5 and 7 and the cervical ASs (CeASs) from day 6 to 8. In quick succession, between days 7 and 12, the clavicular ASs (ClASs), the cranial thoracic ASs (CrTASs), and caudal tho-

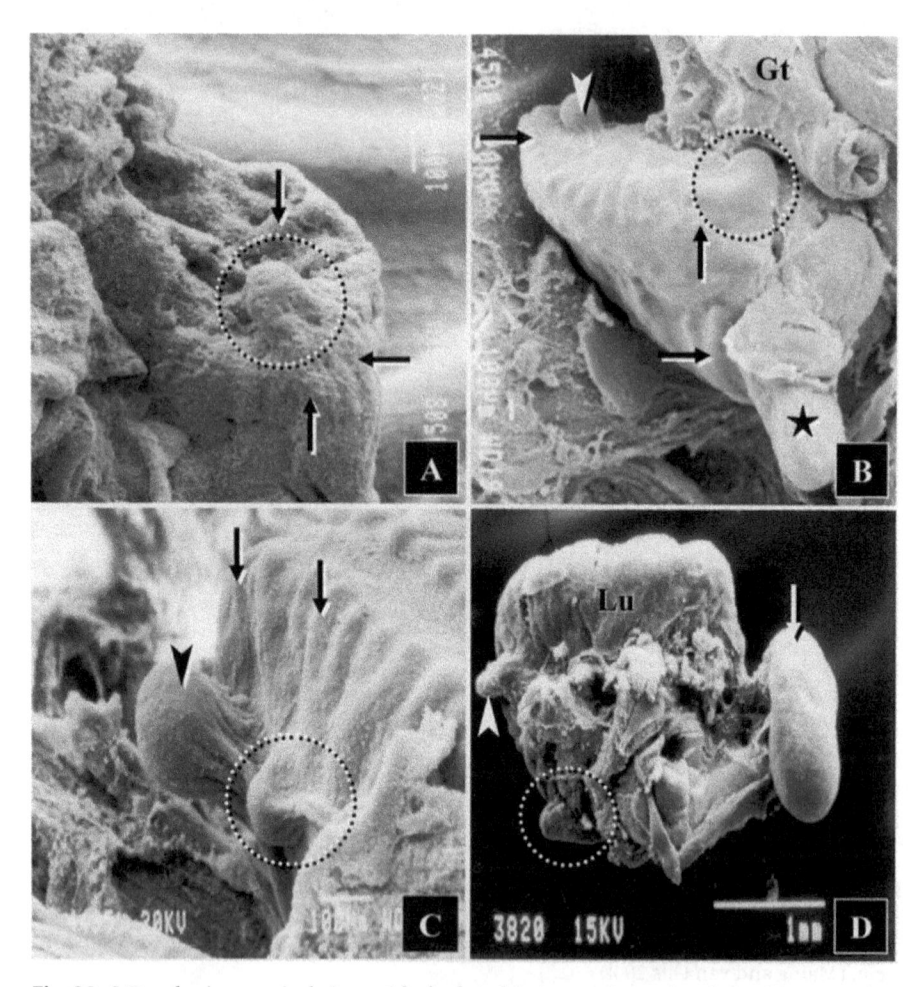

Fig. 22. A Developing cervical air sac (*dashed circle*). *Arrows* Secondary bronchi converging onto the air sac. Day 6. **B** Developing cervical (*arrowhead*), abdominal (*star*), and caudal thoracic (*dashed circle*) air sacs. *Arrows* Secondary bronchi. Day 7. **C** Close-ups of developing cervical (*arrowhead*) and interclavicular (dashed circle) air sacs. *Arrows* Secondary bronchi converging onto the formative air sacs. Day 6. **D** Views of abdominal (*arrow*), cervical (*arrowhead*), and craniothoracic (*dashed circle*) air sacs. *Lu* Lung. Day 8. (Maina 2003a)

racic ASs (CaTASs) appear. The ClASs and the CeASs arise from the cranial edge of the lung, the CrTASs and the CaTASs from the ventral aspect, and the AASs from the caudal edge. Sites where the ASs connect to the PB, SB and PR are called ostia (Chap. 3.5.2). Arising from the terminal end of the developing IPPB (Figs. 7 and 8), the AASs pierce the horizontal septum and extend into the coelomic cavity where they interface with the surrounding visceral structures and in some species radiate to pneumatize adjacent bones. By day 15, the CeASs have extended up the neck and particularly pneumatized the cervical vertebrae. In the domestic fowl, among the entire ASs, from day 16, the AASs are the largest. At hatching, except for the ClASs, the other ASs are paired.

2.4
Pulmonary Vasculature

During embryogenesis, the cardiovascular system is the first to appear. Coordinated growth and specialization of the circulatory system are critical for normal maintenance, growth, differentiation, and function of developing organs and tissues (e.g. Ruberte et al. 2003). Each vascular bend is specialized to meet the specific functional requirements of the organ that it supplies blood to. This makes the vascular system the single most structurally diverse organ system in the body. The morphogenesis of the pulmonary circulatory system, one of the many functionally distinct vascular beds formed during development, has been studied over a long period of time. The growth and branching of the epithelium that are destined to give rise to the airways correlate closely with that of the blood vessels, producing an architectural design that grants optimal exposure of air and blood. The factors/mechanisms involved in the development of the vasculature must provide organizational information as well as growth signals. The close anatomical patterning between the airway and the vascular systems in the developing lung (Fig. 21) has led to the hypothesis that cellular interactions between the epithelium and the mesenchyme control blood vessel formation.

The earliest observation on microvascular growth was ostensibly made in 1853 by Meyer (cited by Hudlicka and Tyler 1986) who discerned spindle-shaped structures arising from blood capillaries in tails of tadpoles, which ultimately became hollow/tubular. Describing the mechanism of formation of vascular plexi from isolated angioblast clusters within the area pelucida of the early chick embryo, Florence Rena Sabin (Sabin 1920) was the first person to empirically characterize vascular morphogenesis. Subsequently, key advances in the study of vascular formation occurred with development of antibodies specific to several angioblast endothelium-specific molecules, QH1 (Pardanaud et al. 1987), SCL/TAL-1 (Kallianpur et al. 1994; Drake et al. 1997), and platelet endothelial cell adhesion molecule-1 (PECAM; Newman and Albeida 1992), as well as quail/chick chimeric analyses (e.g. Poole and Coffin 1991). The mesodermal origin of the blood vessels was first suggested by Hahn

(1909) and Miller and McWhorter (1914). The origin of blood cells and the mechanism by which they enter the lumina of blood vessels have been of long-standing interest. The first step in the formation of the blood vessels (vasculogenesis) is the formation of a subpopulation of motile mesenchymal cells into hemangioblast clusters or blood islands (e.g. Noden 1989; Risau and Flamme 1995; de Mello et al. 1997). The endothelial cells and hematopoietic cells originate from a common mesenchymal precursor cell, the hemangioblast (e.g. Asahara et al. 1997; Choi et al. 1998).

2.4.1
Hematogenesis

Arising from a cytoarchitecturally rather homogeneous group of primordial mesenchymal cells in the embryonic lung of the domestic fowl, hematogenetic cells become conspicuous on day 5 of development (Maina 2003a, 2004a). Thereafter, the cells differentiate and transform into committed erythrogenetic cells (erythroblasts) and leukogenetic cells (leukoblasts). Some of the mesenchymal cells undergo apoptosis: they round-up, disintegrate, the filopodia fold-up, the cytoplasm granulates, and the nucleus becomes peripherally displaced (Fig. 23A,B). In other cells, haemoglobin accumulates in the cytoplasm (Fig. 23C,D). Such cells are immediately surrounded by adjacent mesenchymal cells (Fig. 23D). The leukogenetic cells are spherical (Fig. 24A,B), the nucleus is eccentrically located, and large electron dense inclusion bodies exist in the cytoplasm (Fig. 24C,D). By day 8, many erythrocytes are scattered in the mesenchymal stroma of the developing lung (Fig. 25). Relatively few leukocytic cells are, however, evident. Erythrogenetic cells appear to induce adjacent mesenchymal cells to differentiate into angiogenetic (blood vessel-forming) cells (Fig. 26). The embryonic erythroblasts mitotically divide in the mesenchymal tissue stroma (Fig. 25C,D). After surrounding the erythroblasts, the immediate angiogenetic cells form endothelial cells that line the formative blood vessels (Figs. 26–28), with the outer peripheral cells forming the vessel wall. Cytoarchitecturally, angioblasts possess filopodial extensions and large nuclei with scattered nucleoli (Fig. 29). In some parts of the developing lung, adjacent mesenchymal cells differentiate into arrays of erythrocytes (Fig. 30A–C). The cellular groupings are then surrounded by angioblasts that form an endothelium (Fig. 30B–D). Interestingly, leukocytes do not form cellular continuities nor do they induce mesenchymal cells to differentiate into angioblasts and surround them (Fig. 24A,D).

2.4 Pulmonary Vasculature

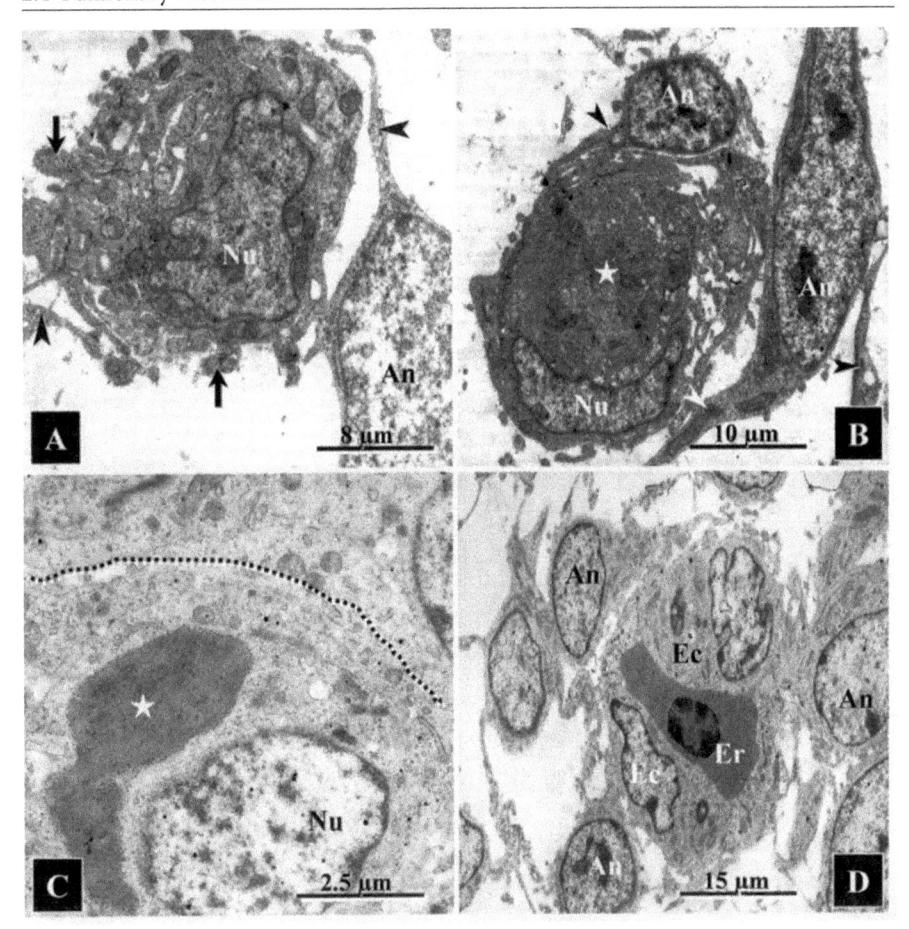

Fig. 23. A A mesenchymal cell undergoing apoptosis: filopodia shorten (*arrows*) and the cytoplasm granulates. *An* Angioblast contacting the cell through filopodia (*arrowheads*); *Nu* nucleus. Day 4. **B** A mesenchymal cell at an advanced stage of apoptosis with dense, granular cytoplasm (*star*) and reduced eccentrically located nucleus (*Nu*). *An* Angioblasts contacting the cell through filopodia (*arrowheads*). Day 5. **C** A mesenchymal cell and accumulating haemoglobin (*star*). *Nu* Nucleus; *dashed line* top edge of the transforming cell. Day 5. **D** A mesenchymal cell with accumulated hemoglobin (erythroblast, *ER*) surrounded by angioblasts that ultimately form endothelial cells (*Ec*). *An* Angioblasts with conspicuous filopodia. Day 6. (Maina 2003a)

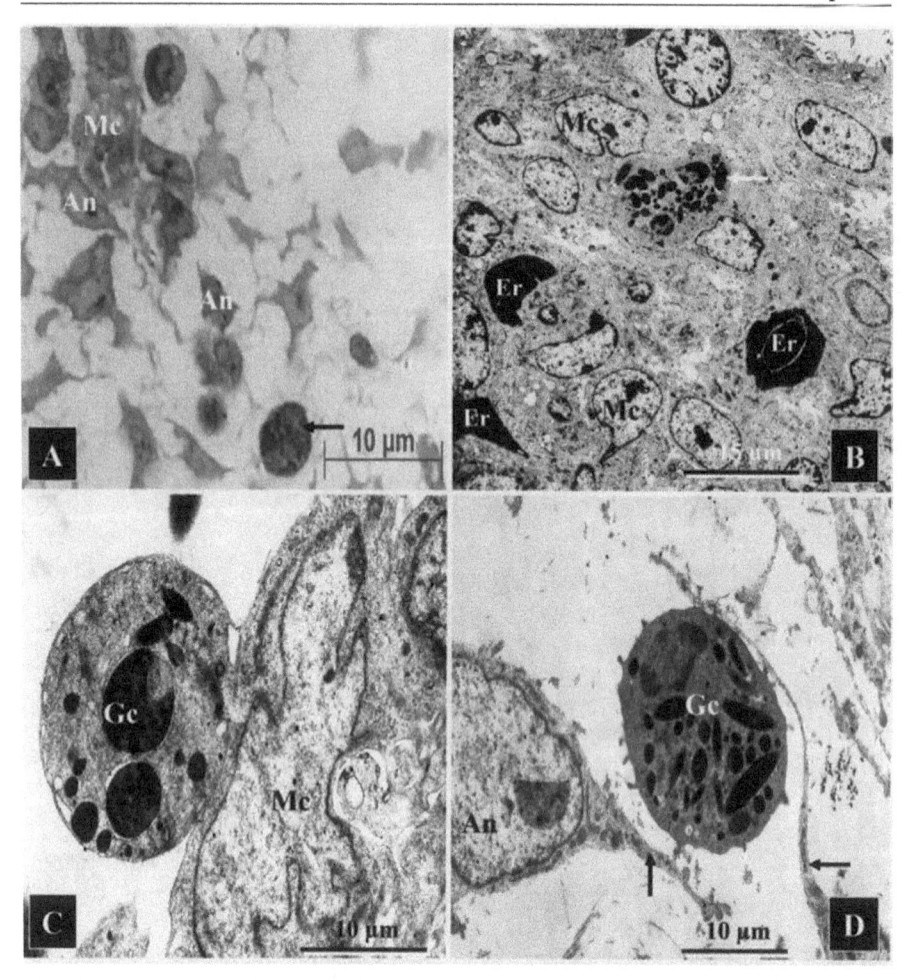

Fig. 24. A A forming granular leukocyte (*arrow*) lying in the midst of angioblasts (*An*) and undifferentiated mesenchymal cells (*Mc*). Day 6. **B** Granular leukocyte (*arrow*) and erythrocytes (*Er*) developing among the mesenchymal cells (*Mc*). Day 14. **C** A granular leukocyte (*Gc*) budding from a mesenchymal cell (*Mc*). Day 11. **D** A granular leukocyte (*Gc*) contacted by angioblasts (*An*) through long filopodial processes (*arrows*). Day 10. (Maina 2003a)

2.4 Pulmonary Vasculature

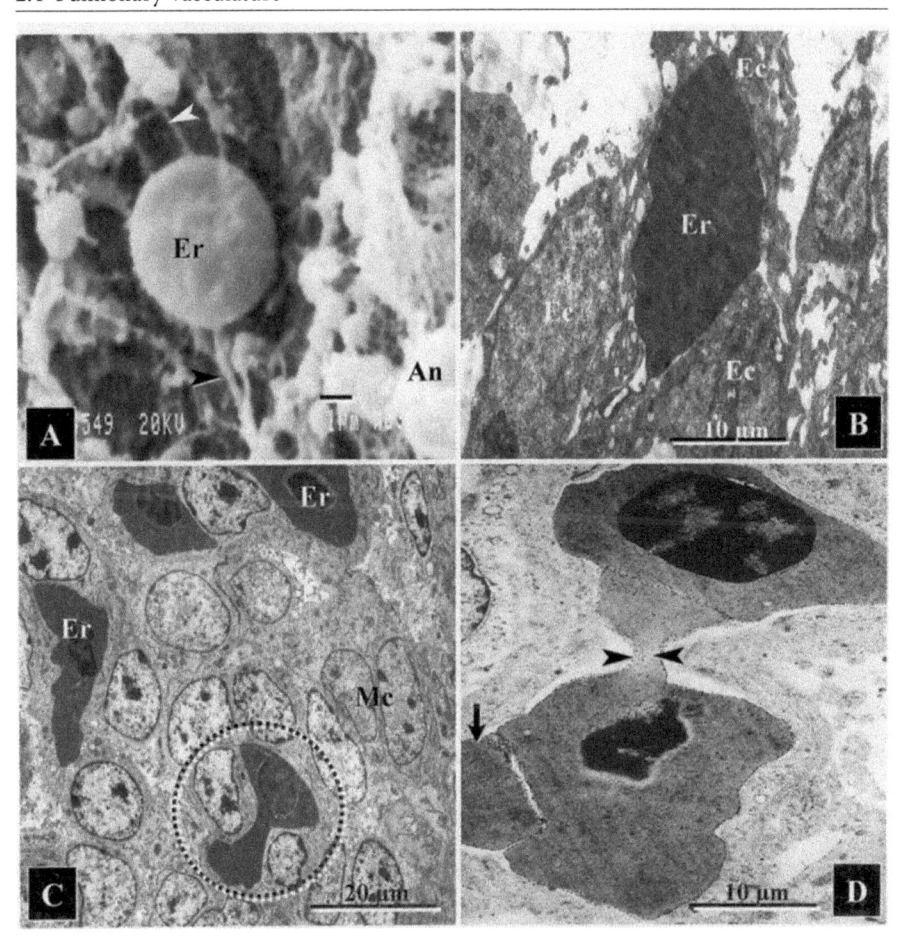

Fig. 25. A, B Erythrocytes (*Er*) surrounded by angioblasts (*An*) which on contacting the erythrocyte form endothelial cells (*Ec*). Day 6. **C, D** Erythrocytes (*Er*) form from mesenchymal cells (*Mc*) and appear to divide (*dashed circle*, C; *arrowheads*, D). The *arrow* in **D** shows an erythrocyte that has separated from another. Day 7. (Maina 2003a)

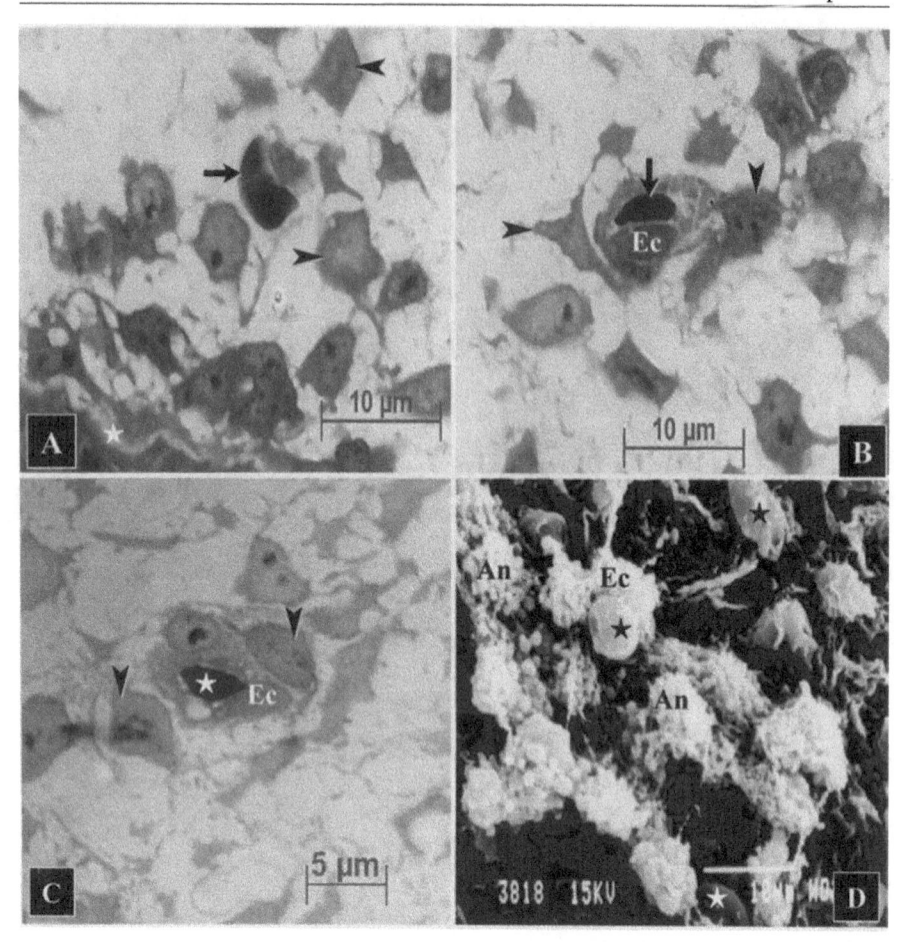

Fig. 26. A An erythrocyte (*Er*) surrounded by angioblasts (*arrowheads*). *Star* Basement lamina of the epithelial cells that surround a parabronchus. Day 9. **B** Erythrocyte (*arrow*) surrounded by an endothelial cell (*Ec*). *Arrowheads* Angioblasts. Day 9. **C** Erythrocytes (*star*) surrounded by an endothelial cell (*Ec*). *Arrowheads* Angioblasts. Day 9. **D** Erythrocytes (*stars*) lying in close proximity to endothelial cells (*Ec*). *An* Angioblasts. Day 11. (Maina 2003a)

2.4 Pulmonary Vasculature

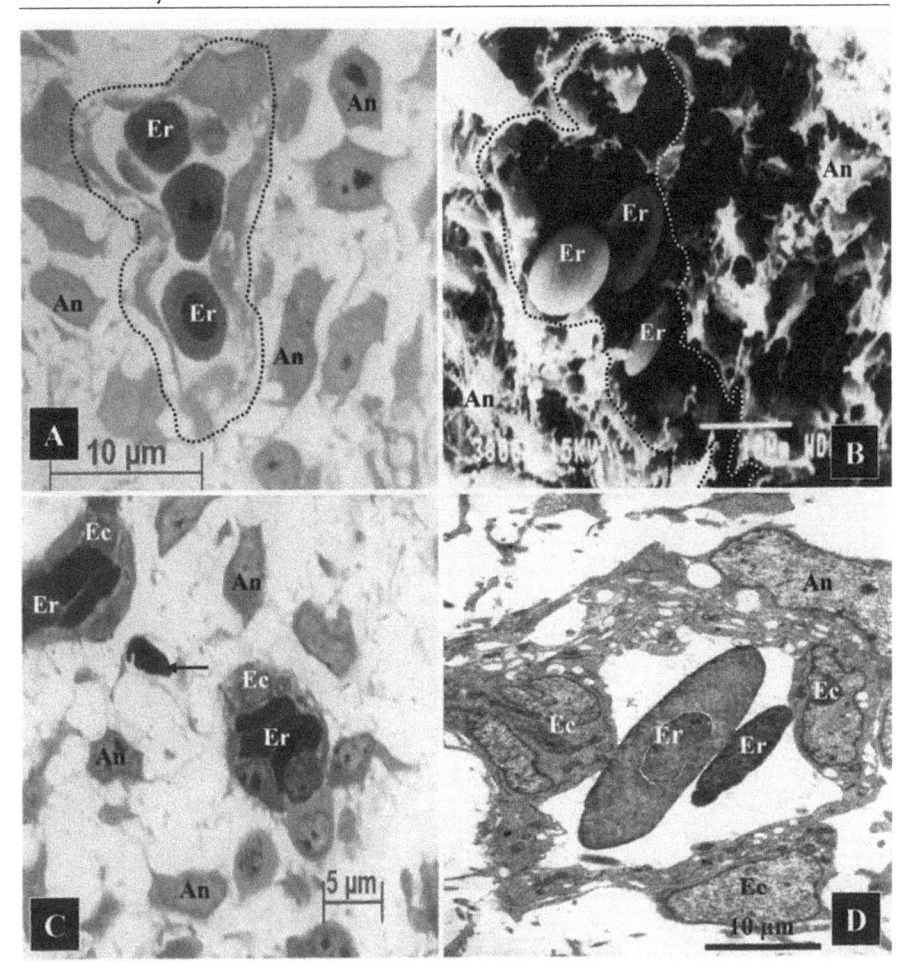

Fig. 27. A, B Blood vessels (*circled*) containing erythrocytes (*Er*) and angioblasts (*An*) lying in close proximity. Day 7. **C** Erythrocytes (*Er*) with endothelial cells surrounding them. *An* Angioblasts; *arrow* an erythrocyte. Day 8. **D** Formative blood vessel with endothelial cells (*Ec*) surrounding erythrocytes (*Er*). Day 8. (Maina 2003a)

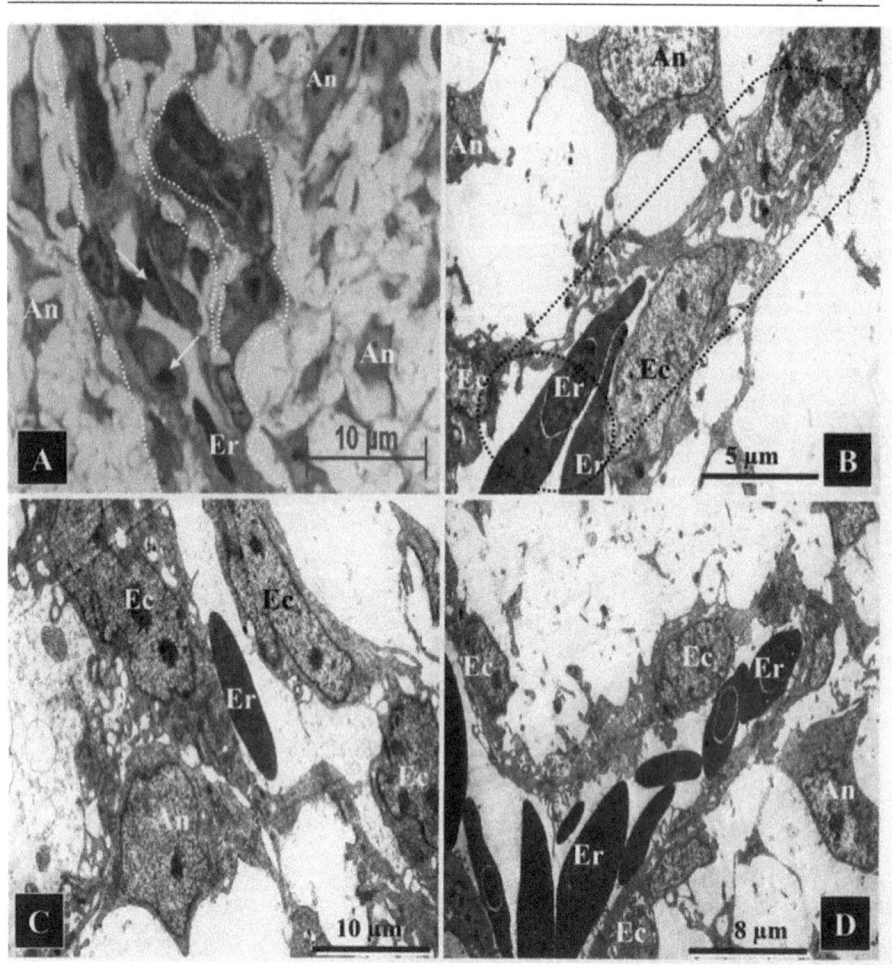

Fig. 28. A Blood vessels developing from aggregation of angioblasts (*An*) that convert into endothelial cells (*arrows*; enclosed *by dashed lines*). *Er* Erythrocytes. Day 14. **B–D** The arrangement of the angioblasts (*An*) determines the direction along which the forming blood vessel elongates (*dashed area*, **B**). *Er* Erythrocytes; *Ec* endothelial cell. Day 14. (Maina 2003a)

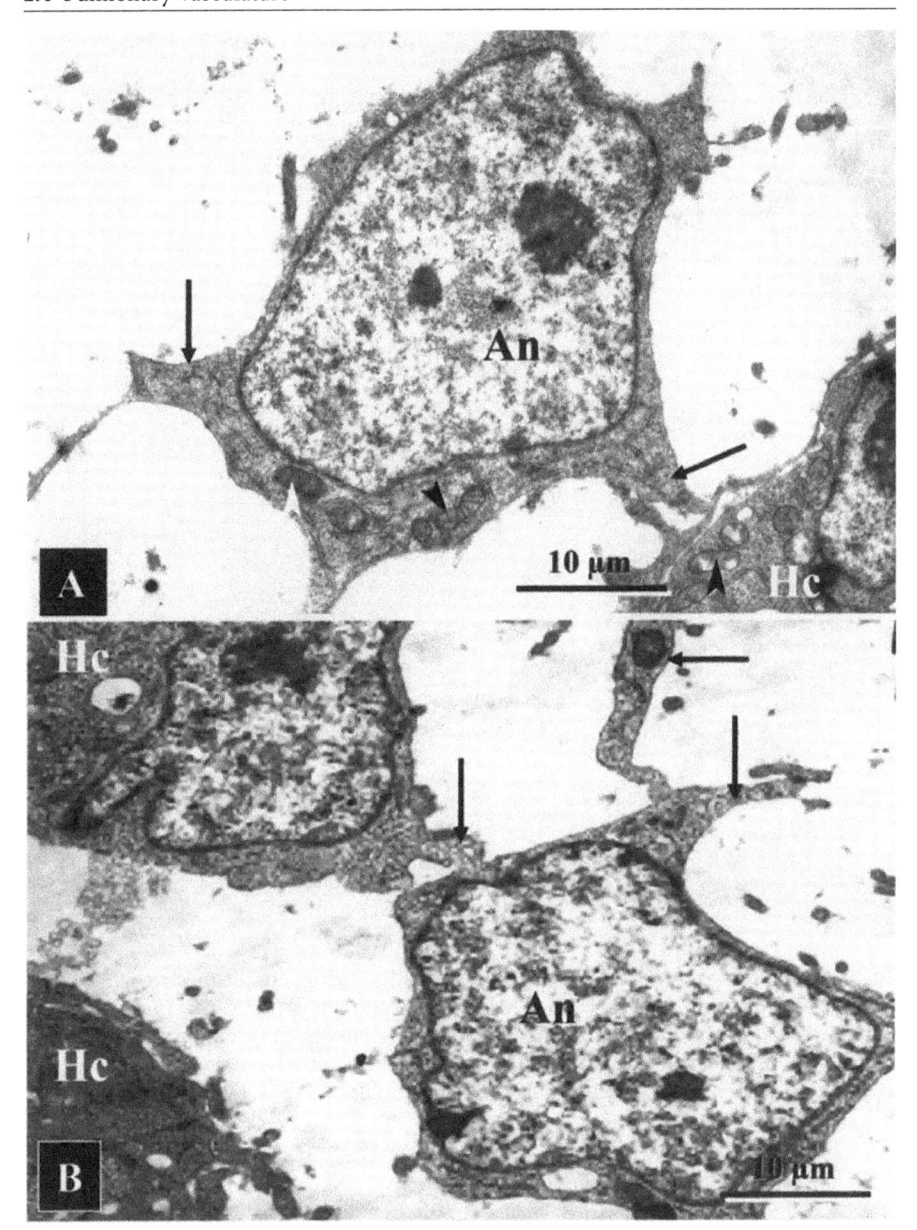

Fig. 29. A, B Angioblasts (*An*) with characteristic filopodia extensions (*arrows*). During the development of the lung, some of the cells (*Hc*) undergo apoptosis. During the process, the filopodia blunt and the cytoplasm granulates. Day 8. (Maina 2003a)

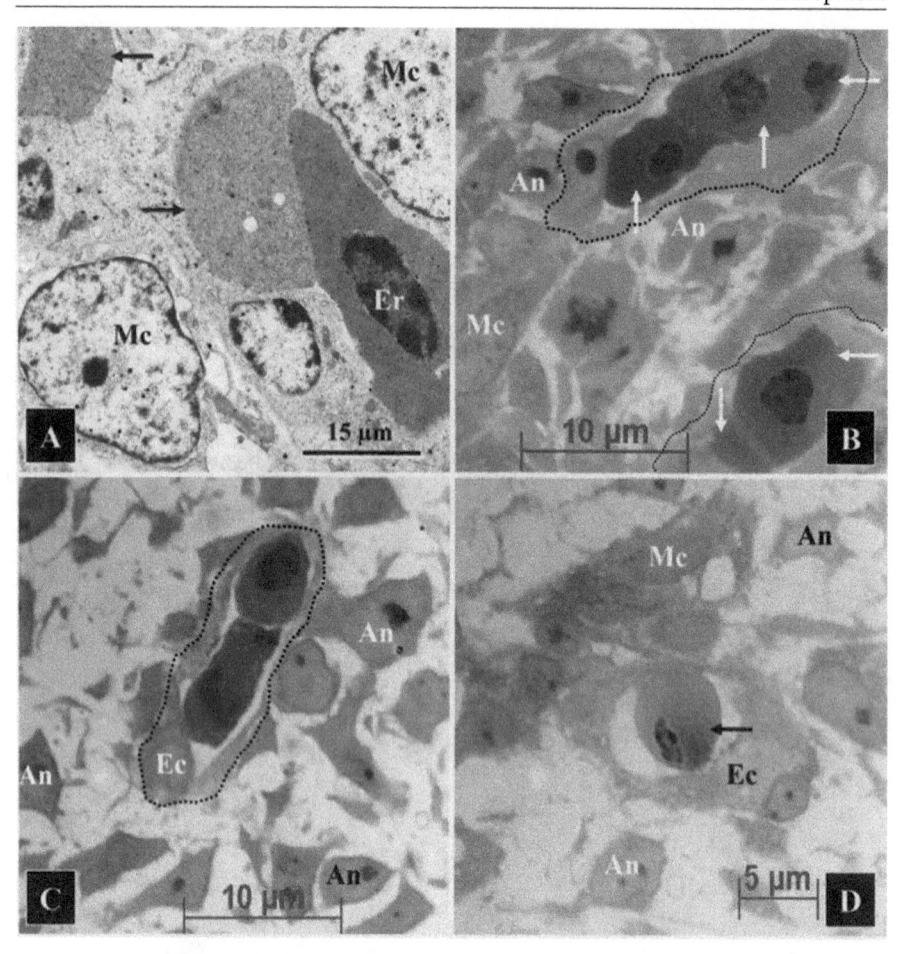

Fig. 30. A Erythrocytes (*Er*) forming by transformation (*arrow*) of the mesenchymal cells (*Mc*) of the developing lung. Day 9. **B, C** Arcades of erythrocytes (*arrows*, **B**) that change to blood vessels (*dashed areas*) once surrounded by angioblasts (*An*) that form endothelial cells (*Ec*). *Mc* Mesenchymal cells. Day 9. **D** An erythrocyte (*arrow*) contained in a blood vessel. *Ec* Endothelial cell; *Mc* mesenchymal cell; *An* angioblast. Day 11. (Maina 2003a)

2.4.2
Vasculogenesis and Angiogenesis

Whilst the organization of the pulmonary circulation in the mature avian lung has been well studied (e.g. Abdalla and King 1975, 1976a,b, 1977; West et al. 1977; Maina 1982a, 1988), and reviewed by Abdalla (1989), the mechanism by which the vascular system is assembled is largely uncertain. Initially, it was postulated by His (1900), among others, that all blood vessels within the embryo were derived from extraembryonic precursor cells. Through programmed budding, branching, and extension, the primordial blood vessels grew and proliferated throughout the whole embryo. That view was, however, later rejected by Evans (1909) and Reagan (1916), among others, who demonstrated that systemic blood vessels originated intraembryonically – not by invasion from the vascular extraembryonic yolk sac. Subsequent studies, e.g., using electron microscopy (e.g. Gonzalez-Crussi 1971; Hirakow and Hiruma 1981) and immunostaining with antibodies specific to primitive hematopoietic and vascular precursor cells (e.g. Coffin and Poole 1988; Pardanaud et al. 1987), further showed that the endothelium of intraembryonic blood vessels does not form by extension of extraembryonic vessels into the embryo but rather that extraembryonic vascularization heralds that of the embryo. Coffin and Poole (1988), Poole and Coffin (1989), and Pardanaud et al. (1989) observed that the cellular mechanisms that occur in vasculogenesis and angiogenesis are different and may be regulated by different molecular factors. Used by His (1900) to describe the nascent mesenchymal cells committed to the endothelial lineage, the word 'angioblast' was adopted by Lillie (1918) to describe the layer of mesoderm in the area opaca of the chick embryo that gives rise to blood islands while Bremer (1912) and Sabin (1920) suggested that the term 'angioblast' should describe the cell type intermediate between mesoderm and endothelium. Presently, there is no single marker available to differentiate an angioblast from an endothelial cell (Healy et al. 2000). It is as a precursor of an endothelial cell (yet incorporated to the endothelial tissue) that the term angioblast is currently used (e.g. Noden 1989; Maina 2004a).

Local environmental conditions together with proximity to certain cellular elements such as the endoderm greatly affect vasculogenesis (e.g. Davis and Bayless 2003). Hirakow and Hiruma (1981) observed that, in the chick embryo, the pattern of vasculogenesis varies according to the developmental stage and the particular body region. Unlike those derived from the somatopleuric mesoderm, intraembryonic angioblasts derived from the splanchnic mesoderm can produce hematogenic cells (e.g. Pardanaud and Dieterlen-Liëvre 1993; Pardanaud et al. 1996). Since the pioneering studies of Sabin (1917, 1920) and Danchakoff (1918), the question of the origin(s) of the vasculoendothelial cells (VECs) and the hematogenetic cells (HGCs), especially erythrocytes, has remained elusive. Moreover, clear categorization of the various angiogenetic (vasoformative) progenitor cells is lacking (e.g. Downs 2003).

In the developing avian lung, the differentiation of the mesenchymal cells into hematogenetic and angiogenetic cells starts as early as day 5 of development (Maina 2003b, 2004b). With a clear definition of the various vasculogenetic cells lacking in the literature, Maina (2003a,b; 2004a) called the committed cells 'angioblasts' and the first of such cells to surround the formative erythrocytes 'endothelial cells'. Investigators like Murray (1932), Pèault et al. (1983), and Lacaud et al. (2001) suggested that a common precursor cell, 'a hemangioblast', produced both VECs and HGCs while Pardanaud et al. (1989, 1996) and Pardanaud and Dieterlen-Lièvre (1993) observed that, in visceral organs, the cells originated differently. Lacaud et al. (2001) considered the hemangioblast to be the precursor cell that produces both the endothelial and blood cells. In the embryo of the eel, *Fundulus heteroclitus*, Stockard (1915) remarked that *"vascular endothelium, erythrocytes, and leukocytes although all arising from mesenchyme are really polyphyletic in origin"*. In the embryonic avian lung, the inaugural blood cells, i.e. erythroblasts and leukoblasts, arise from cytoarchitecturally similar mesenchymal cells (Maina 2003b; 2004a). In complete contrast, in the embryo of *Fundulus heteroclitus*, Danchakoff (1918) observed that *"erythrocytes develop within the vessels, leukocytes outside of them"*. In the embryonic avian lung, erythrocytes form early and are seemingly vital for vasculogenesis and subsequently angiogenesis (Maina 2003b, 2004a). In complete contrast to the developing avian lung, interestingly, in teleost hybrids, Reagan and Thorington (1916) noted that *"leukocytes are never developed in embryos that have never possessed a circulation"*.

In the embryonic lung of the domestic fowl, vasculogenesis occurs diffusely from about day 7. With conspicuous filopodial extensions, the characteristically stellate angioblasts (Fig. 29) approach and surround the erythroblasts (Figs. 26 and 27C). Gradually, the angioblasts connect and form blood vessel walls (Figs. 27 and 28): angioblasts that lie next to erythrocytes form the endothelial lining while the outer ones contribute to the formation of the vessel wall. Endothelial cell density within a certain area of an embryonic vasculature can be used to distinguish between a small diameter capillary-like vascular network (low endothelial cell density) and a large diameter, presinusoidal network (high endothelial cell density; LaRue et al. 2003). Interestingly, in the developing avian lung, the granular leukocytes are never surrounded by angioblasts (Fig. 24A,D). From day 9, the initially scattered vascular units begin to connect and surround the developing PR (Fig. 31A). The interparabronchial blood vessels (IPRBVs) are located midway between the ETs of adjacent PR (Fig. 31B). In the ET, the BCs form through transformation of the mesenchymal cells into erythrocytes (Fig. 31C). Subsequently, the formative vascular units connect after substantial remodeling of the parenchymal stroma, particularly by apoptosis (Fig. 31D). Presence of erythrocytes in the lumina of the formative PR (Maina 2003a) indicated pluripotentiality of the epithelial and mesenchymal cell differentiation. Epithelial cells may transform into hematogenic cells, especially erythroblasts (Maina 2003a). The

2.4 Pulmonary Vasculature

arrangement and orientation of angioblasts around a formative vascular unit set the direction along which a blood vessel sprouts (Fig. 28). Endothelial cells appear to transform into erythrocytes, detach, and fall into the vessel lumen (Maina 2003a), a feature that further illustrates the plasticity of the vasculogenetic primordial cells of the developing avian lung bud. The IPRBVs (Fig. 31A,B) are well formed by day 10. Complete pulmonary circulation, i.e. from the heart through the ET of the lung, is not, however, established until

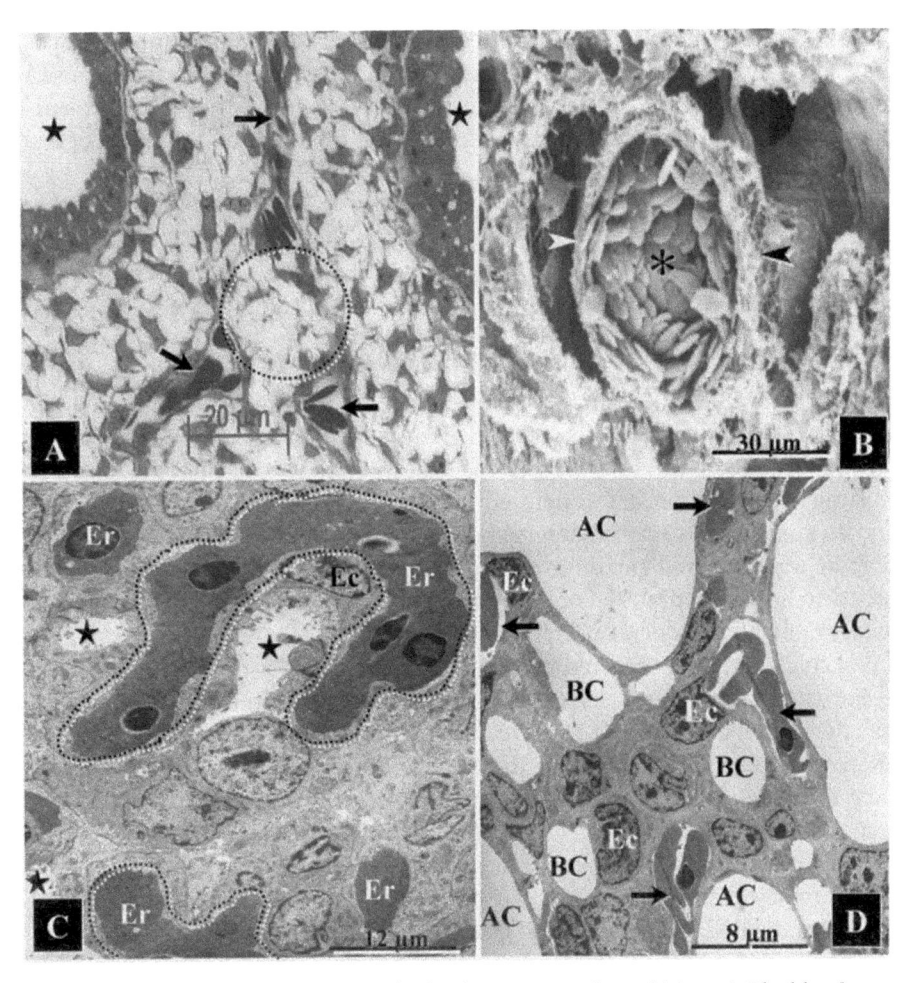

Fig. 31. A Blood vessels (*arrows*) developing between parabronchi (*stars*). The blood vessels are about to fuse (*dashed circle*). Day 14. **B** An interparabronchial blood vessel with a thick wall (*arrowheads*) and lumen packed with erythrocytes (*asterisk*). Day 18. **C** Continuity of erythrocytes (*Er; dashed areas*) forming around air capillaries (*stars*). *Ec* Endothelial cell. Day 18. **D** Blood capillaries (*BC*) developing in the gas exchange tissue by intracytoplasmic lumenization of the endothelial cells (*Ec*). *Arrows* Erythrocytes; *AC* air capillaries. Day 19. (**A** from Maina 2003b; **B, C** from Maina 2004a; **D** from Maina 2003a)

after day 18 of embryonic development when the BC are well formed (Figs. 15D, 16C,D, 17 and 18). Since before a functional pulmonary circulatory system is formed and the embryo has had access to air the earliest erythrocytes cannot play a role in gas exchange, it is implicit that the role of de novo erythropoiesis is that of initiating and regulating vasculogenesis and possibly angiogenesis.

Developmentally, the avian pulmonary vasculature inaugurates in a scattered manner. The discrete vascular units steadily connect and ultimately link up with the systemic circulation across the heart through the pulmonary arteries and veins. In the mature lung, vascular architecture displays cross-current, counter-current, and multicapillary serial arterialization arrangements (Chap. 3.8), means by which air and blood are presented and exposed to each other in the ET. The fabrication of the vasculature is spatially and temporally well coordinated: it forms by piecemeal interconnection of independently formed vascular units.

2.5
Blood–Gas Barrier (BGB)

In certain consequential ways, compared with other organs, the vertebrate lung is structurally and functionally unique: it is the only organ in the body that accepts the total cardiac output, a measure that increases several-fold during strenuous exercise; during life, the pulmonary vasculature tolerates shifting tensions that are generated by contractions of the cardiac muscle, and, in contrast to gas exchangers such as the gills that are stabilized by support of water, a fluid medium of which the density is equivalent to that of blood, in lungs, air, material of lower specific density balances blood, a fluid of measurably higher density. The BGB of the vertebrate lung hence encounters a real challenge of maintaining structural integrity.

For transition from water to land, one of the most important adaptations was the switch from water to air breathing (e.g. Little 1990; Graham 1997). For efficient gas exchange by passive diffusion, a thin and extensive BGB is necessary. Trade-offs and compromises were invoked in the process of producing a structure of which the properties were totally at variance: thin to offer efficient flux of respiratory gases and adequately strong to withstand stress during ordinary conditions of operation. The three-ply (tripartite laminated) design of the BGB in the vertebrate lung (Chap. 3.3) has apparently been conserved for about 400 million years (e.g. Pough et al. 1989). Body designs that have remained constant for a long evolutionary period have been termed 'Bauplans' (='blue-prints' ='frozen cores'; e.g. Wagner 1989). The rarity of such designs in biology bespeaks of the importance and the immense material cost of establishing and supporting such structures.

Recently, Schittny and Burri (2004) stated that *"it is only poorly understood what governs the formation of the air–blood barrier"*. In the developing lung

2.5 Blood–Gas Barrier (BGB)

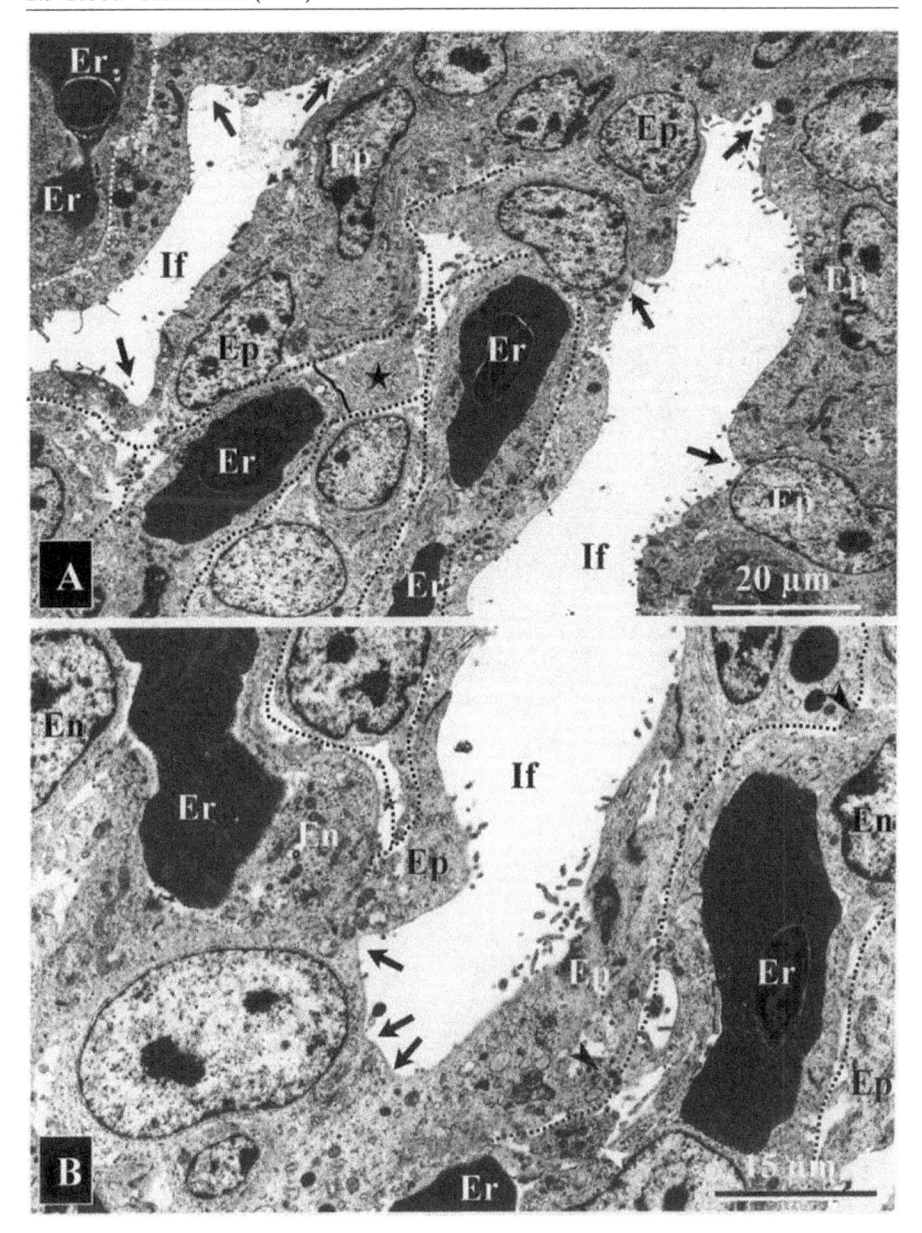

Fig. 32A,B. Thinning of the BGB by extension of air capillaries into the mesenchymal tissue stroma (*arrows*). *Dashed lines* Intercellular spaces along which epithelial (*Ep*) and endothelial (*En*) cells fuse; *Er* erythrocytes; *star* a wide intercellular space between an endothelial and an epithelial cell; *arrowheads* site where endothelial and epithelial cells have begun to fuse forming common basement lamina. Day 16. (Maina 2004b)

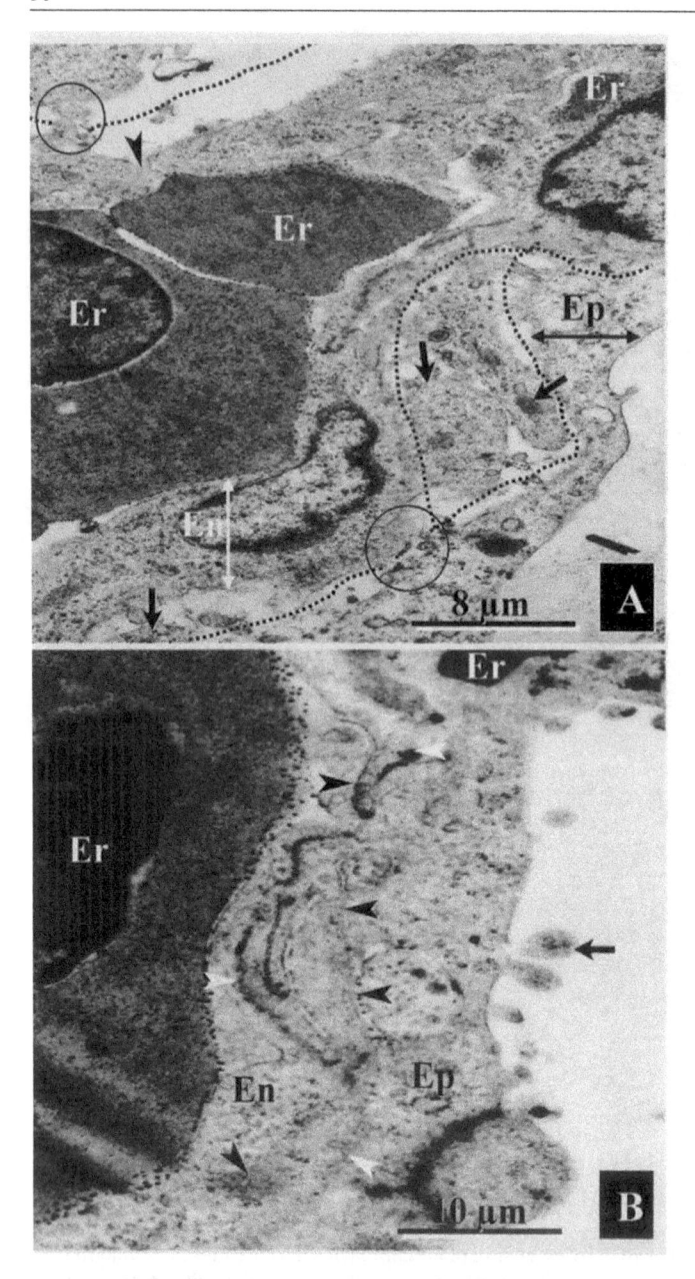

Fig. 33. A Formation of thin BGB by approximation of the endothelial (*En*) and epithelial (*Ep*) cells after disintegration of mesenchymal (interstitial) cells (*arrows*). *Dashed lines* Space between epithelial and endothelial cells; *arrowhead* intercellular junction between endothelial cells; *Er* erythrocytes; *circle* a point where endothelial and epithelial cells have fused. Day 17. **B** Development of thin BGB showing deposition of extracellular matrix (*arrowheads*), material that comes to constitute a common extracellular space. *Er* Erythrocytes; *En* endothelial cell; *Ep* epithelial cell; *arrow* microvilli. Day 17. (Maina 2004b)

2.5 Blood–Gas Barrier (BGB)

of the domestic fowl, originating from the infundibulae that in turn arise from the atria (Figs. 14, 15 and 16A,B), the AC project profusely into the periparabronchial tissue from day 18. As the AC extend deeper into the ET, the BC come closer to the forming air spaces. Progressively, the AC and BC intertwine in three dimensions. A thin BGB forms by means of cell transformation, translocation, and breakdown (Maina 2003b, 2004b). In the mammalian lung, canaliculi form by apoptosis (Schittny et al. 1998; Bruce et al. 1999; Stiles et al. 2001). In the rat lung, the thinning of the alveolar septa results in reduction of the absolute mass of the interstitial tissue in spite of a 25 % decrease in lung volume (Schittny and Burri 2004). The absolute number of fibroblasts is reduced by 10–20 % and the epithelial cells by more than 10 % (Kaufman et al. 1974). Interstitial (mesenchymal) cells that are located between endothelial

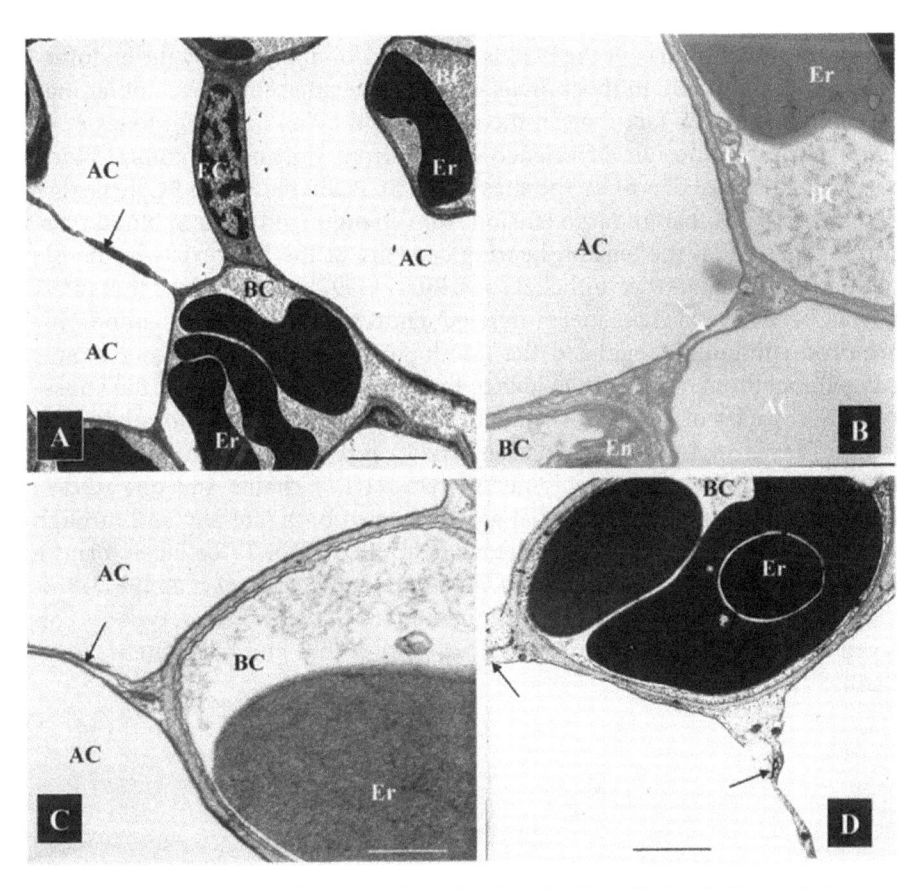

Fig. 34. A Exchange tissue of the avian lung showing blood capillaries (*BC*) containing erythrocytes (*Er*) and air capillaries (*AC*). *EC* Perikaryon of an epithelial cell; *arrow* area where an epithelial cell has disintegrated. Day 26. **B–D** In the sites where air capillaries lie next to each other (*arrows*), an extracellular matrix space is lacking (*arrows*). *Er* Erythrocyte; *BC* blood capillary; *En* endothelial cell; *AC* air capillary. Day 26. *Scale bar* **A**, 10 µm; **B–D** 5 µm

and epithelial cells are displaced while others disintegrate (Figs. 32 and 33): gradually, endothelial and epithelial cells approximate until they contact. Subsequently, the cells deposit extracellular matrix between them (Fig. 33). In transgenic mice in which the sequence coding for nidogen-binding site γ1III4 within the laminin-γ1 chain (*Lamc1* gene) was selectively deleted by gene targeting, in large parts of the BGB the BL was disrupted or missing and the epithelial and endothelial cells failed to contact (Willem et al. 2002). By day 21 (hatching), in the domestic fowl, a thin BGB has formed and the lung is structurally well prepared for gas exchange: the BC and the AC are well formed (Fig. 34A) and, in the sites where BC and AC lie adjacent to each other, the BGB adopts a three-ply construction, i.e. epithelial and endothelial cells contact across an extracellular matrix space (basement lamina; Fig. 34B–D). In those areas where AC lie next to each other (Fig. 34A), the perikarya of the type 1 epithelial cells disintegrate, leaving sites where the cytoplasmic flanges of the epithelial cells contact directly (Fig. 34B-D): it is tacit that in the avian lung, the extracellular matrix of the BGB is made and/or deposited by the endothelial cells. Lack of a BL in those areas where AC lie adjacent to each other may be explained by the fact that, in the rather rigid avian lung (e.g. Jones et al. 1985), such sites do not experience tension from shifting intramural blood pressure. Presence of a BL in the sites of the BGB where AC and BC lie next to each other, parts that undergo tension from changing intramural blood pressure, suggests that the tension-bearing capacity of the BGB exists in the BL, not in the epithelial or endothelial cells. Burri (1997) hypothesized that interaction between the mesoderm-derived endothelium and the endoderm-acquired epithelium regulated the development of the BGB. There is now compelling, though indirect, evidence that in the vertebrate lung, the stress-bearing capacity of the BGB is contributed by collagen type IV located in the lamina densa of the extracellular matrix (e.g. West and Mathieu-Costello 1999). Essentially, a heteropolymer of two $\alpha 1(IV)$ chains and one $\alpha 2(IV)$ chain (e.g. Martin et al. 1988) that are connected both laterally and through their terminal domains (e.g. Yurchenco et al. 1986), type IV collagen forms a dense molecular network in the basement lamina (e.g. Timpl et al. 1981). Providing high tensile strength with minimal increase in thickness, type IV collagen is the ideal material for construction of a thin and adequately strong BGB.

2.6
Molecular and Genetic Aspects in Lung Development

2.6.1
General Observations

While plentiful data now exist on the development of the mammalian lung (e.g. Cardoso 2000; Warburton et al. 2000; Alsberg et al. 2004; Tuyl et al. 2004), little work has been done to identify and categorize the genes and molecular factors that are involved in the morphogenesis of the avian lung. Given the greater morphological complexity of the avian lung (Chap. 3), it is implicit that the molecular factors and the genetic interplay involved in its development should be more intricate. Quite possibly, there are species-specific differences in the reactions of the epithelial and the mesenchymal components of developing lungs to similar and different inductive cues. A general review of pulmonary development is given here to highlight the critical lack of data on the avian respiratory system.

From the early stages of pulmonary development, progressive structural transformation is controlled by suites of genetically conserved intercellular signaling pathways (e.g. transcriptional factors), soluble peptide GFs, and insoluble extracellular matrix (ECM) molecules (e.g. Healy et al. 2000). Closely coordinated signals regulate cell proliferation, growth, motility, and apoptosis (e.g. Affolter et al. 2003; Warburton et al. 2003; Alsberg et al. 2004; Tuyl et al. 2004): an organ primordium (anlage) that is formed within a diffuse embryonic domain changes into recognizable morphological patterning. The progenitor cells proliferate, differentiate, and diversify into different lineages and appropriate organ-specific stem cells in different cell environments. At maturity, the genetically programmed stem cells produce new progenitors that typically differentiate along the same tissue-specific lineage pathways (e.g. Brazelton and Blau 2004). Differences in the growth rates of the epithelial cell are caused by complicated interplay between soluble, insoluble, and physical signals from local microenvironments. From simple outpouchings of the foregut endoderm, epithelial cell cords (formative airways) bifurcate and lengthen in a proximal-distal axis, giving complex three-dimensional architecture (e.g. Bernfield 1977).

While the molecular mechanisms involved in the development of the bronchial system are well known (e.g. Perl and Whitsett 1999; Alsberg et al. 2004; Tuyl et al. 2004), those involved in the development of the pulmonary vascular system are relatively little known. Members of the vascular endothelial growth factor (VEGF) and its flk-1/KDR receptor (e.g. Ferrara 2000; D'Angio and Maniscalco 2002), angiopoetin and the emprin family (Dumont et al. 1994; Hall et al. 2002) have been implicated in the development of the pulmonary system. Mice with an inactivated flk-1 receptor or VEGF gene die in utero (e.g. Ferrera 2000), while knockout mice lack yolk-sac blood-islands and there are no organized blood vessels at any stage of development (Shalaby et al. 1997). The lethality of deletion of a single allele illustrates the critical importance of VEGF in embryonic vascular development (Carmeleit et al. 1996). The *Gli* family of transcription factors (*Gli 1–3*) are transducers of sonic hedgehog (*Shh*) signaling located in the foregut mesoderm (Hui et al. 1994). In the course of lung morphogenesis, the genes are expressed in overlapping but distinct areas of the lung mesenchyme (Grindley et al. 1997). *Gli2*[1-] and *Gli3*[1-] double mutant mice die typically by day 10.5 without evidence of primitive lung or trachea formation (Motoyama et al. 1998). *Shh.Shh*[1-] null mutants form lungs but branching is severely impaired (Litingtung et al. 1998; Pepicelli et al. 1998).

Genetic studies have shed light on the mechanisms underlying mammalian lung cell lineage diversification. For example, the bHLH genes *Ascl1* (*Mash 1*) and *Hes1* promote the development and production of the neuroendocrine cell lineage (Borges et al. 1997; Ito et al. 2000). The three-dimensional pattern of expression and production of fibroblast growth factor-10 (FGF-10) in discrete regions of the mesenchyme around the developing lung bud (e.g. Park et al. 1998; Cardoso 2000; Kaplan 2000) may determine how, where, and when new buds develop. Deletion or mutation of FGF-10 (e.g. Bellusci et al. 1997) or its receptor FGFR-2 inhibits lung branching (Peters et al. 1994). No bifurcation of the lung epithelial cell cords occurs in the absence of FGF-10 (Sekine et al. 1999). Expressed in the mesenchyme, FGF-10 controls the expression of bone morphogenetic protein-4 (BMP-4) at the growing (terminal) epithelial bud (Bellusci et al. 1996a; Weaver et al. 2000; Hyatt et al. 2002). BMP-4 belongs to the tumor growth factor-β (TGT-β) superfamily and controls cell differentiation and proliferation at the epithelial buds (Hyatt et al. 2002). Stimulation of the growth of the epithelial cell tubules is mediated by an array of molecular factors that include endothelial growth factor (EGF or TGT-α), hepatocyte growth factor (HGF), and FGF-7: in a loop that controls the developmental process, TGF-β1 suppresses the effects of these molecular factors (e.g. Cardoso 2000; Desai and Cardoso 2002). Moreover, BMP-4 (Bellusci et al. 1996b), transforming growth factor-β1 (TGF-β1) (Serra and Moses 1995) and *Shh* (Bellusci et al. 1996b, 1997), factors produced by the lung epithelial cells, all inhibit FGF-10 production in the mesenchyme and suppress epithelial cell growth (Hogan 1999; Lebeche et al. 1999; Cardoso 2000). Upregulation of these factors in the highly proliferative regions of the lung may stop, slow down growth, induce quiescence, and promote lung bud maturation.

In addition to the above-mentioned molecular factors and genes, extracellular matrix proteins including collagens, elastin, laminin-isoforms, fibrillins, and nidogens as well as their receptors including integrins and dystroglycan are significantly involved in lung development (Durbeej and Ekblom 1997; Wendel et al. 2000; Coraux et al. 2002; Gloe and Pohl 2002). By modulating proliferation and/or migration of the epithelial cells in the respiratory buds and establishing branching focal points, laminin-5 may play multiple roles during branching morphogenesis (Coraux et al. 2002). Considering the greater morphological complexity of the avian lung (Chap. 3), it is conceivable that more intense dynamics of the molecular and genetic factors are involved in its fabrication.

Some of the studies that have touched on aspects of the genetic and molecular mechanisms of the development of the avian lung are those by Goldin and Opperman (1980) and Hacohen et al. (1998) who investigated the stimulation of DNA synthesis in the embryonic chick lung and trachea by the epidermal growth factor (EGF); Chen et al. (1986) examined the expression and distribution of cell-to-cell adhesion molecules (fibronectin and laminin) on the embryonic chick lung cells; Muraoka et al. (2000) investigated the expression of nuclear factor-kappaβ on epithelial growth and branching of the embryonic chick lung; Stabellini et al. (2001) looked into the roles of polyamines and transforming growth factor beta-1 (TGF-β1) on branching morphogenesis during the development of the embryonic chick lung; and, recently, in the chicken embryo, Sakiyama et al. (2003) investigated the effect of the Tbx-4-Fgf-10 system on the separation of the lung bud from the oesophagus. Recent studies by Maina and Madan (2003) and Maina et al. (2003) have shown that basic fibroblast growth factor-2 (bFGF-2) is widely expressed in the epithelial and mesenchymal cells of the developing avian lung from very early stages. Expression and upregulation of the GF in different areas of the developing lung appear to regulate the rate of growth, the trajectories followed, the areas appropriated, the three-dimensional symmetry, and the interconnections between the airways. Expressed and maintained from very early to late stages of development, implicitly, FGF-2 should play an important role in the formation and growth of the avian lung.

2.6.2
Fibroblast Growth Factors (FGFs)

FGFs are a family of some 23 gene-encoding proteins, i.e. functional regulators of development (e.g. Gospodarowicz 1991; Cardoso 2000; Ornitz and Itoh 2001). Six FGFs, namely 1, 2, 7, 9, 10, and 18, are expressed in the lung (e.g. Gonzalez et al. 1990; Fu et al. 1991; Colvin et al. 1999). Four FGF receptors (FGFRs) are expressed in the lung (Weinstein et al. 1998). Particularly important in the development of complex, heterogeneous organs that form by budding and branching, e.g. lungs (e.g. Cardoso 2001), glands (e.g. Goldin 1980), kidneys (e.g. Bard 2002), and the tracheal system of insects (e.g. Franch-Marro and Casanova 2002), FGFs were the first angiogenetic GFs to be sequenced (Klagsbrun 1989). The GFs are generally produced by the pulmonary mesenchyme while their receptors are present in the lung epithelium. However, exceptions are FGF-1 and FGF-2 which are expressed in the fetal pulmonary epithelium and mesenchyme (Han et al. 1992; Gonzalez et al. 1996). Of the FGFs expressed in the lung, only FGF-10 has been shown to be utterly necessary for the initiation of lung development (Hyatt et al. 2002). FGF-1 and FGF-7 induce different patterns of pulmonary growth and development (Cardoso et al. 1997). FGF-2 is a particularly highly conserved GF that is highly instructive in the growth and development of many different organs and tissues as well as induction of the mesoderm (e.g. Kessler and Melton 1994; Bikfalvi et al. 1997; Le and Musil 2001). Together with FGF-receptor (FGF-R), FGF-2 is reported to be involved in the morphogenesis of the mammalian (Han et al. 1992) and avian (Maina et al. 2003) embryonic lungs and is a potent mitogen for type II pneumocytes (Tanswell et al. 1999). The GF has been associated with compensatory lung growth after damage from exposure to 95 % oxygen (Jankow et al. 2003). FGF-2 is a rather peculiar GF: although they display certain vascular and hematological deficiencies, FGF-2 knockout mice are reported to be morphologically normal (Zhou et al. 1998). In spite of the fact that it is present in the extracellular matrix (e.g. Vlodavsky et al. 1987), interestingly, FGF-2 does not appear to have a signaling peptide (e.g. Abraham et al. 1986). On that basis, questions have been raised whether the GF is secreted from cells under normal physiological conditions. Investigators like Park et al. (1998) have mostly attributed the morphogenesis of the airways in the mammalian lung to FGF-7 and FGF-10: very little of it has been ascribed to FGF-2.

The diffuse pattern of expression and distribution of FGF-2 in the rat fetal lung, i.e. localization in the airway epithelial cells, their BL, and the extracellular matrix (Han et al. 1992), corresponds with that observed in the avian lung (Maina and Madan 2003; Maina et al. 2003). In the chick embryo lung, the basic FGF-2 (bFGF-2) is mainly expressed in the epithelial cells and to a lesser extent in the mesenchymal cells that surround the formative airways (Figs. 35 and 36). In the epithelial cells themselves, bFGF-2 is upregulated in the apical aspects and in the mesenchymal cells overlying the BL (Fig. 36B-D). A Tbx4-FGF-10 system is implicated in the inauguration of the chick embryo lung bud and its separation from the foregut (Sakiyama et al. 2003). In the mouse lung, a key factor that elicits lung bud induction is activation of FGF signaling in the foregut endoderm by local expression of FGF-10 in the mesoderm (Bellusci et al. 1997). FGF-10 is the major driving force for budding during bronchial bifurcation (e.g. Park et al. 1998): its signaling is mediated by the receptor FGFR-2IIIB whose expression is maintained in the respiratory epithelium once lung development starts (e.g. Xu et al. 1998). During branching morphogenesis, in a dynamic stereotypical fashion, FGF-10 diffuses from the mesenchyme to activate the FGFR-2IIIb in the adjacent epithelium in the distal lung, initiating a regular pattern of bifurcation (Bellusci et al. 1997). The on-and-off pattern of expression of FGF-10 in the lung is analogous to that described during the patterning of *Drosophila* trachea (Sutherland et al. 1996). In the tracheal system, local expression of *branchless* (the invertebrate homologue of FGF) directs tracheal epithelial budding to proper positions during branching morphogenesis (Metzger and Krasnow 1999). Mechanisms that prevent diffuse distribution of FGF-10 signals are important in preserving FGF-10 spatial gradients. In the lungs of $Shh^{-/-}$ mice where FGF-10 expression is not focal (as in the wild type), branching morphogenesis is severely

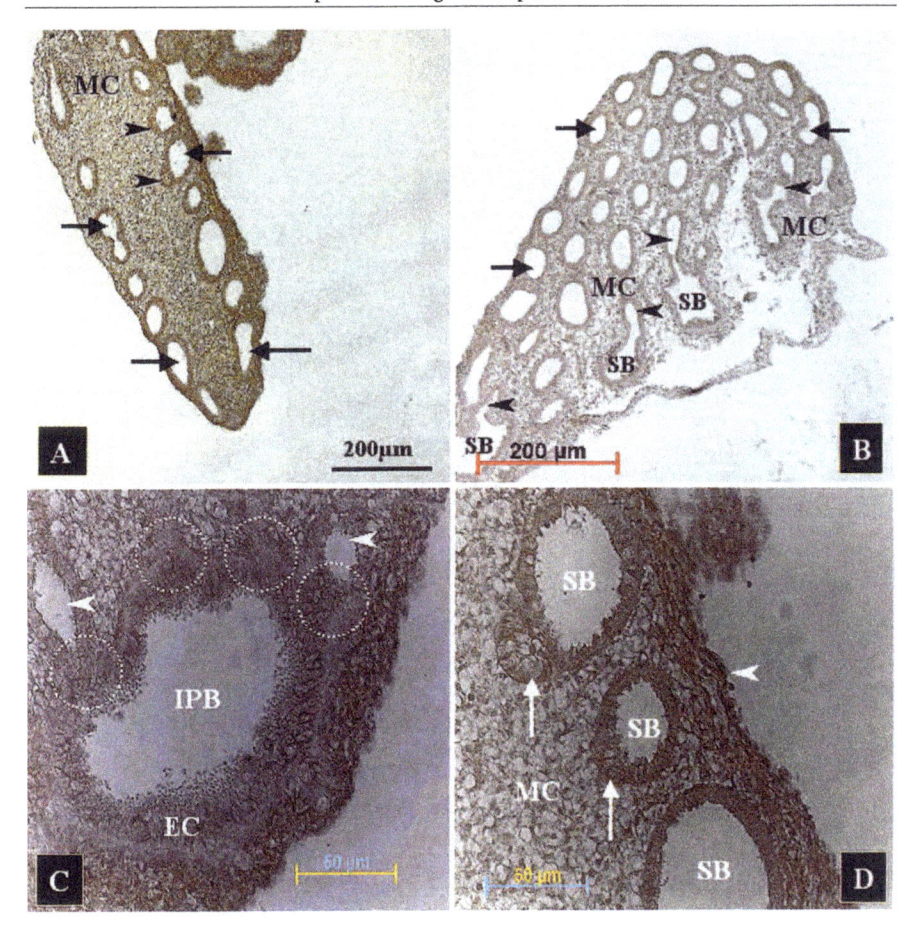

2.6 Molecular and Genetic Aspects in Lung Development

Fig. 35. Immunocytochemical study of expression of fibroblast growth factor-2 of developing lung of the domestic fowl, *Gallus gallus* variant *domesticus*. **A** Longitudinal section of the developing lung showing secondary bronchi (*arrows*) giving rise to parabronchi (*arrowheads*). *MC* Mesenchymal cells. Day 5. **B** Transverse section of the lung showing secondary bronchi at the periphery of the lung (*arrows*) and some lying deep in the lung (*SB*). *MC* Mesenchymal cell stroma. Day 5. **C** Intrapulmonary primary bronchus (*IPB*) giving rise to secondary bronchi (*dashed circles*). *Arrowhead* Lumina of the secondary bronchi. Day 5. **D** Periphery of the lung showing secondary bronchi (*SB*) with parabronchi (*arrows*) budding from them. *MC* Mesenchymal cell stroma. Day 7.5. (Maina et al. 2003)

impaired (Pepicelli et al. 1998). FGF-10 knockout mice develop a normal trachea but lungs do not form (Min et al. 1998; Sekine et al. 1999). Unlike FGF-7, which in the presence of other soluble factors can induce the trachea to transdifferentiate into distal lung, FGF-10 cannot do so (Shannon et al. 1999). This indicates that tracheal and lung morphogenesis is coordinated by independent events. High levels of FGF-9 have been reported in the mesothelial layer

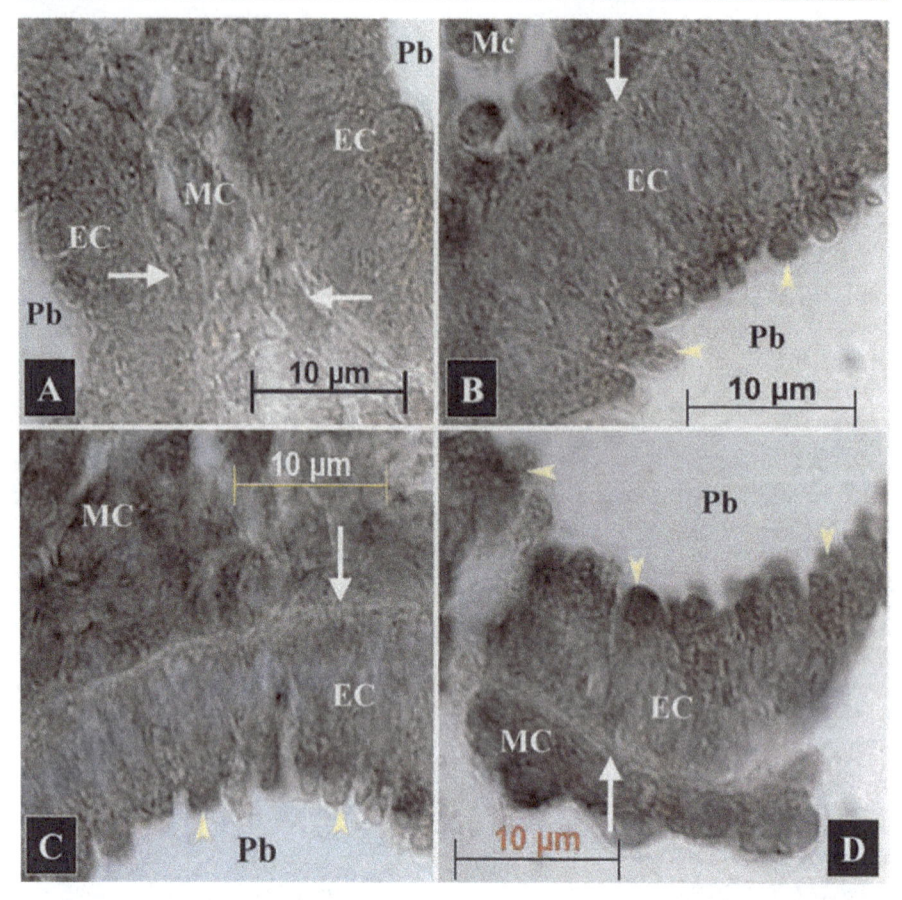

Fig. 36A-D. Expression and upregulation of fibroblast growth factor-2 (FGF-2) in the mesenchymal cells (*MC*) and epithelial cells (*EC*) of the developing lung of the domestic fowl, *Gallus gallus* variant *domesticus*. *Arrows* Basement lamina; *arrowheads* FGF-2 upregulated in the apical parts of the epithelial cells. Day 7. (Maina et al. 2003)

and epithelium of the embryonic gut and lung at day 10.5 (Colvin et al. 1999). It has been suggested that FGF-9 may diffuse to the mesenchyme to activate FGFR-1 signaling (Szebenyi and Fallon 1999) and presumably regulate expression of mesenchymal genes, including FGF-10 (Arman et al. 1999). Mice deficient in the FGF-9 gene have hypoplastic lungs and thin mesenchyme (Colvin et al. 2001). In the rat, deletion of the FGF-18 gene has no apparent effect on lung development (Liu et al. 2002; Ohbayashi et al. 2002). In the development of the mammalian lung, FGF-7 is a more potent morphogen compared to FGF-10 (e.g. Tichelaar et al. 2000) and is expressed very early in development (Park et al. 1998). Deposition of matrix components has been observed to generate dichotomous branching (Hilfer 1996; Hogan et al. 1997).

There is urgent need to find out whether these and other signaling molecules and genes are expressed in the developing avian lung and if they are to know the specific roles that they play in its development.

2.6.3
Vascular Endothelial Growth Factor (VEGF)

Airway epithelial cells synthesize VEGF during the development of the lung and deposit it into the subendothelial matrix while pulmonary endothelial cells synthesize appropriate receptors (e.g. Acarregui et al. 1999; Miquerol et al. 1999). Matrix-associated VEGF stimulates endothelial cell differentiation, migration, and proliferation that results in development of a primitive vascular plexus in the mesenchyme directly surrounding the airways. These functions are mediated through binding of high-affinity cell receptors and matrix-binding sites (e.g. Yamaguchi et al. 1993; Soker et al. 1997). By stimulating neovascularization at the leading edge of branching airways, VEGF synchronizes airway branching with blood vessel formation (Healy et al. 2000). VEGF is a dimeric, heparin-binding glycoprotein that is an endothelial cell-specific mitogen capable of inducing cell proliferation and chemotaxis (e.g. Fong et al. 1995; Ferrara et al. 1996; Shalaby et al. 1997; Ferrara 1999). By differential mRNA splicing, the VEGF gene gives rise to at least five protein isoforms named $VEGF_{122}$, $VEGF_{145}$, $VEGF_{164}$, $VEGF_{188}$, and $VEGF_{206}$ that have different affinities for heparin sulfate as well as for the receptors, VEGFR-1 (flt-1), VEGF-2 (flk-1/KDR), and neuropilin-1 (Ferrara et al. 1992; Shima et al. 1996; Larrivée and Karsan 2000). $VEGF_{122}$ does not bind to heparan sulfate and is freely diffusible; $VEGF_{188}$ is heparin binding and is primarily associated with the cell surface and extracellular matrix; while $VEGF_{164}$ has intermediate properties (e.g. Park et al. 1993; Ferrara and Davis-Smyth 1997). The existence of multiple VEGF ligands and receptors suggests specific and probably redundant regulation of vascular development (e.g. Ng et al. 2001; Tomanek et al. 2002). The import of VEGF in vascular development is evidenced by the fact that mice lacking a single allele of the VEGF gene are embryonic lethal (Carmeliet et al. 1996; Ferrara et al. 1996). Interestingly, VEGF-stimulated vessel growth cannot be explained solely on the basis of its mitogenic and chemoattractant effects on endothelial cells (EC). VEGF-deficient embryos contain vascular EC that fail to form blood vessels (e.g. Carmeliet et al. 1996; Ferrara et al. 1996): this suggests that VEGF is involved in tubulogenesis. Produced in the pulmonary epithelium, especially by the type II cells (Maniscalco et al. 1995), $VEGF_{188}$ may mediate assembly and stabilization of the highly organized vessel networks that come to surround the alveoli.

While their structure might appear simple and the growth by accretion of cells may seem deceptively easy, normal vascular development is a highly complex, well-coordinated process that involves physical and chemical stimulators and inhibitors, and multiple gene activity and signaling molecules (e.g.

Burri and Tarek 1990; Risau 1997; Cleaver and Krieg 1999; Yancopoulous et al. 2000; Gerritsen et al. 2003). The molecular events involved in endothelial cell alignment into tube-like structures with patent lumens have eluded scientists for decades (e.g. Taichman et al. 2002; Gerritsen et al. 2003; Tuyl et al. 2004). In an in vitro three-dimensional gel environment, expression of as many as 1000 different genes is upregulated during endothelial tubulogenesis (Gerritsen et al. 2003; Alsberg et al. 2004; Tuyl et al. 2004). Although remarkable progress has been made recently towards understanding the effects and roles of biochemical characteristics of angiogenetic factors, it is becoming quite clear that the signaling pathways are very intricate and our knowledge of GF interplay in angiogenesis is far from lucid. What is certain is that the process of endothelial differentiation, accretion, and juxtaposition into a network of cylindrical structures requires a well-integrated program of gene expression (Pepper 1997; Tomanek and Schatteman 2000; Gerritsen et al. 2003).

Members of the VEGF (Neufeld et al. 1999; Ng et al. 2001; D'Angio and Maniscalco 2002), angiopoetin, and emprin families (Hall et al. 2002) have been implicated in the vascularization of the pulmonary system. Other factors that are involved include local hormones, extracellular matrix elements, hemodynamic forces (that cause shear stress, stretch, or deformation of endothelial cells and modification of the BL), hypoxia, hyperoxia, ischemia, intercellular interactions with the surrounding and supporting mesenchymal cells, and growth and other molecular factors and their signaling pathways, and facilitators like the vasodilator nitric oxide (e.g. Madri et al. 1988; Levy et al. 1996; Stone et al. 1996; Li et al. 1997; Maniscalco et al. 1997; Seko et al. 1999; Ferrara and Gerber 1999; Nicosia and Villaschi, 1999; Akeson et al. 2000; Hudlicka and Brown 2000; Agraves et al. 2002). Disturbances in the interactions of the myriad factors and agents result in abnormal vessel formation (e.g. Auerbach and Auerbach 1997). Extravagant vascular growth may result in pathological states such as cancer and chronic inflammatory disorders while paucity of blood vessels may impair processes like atherosclerosis, diabetic retinopathy, and wound and burn healing or organ repair. The interactions between endothelial cells and adjacent mesenchymal cells determine the phenotypic heterogeneity of endothelial cells (e.g. Carmeliet 2000). It has been suggested that the structure of endothelial cells is specifically suited for the tissues that blood vessels form in and serve (e.g. LeCouter et al. 2002). If the premise is correct, vascular growth and development should entail determinate qualitative and quantitative adjustments that produce vital structural and functional properties. Regardless of developmental stage, location, or species in which they form, networks of blood vessels formed during vasculogenesis share similar geometric properties like mean blood vessel diameters and avascular space diameters (LaRue et al. 2003).

Mature blood vessels are comprised of endothelial cells and supporting cells, e.g. pericytes, connective tissue, and smooth muscle cells (e.g. Risau and Flamme 1995). Blood vessels basically form through two mechanisms: 'vasculogenesis' involves de novo differentiation of progenitor cells (angioblasts) from uncommitted mesoderm and their coherence into endothelial cords leading to formation of rudiment blood vessels, while 'angiogenesis' entails growth and sprouting of new blood vessels from preexisting ones (e.g. Poole and Coffin 1989; Risau and Flamme 1995; Risau 1997; Poole et al. 2001). The patterning of blood vessels is modified by the process of 'intussusceptive growth' that entails formation of endothelial cell pillars that separate blood vessels (e.g. Caduff et al. 1986; Burri and Tarek 1990; PÈrez-Aparicio et al. 1996). Both vasculogenesis and angiogenesis require sequential changes of endothelial cell gene expression and functions to allow cells to proliferate, migrate and assemble into vascular channels (e.g. Hanahan 1997; Petrova et al. 1999).

2.6.4
Wnt Genes and Signaling

Wnts are secreted proteins that bind to cell membranes or extracellular matrix once they are released (e.g. Nusse and Varmus 1992). Based on capacity to mediate cell-to-cell interactions, set cell planar polarity, and determine fates of cells, Wnt signaling is significantly involved in the patterning of embryos and in many developmental and pathological processes (e.g. Wodarz and Nusse 1998; Pfeifer and Polakis 2000; Okubo and Hogan 2004). Receptors for Wnt proteins, called "Frizzled" (Fzd), are members of a seven transmembrane protein family (e.g. Cadigan and Nusse 1997). Wnt–frizzled interactions activate intracellular pathways and cause β-catenin accumulation or an increase in intracellular Ca^{2+} concentration. Based on the different intracellular events that are activated after Wnt–Fzd binding, Wnt signaling has been divided into the canonical (Wnt/β-catenin pathway) and the noncanonical pathways (e.g. Dale 1998). The canonical pathway of Wnt signaling is initiated by the interaction between extracellular Wnt ligands and their receptors, resulting in stabilization of β-catenin which then interacts with nuclear T-cell factor/lymphoid enhancer factor (TCF/LEF) and transcription factors to modulate the activity of target genes (Miller et al. 1999). Less well known, the noncanonical pathway requires the binding of specific Wnts to specific Fzds that result in intracytoplasmic release of Ca^{2+} and FAK pathway activation (Cohen et al. 2002). Not only is β-catenin the key component of the canonical Wnt signaling pathway, but it is also important in the formation of adhesive intercellular junctions of endothelial cells that regulate cell function (e.g. Gory-Faure et al. 1999). Wnt signaling participates in angiogenesis by regulating endothelial cell growth and function (Cheng et al. 2003).

There are several soluble protein antagonists of Wnt signaling. These include secreted frizzled related protein (sFRP), Wnt inhibitory factor (WIF) and Dickkopf (Dkk; e.g. Moon et al. 1997a,b; Hsieh et al. 1999; Fedi et al. 1999). Associated with regulation of epithelial cell proliferation, Wnt signaling has been localized in the lungs of developing mice (Okubo and Hogan 2004). By in

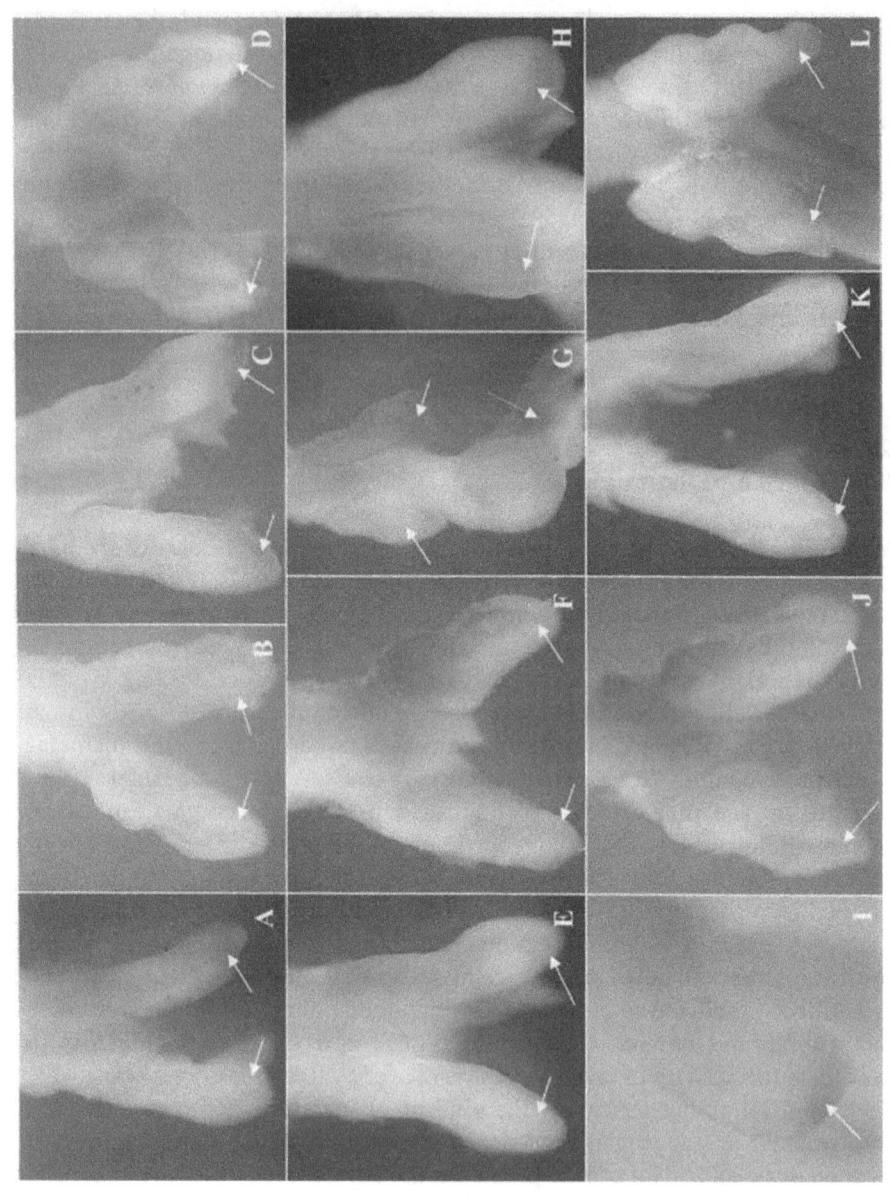

Fig. 37. Expression of wnt genes during stages 26 and 29 of the development of the lung of the domestic fowl, *Gallus gallus* variant *domesticus*. wnt 1 is not expressed at stages 26 and 29 (**A, B**); wnt 2 is expressed in the growing tips of the lung at stage 26 (**C**) but not at stage 29 (**D**); wnt 3a is not expressed at stages 26 (**E**) and 29 (**F**); wnt 4 is expressed in the distal parts of the lung and the gut (*dashed arrow*) at stage 26 (**G, H, I**) but not at stage 29 (**J**); and wnt 5a is not expressed at stages 26 (**K**) and 29 (**L**). (J.N. Maina and R.G. Macharia, unpubl. observ.)

2.6 Molecular and Genetic Aspects in Lung Development

situ hybridization, Ishikawa et al. (2001) found that Frz5 mRNA was expressed in the yolk sac, eye, and lung bud during mouse embryogenesis. A number of Wnt ligands and receptors, e.g. β-catenin, Tcf1, Tcf4, and Lef1, are expressed in the endoderm and distal epithelium of the developing lung (Tebar et al. 2001). For example, Wnt7b is transcribed in the distal endoderm during branching morphogenesis while Wnt2 is expressed in the adjacent mesoderm (Weidenfeld et al. 2002). Transcription factors of the TCF/LEF family are expressed in the developing lung, both in the endoderm and mesoderm (Tebar et al. 2001). Conditional deletion of the β-catenin component of the Wnt-signaling pathway has indicated an unequivocal role for Wnt signaling in mouse lung endoderm development (Okubo and Hogan 2004). In the avian lung, our preliminary observations (Maina and Macharia, unpubl.) show that between stages 26 and 30 of embryogenesis (Hamburger and Hamilton 1951), Wnt2, 4, and 7a are expressed but Wnt 6 is not. The expression mainly occurs on the distal (leading=growing) tips of the lung buds (Fig. 37). The Wnts appear to be localized at different time points, with expression apparently ceasing at stage 30. This may mark a turning point between complete organogenesis and start of local expansion of the target cells.

3
Qualitative Morphology

> *The avian lung exhibits a special architecture and upon our understanding of this architecture will depend our conception of its physiology.* Locy and Larsell (1916a)

3.1
General Observations

Few organs have attracted and sustained as much fascination and interest as the avian respiratory apparatus, the lung–air sac system, and yet remained as stubbornly inexplicable. Hans-Reiner Duncker (1974) pointed out that *"the avian respiratory tract has been investigated by scientists as long as they have been studying comparative anatomy"* while Donald Farner (1970) remarked that *"historically, the avian respiratory system is highly ranked among the controversial organ-systems"*. After the unremitting study of the functional morphology of the avian respiratory system for at least the last four centuries (e.g. Coitier 1573), Peter Scheid (1990) conceded that *"we have then to admit that we cannot decide whether the lung structure of birds has evolved out of functional needs or simply out of structural constraints with no significance for the higher efficiency"*. Taking into account the elementary tools used and the less explicative techniques applied, studies by many of the early investigators are impressive for their meticulous attention to details and interpretative skills. Appropriately calling the nineteenth century avian morphologists *"astute scholars"*, Berger (1960) stated that *"any thorough anatomical study (of birds) must begin with a careful analysis of their papers"*. We must, however, add that it is important that the works now be corroborated and subjected to modern research methods. Where equivocal, the conclusion(s) must be revised or rejected. For instance, the one time intuitively appealing hypothesis that 'anatomical (sphincters) valves' directed the air flow in the avian lung (e.g. Dotterweich 1930; Vos 1934) is now totally unsustainable. Based on new physiological and morphological findings (e.g. Banzett et al. 1987, 1991; Wang et al. 1988; Maina and Africa 2000; Maina and Nathaniel 2001), a new insight has emerged: among other structural features, syringeal constriction, bronchial size and geometry, and presence of a segmentum accelerans (SA), may generate inspiratory aerodynamic valving (IAV), means by which air is accelerated downstream of the extrapulmonary primary bronchus (EPPB) past the openings of the medioventral secondary bronchi (MVSB) into the intrapulmonary primary bronchus (IPPB; Sect. 3.11). For want of brevity here, only the main structural features and functional concepts of the avian lung will be highlighted. A comprehensive review of the structure and function of the lung-air sac system of birds was given in Seller (1987), King and McLelland (1989) and Maina (1996, 2002a): further details can be found there.

3.2
Lung

In certain aspects that bear strongly on its function, the structure of the avian respiratory system is exceptional: the lung is practically rigid and inflexible; unlike the mammalian lung that is tidally (bidirectionally=in-and-out) ventilated, the exchange tissue (ET) of the avian lung, specifically the paleopulmonic parabronchi (PPPR; Sect. 3.7), is ventilated unidirectionally and continuously by synchronized action of the air sacs (ASs); the gas exchanger (the lung) has been totally disengaged from the ventilator (the air sacs); compression of the avian lung does not cause significant collapse of the ACs (Macklem et al. 1979); and capacious and transparent, the ASs are totally avascular and play no direct role in gas exchange (e.g. Magnussen et al. 1976). Although deeply marked by the vertebral ribs on the dorsolateral part, i.e. the costal and vertebral surfaces (Fig. 38), in complete contrast to the mammalian lung, the avian lung is never divided into lobes (Fig. 39). From about one-fifth to one-third of the volume of the avian lung is contained between the ribs (e.g. King and Molony 1971). In the more derived birds, where the neopulmonic parabronchi (NPPR) are well developed (Sect. 3.6), the avian lungs are small and fairly wedge-shaped. Such lungs essentially have three surfaces: the dorsolateral surface that contacts the ribs, i.e. the thoracic wall, and is called the costal surface; the dorsomedial surface that contacts the vertebrae and is called the vertebral surface; and the ventromedial surface that is in contact with the tissue of the horizontal septum and is termed the septal surface. Scattered on the septal surface are ostia, openings that connect the bronchi, i.e. PB, SB, and PR, to the ASs (Figs. 40 and 41). With the lungs displaced to the dorsal part of the coelomic cavity (where they are firmly attached to the vertebral ribs; Figs. 38 and 39) and a diaphragm lacking, the liver rather than the lungs, as is the case in mammals, surrounds the heart. In most species of birds, the lungs extend cranially to about the level of the first cervical rib while they terminate caudally near the cranial border of the ilium.

Pertaining to certain general design features that may be purely coincidental or evolutionary (e.g. Perry 1992), the avian respiratory system, the lung-air

3.2 Lung

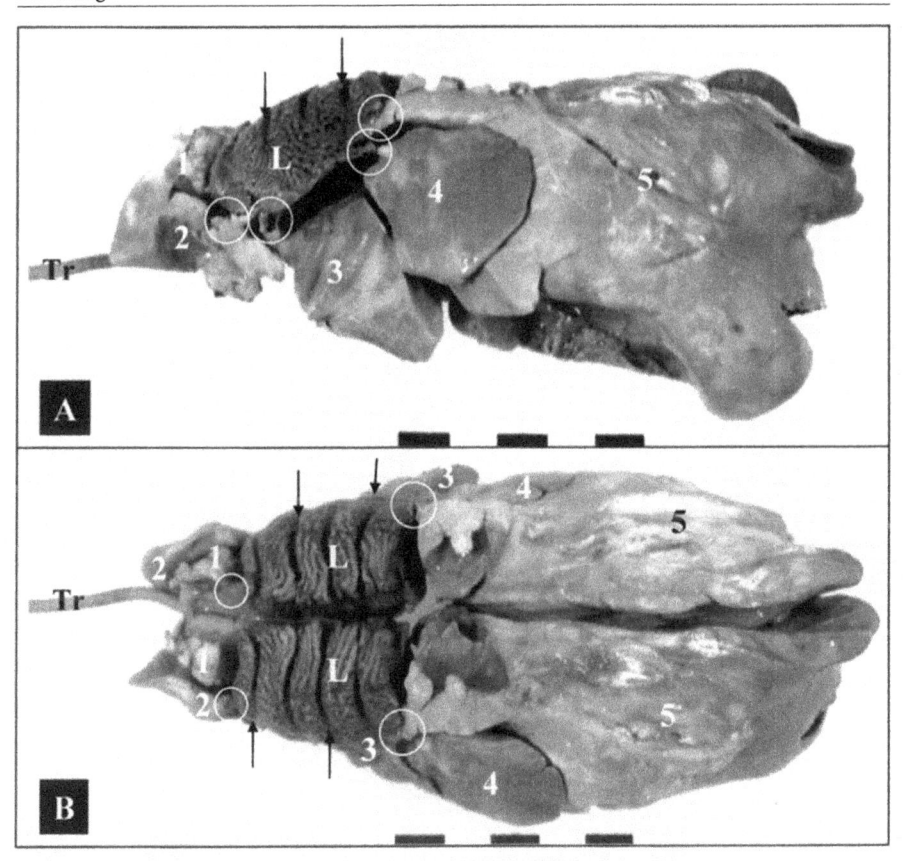

Fig. 38A,B. Latex cast preparations of the lung and the air sacs of the domestic fowl, *Gallus gallus* variant *domesticus*. *Tr* Trachea; *L* lungs; *arrows* costal sulci; *circles* ostia; *1* cervical air sac; *2* interclavicular air sac; *3* craniothoracic air sac; *4* caudothoracic air sac; *5* abdominal air sac. *Scale bars* 1 cm. (**A** from Maina 2002 c; **B** from Maina and Africa 2000)

sac system, resembles the reptilian one (Fig. 42): an anterior compartmented space in which most gas exchange occurs (analogous to the avian lung) and a distended, smooth posterior part (analogous to the ASs of the avian respiratory system) occur (e.g. Maina 1989a; Maina et al. 1999). The African chameleon has some saccular extensions of the lung that branch out into the abdominal cavity (e.g. Patt and Patt 1969; Maina 1998). The high level of gas-exchange efficiency of the avian lung largely arises from the structural uniqueness of the PR and the pattern and relative directions of flow of air and blood, features that form the basis of the cross-current and counter-current gas-exchange systems (Sect. 3.9). Albeit the remarkable specific diversity in the avian taxon, amazingly, the basic structure of the respiratory system in birds is similar. Differences in fine details, particularly regarding the extent of development of the PR, the arrangement of the SB, and the location, connec-

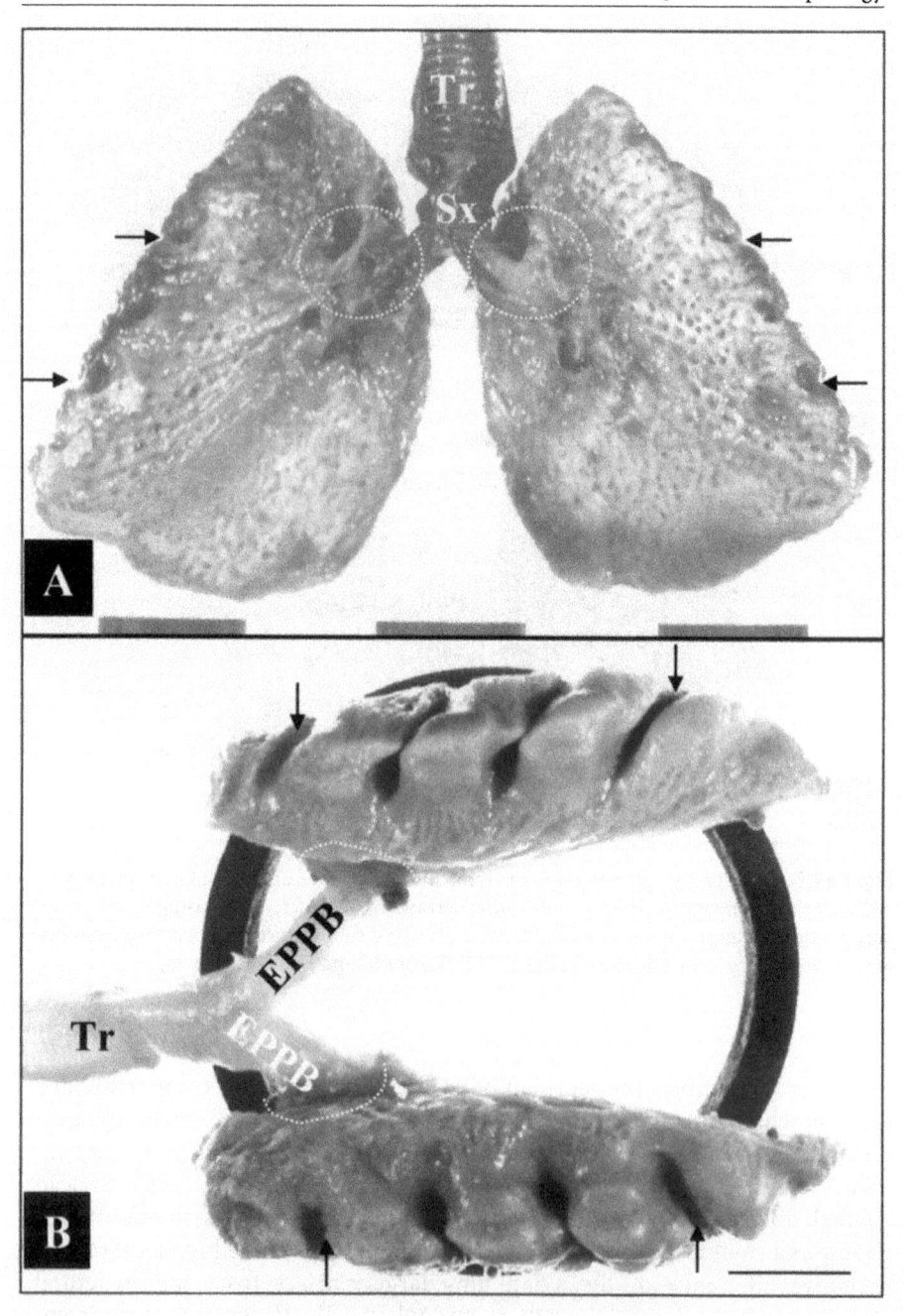

Fig. 39A,B. Medial and dorsal views of the lungs of the domestic fowl, *Gallus gallus* variant *domesticus*, and the ostrich, *Struthio camelus*. *Arrows* Costal sulcae; *Tr* trachea; *Sx* location of the syrinx; *dashed encircled area* hilus; *EPPB* extrapulmonary primary bronchus. *Scale bars* **A**, 1 cm; **B**, 2 cm. (**A** from Maina 2002a; **B** from Maina and Nathaniel 2001)

3.2 Lung

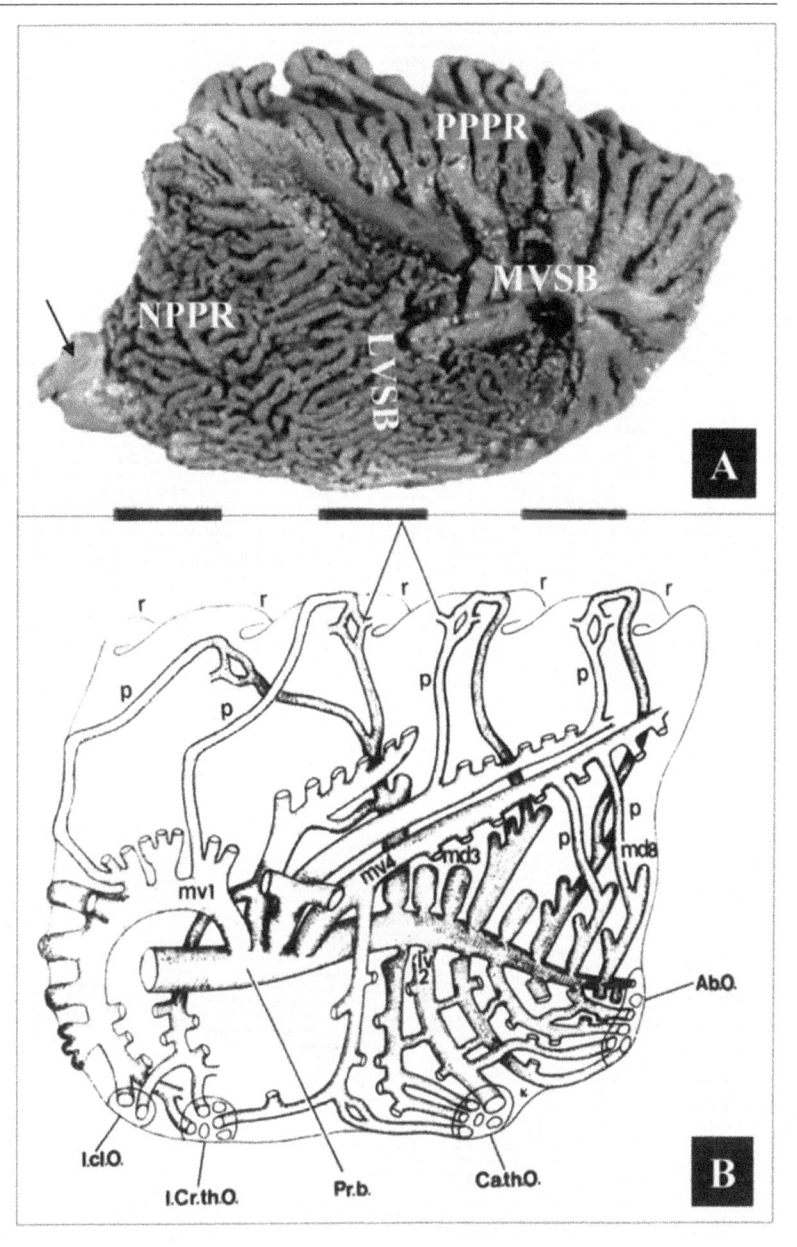

Fig. 40. A Medial view of the lung of the domestic fowl, *Gallus gallus* variant *domesticus*, showing air-conducting passages that include the medioventral secondary bronchi (*MVSB*), lateroventral secondary bronchi (*LVSB*), paleopulmonic parabronchi (*PPPR*) and neopulmonic parabronchi (*NPPR*). *Arrow* Ostium. *Scale bar* 1 cm. **B** The avian lung drawn as transparent to show the airways. *mv* Medioventral secondary bronchi; *md* mediodorsal secondary bronchi; *lv* lateroventral secondary bronchi; *p* parabronchi; *r* costal sulci; *AbO* abdominal ostium; *CathO* caudal thoracic ostium; *Prb* primary bronchus; *ICrthO* cranial thoracis ostium; *IclO* interclavicular ostium. (**A** from Maina 2002c; **B** from King and McLelland 1989, redrawn with permission from the publisher)

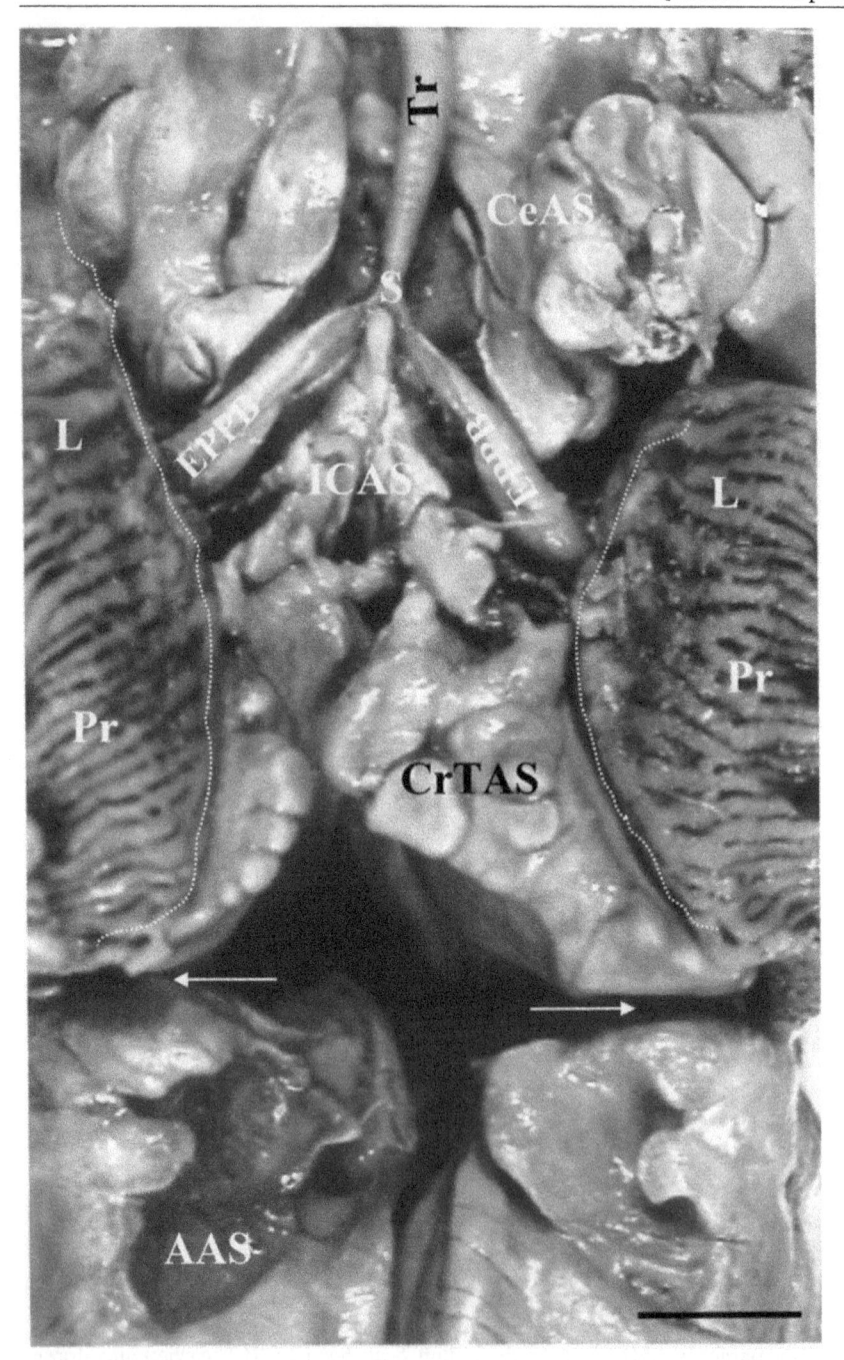

Fig. 41. Ventral view of a cast of the lung and air sacs of the domestic fowl, *Gallus gallus* variant *domesticus*, showing trachea (*Tr*), syrinx (*S*), cervical air sac (*CeAs*), extrapulmonary primary bronchus (*EPPB*), interclavicular air sac (*ICAS*), lung (*L*), primary bromchi (*Pr*), craniothoracic air sac (*CrTAS*), abdominal air sac (*AAS*), oblique septum (*dashed line*), and horizontal septum (*arrows*). *Scale bar* 1 cm. (Maina and Africa 2000)

3.2 Lung

Fig. 42. Schematic diagrams of the lung of the monitor lizard, *Varanus exanthematicus* (**A**), and a mallard duck, *Cairina moschata* (**B**), showing the similarities between the designs of the reptilian and avian respiratory systems. (Reproduced from Duncker 1979b)

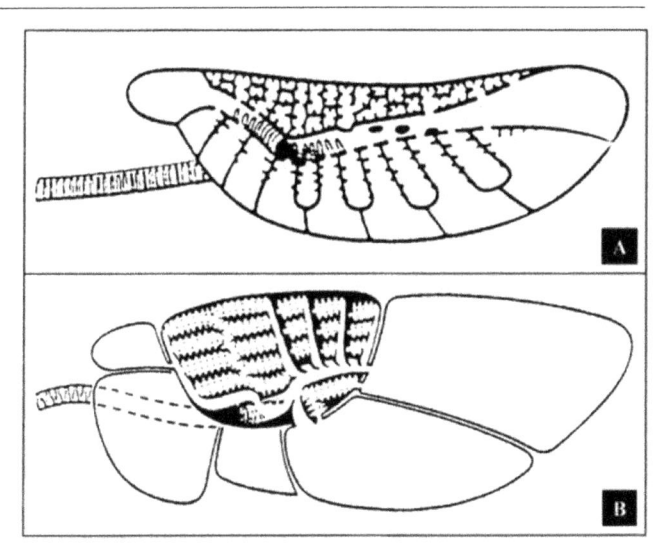

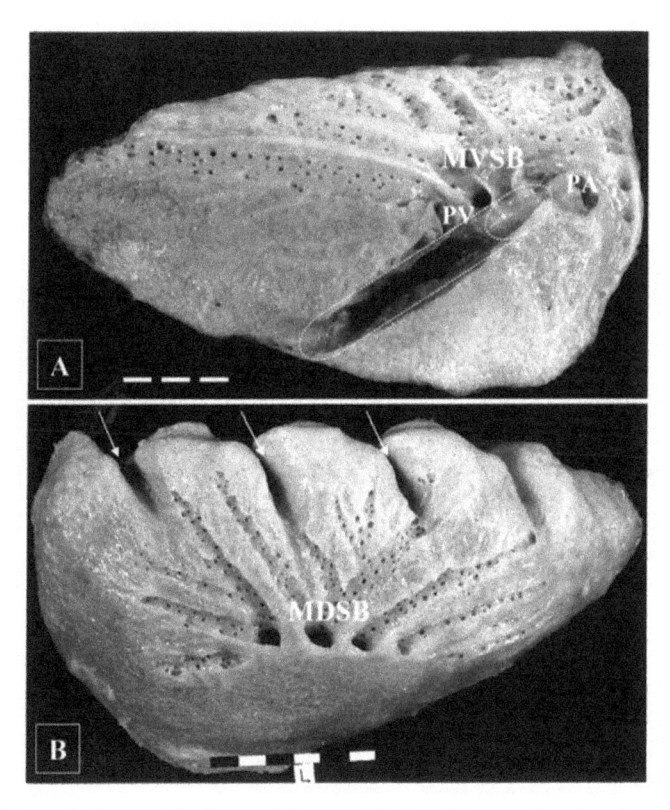

Fig. 43A,B. Medial and lateral views of the lung of the ostrich, *Struthio camelus*, showing medioventral secondary bronchi (*MVSB*), mediodorasal secondary bronchi (*MDSB*), pulmonary artery (*PA*), and pulmonary vein (*PV*). *Arrows* Costal sulci. *Dashed area* (A) Primary bronchus. *Scale bars* 1 cm. (Maina and Nathaniel 2001)

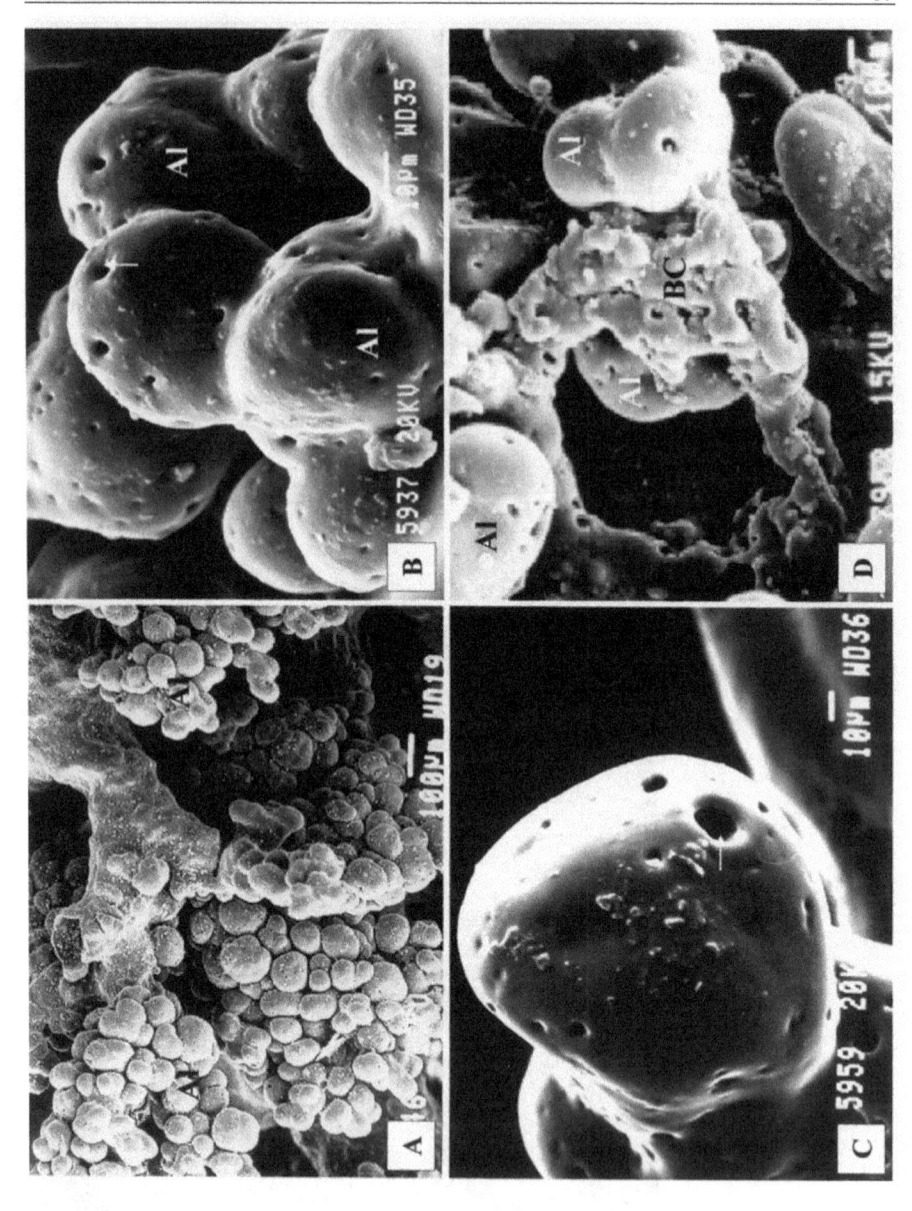

Fig. 44. Cast of a human lung given here to emphasize the structural differences that exist between it and the avian lung. **A** Grape-like arrangement of the alveoli (*Al*). **B, C** Alveoli (*Al*) are spherical, terminal gas-exchange units. *Arrow* Interalveolar pore (pore of Kohn). **D** Alveoli (*Al*) interface with blood capillaries (*BC*) located in the interalveolar septum. (**A, D** from Maina and van Gils 2001)

tion, number of ASs, however, do occur. While the differences may be of some phylogenetic importance, most of them are of little, if any, functional consequence.

3.3 Airway (Bronchial) System

3.3.1 Primary Bronchus (PB)

The airway (bronchial) system of the avian lung comprises of a three-tiered system of air conduits. The IPPB gives rise to four sets of SB as it transits the lung proximal-distally (Figs. 40 and 43). The architecture of the bronchial sys-

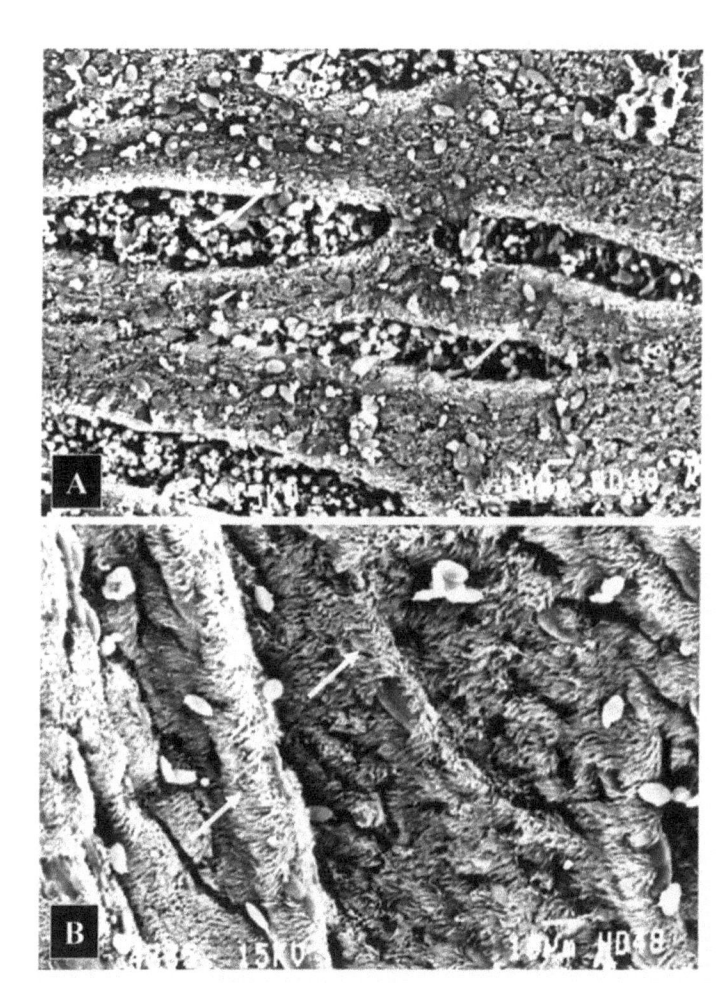

Fig. 45A,B. Different views of the epithelial lining of the primary bronchus of the domestic fowl, *Gallus gallus* variant *domesticus*, showing conspicuous folding (*arrows*) and presence of ciliated epithelial cells. (**A** from Maina and Africa 2000)

tem in the avian lung fundamentally differs from that in the mammalian lung. Unlike in the BRLu where the airway system forms by invariant dichotomous bifurcation, in the avian lung, a continuous bronchial system forms by sprouting of the SB from the IPPB (Figs. 7 and 8). In a hoop-like arrangement, the PR interconnect the SB, establishing continuity of the bronchial system. While the so-called respiratory tree or bronchioalveolar tree of the mammalian lung is blind-ended (Fig. 21), with alveoli forming terminal 'grapefruit-like' arrangements (Fig. 44), in the avian lung the ACs are essentially continuous anastomosing air conduits (Sect. 3.4).

After piercing the horizontal septum, the EPPB enters the lung at the hilus, a site near the junction of the cranial and middle thirds of the lung (Fig. 39). There, it relates to the pulmonary artery (PA) and vein (PV; Fig. 43A). The IPPB passes through the lung in a rather curved manner (Figs. 40B). In transit, it changes in cross-sectional area (e.g. King 1966). The airway terminates on the caudal margin of the lung by entering the abdominal air sac (AAS).

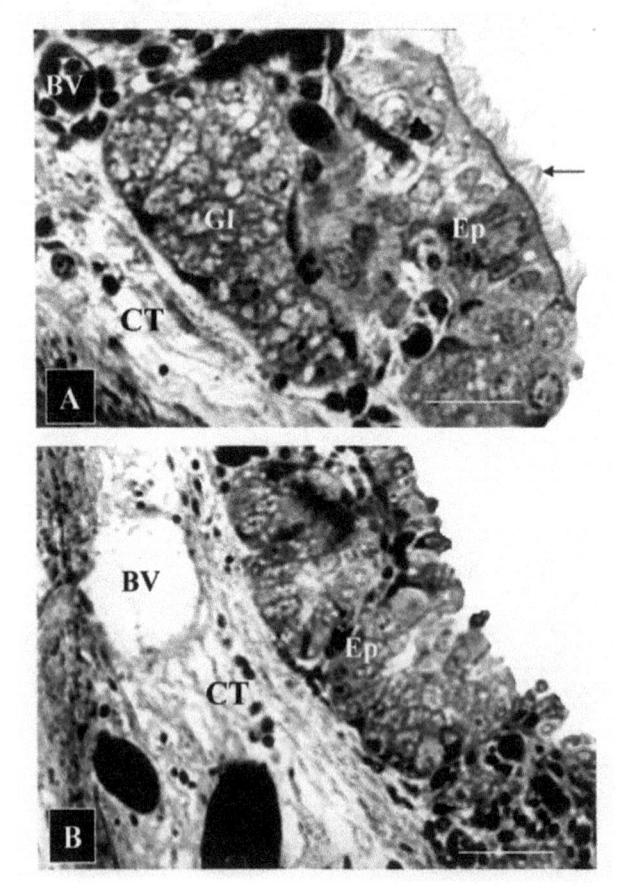

Fig. 46A,B. Epithelium lining the primary bronchus of the domestic fowl, *Gallus gallus* variant *domesticus* (*Ep*), showing ciliated cells (*arrows*). *Gl* Subepithelial gland; *BV* blood vessel; *CT* connective tissue. *Scale bars* 10 µm

3.3 Airway (Bronchial) System

Supported by cartilaginous plates, the IPPB is lined by a pseudostratified simple columnar epithelium with characteristic longitudinal folds that support ciliated cells and mucous-secreting goblet cells (Figs. 45 and 46).

3.3.2
Secondary Bronchi (SB)

The four sets of SB that occur in the avian lung are named according to the area of the lung that they supply with air. In many species, typically four in number, the MVSB arise from the dorsomedial wall of the cranial third of the

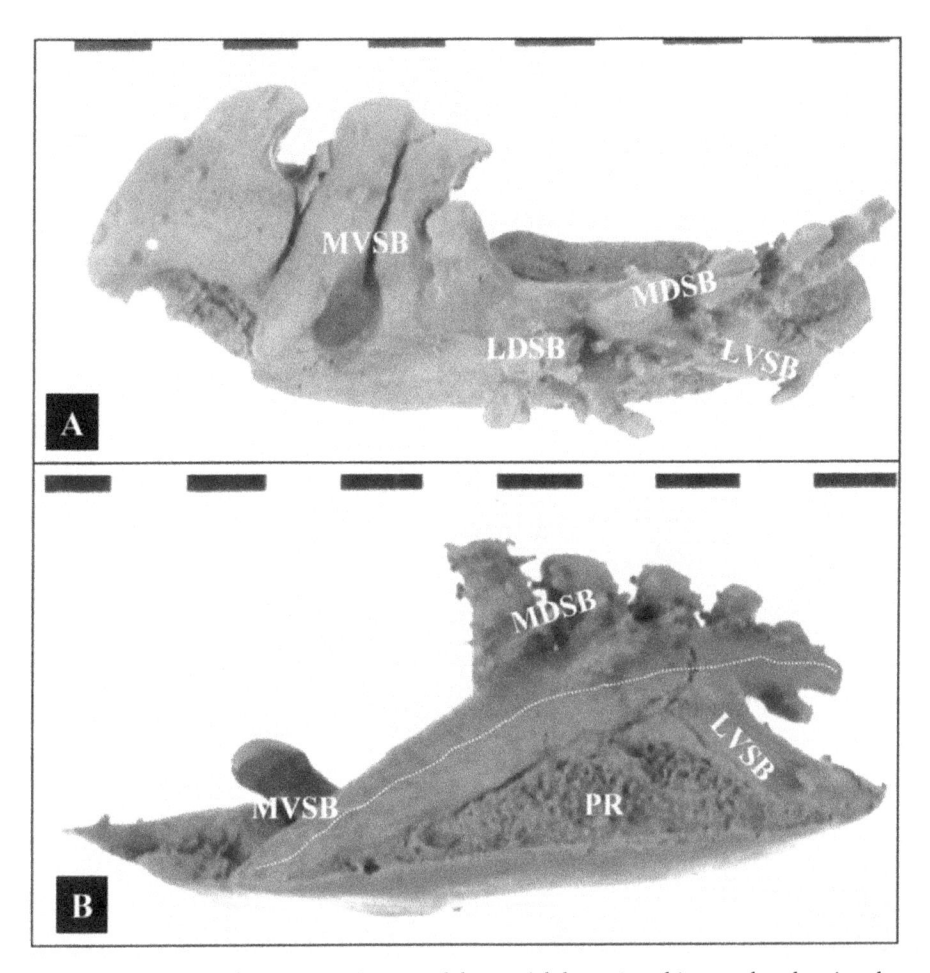

Fig. 47A,B. Casts of the primary airways of the ostrich lung, *Struthio camelus*, showing the origins of the medioventral secondary bronchi (*MVSB*), laterodorsal secondary bronchi (*LDSB*), mediodorsal secondary bronchi (*MDSB*), and laterodorsal secondary bronchi (*LVSB*) arising from the primary bronchus (*dashed line*). PR Parabronchi. *Scale bars* 1 cm. (Maina and Nathaniel 2001)

IPPB (Figs. 40, 43, 47, and 48). The mediodorsal secondary bronchi (MDSB), the lateroventral secondary bronchi (LVSB), and the laterodorsal secondary bronchi (LDSB) originate from the caudal two-thirds of the IPPB (Figs. 40, 43B, 47 and 48B). Seven to ten in number and decreasing in size craniocaudally, the MDSB arise in a line from the dorsal circumferential wall of the IPPB

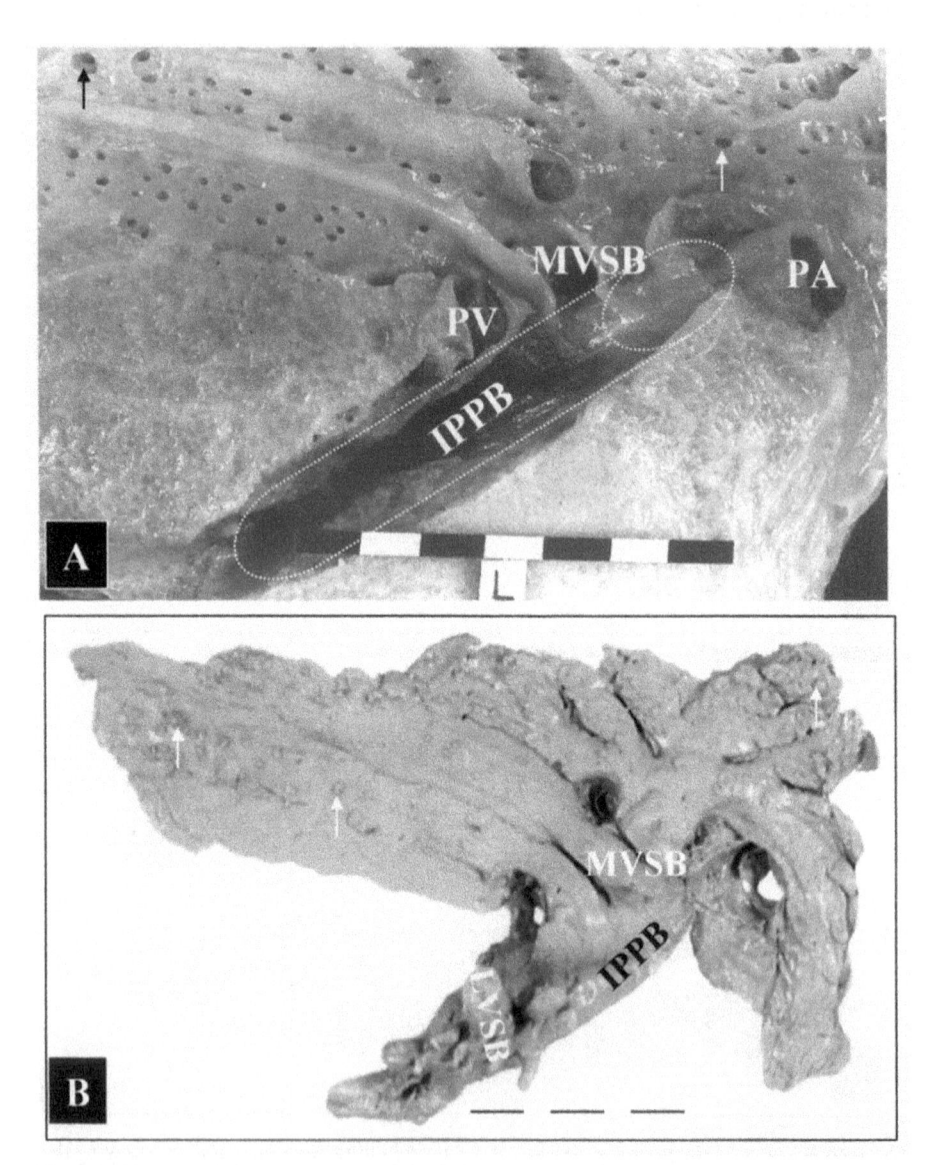

Fig. 48. Close-up (**A**) and cast (**B**) of the hilus region of the lung of the ostrich, *Struthio camelus*, showing parabronchi (*arrows*), medioventral secondary bronchi (*MVSB*), pulmonary vein (*PV*), pulmonary artery (*PA*), intrapulmonary primary bronchus (*IPPB*), and laterodorsal secondary bronchi (*LDSD*). *Scale bars* 1 cm. (Maina and Nathaniel 2001)

(Fig. 47). The LVSB are highly variable in number and size and originate in a row from the IPPB at the same craniocaudal extent as the MDSB, but from the ventral surface (Figs. 40B and 47), i.e. directly opposite the openings of the MDSB. The LDSB arise from the same craniocaudal extent of the IPPB as the MDSB and the LDSB but from the lateral wall of the IPPB (Fig. 47A). Numerous and generally small in size, the LDSB resemble PR in structure. Unlike the other groups of SB that arise in a line, the LDSB have a scattered origin and among SB are the most highly variable in number and size.

The larger SB have a constricted origin and, relative to the course of the IPPB, are angulated (Fig. 49). For a short distance from the IPPB, the SB are lined by an epithelium similar to that which lines the IPPB (Figs. 45 and 46). Mucus glands are, however, lacking in the epithelium that lines the rest of the

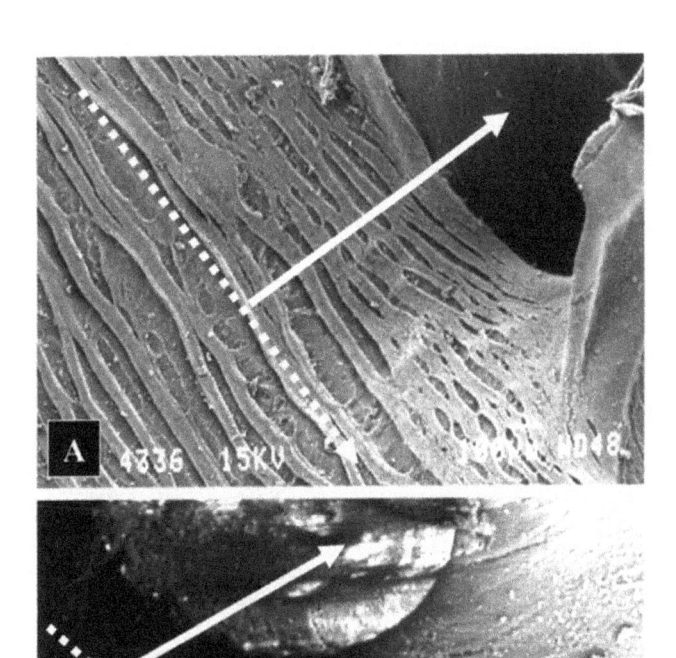

Fig. 49. Openings and angulations of a mediodorsal secondary bronchus (**A**) and a mediodorsal secondary bronchus (**B**) in the domestic fowl, *Gallus gallus* variant *domesticus*. The *continuous arrow* shows the secondary bronchus while the *dashed arrow* shows the primary bronchus. (**A** from Maina and Nathaniel 2001)

SB (Duncker 1971). While in the mammalian lung diametric and histological transition from a bronchus to a smaller order airway occurs gradually, in the avian lung the changes between a SB and a PR may be rather abrupt (del Corral 1995). With much of it consisting of cuboidal (sometimes ciliated) and squamous cells, the epithelium lining some sections of the SB is histologically similar to that of the PR.

3.3.3
Parabronchi (Tertiary Bronchi) (PR)

A short distance from their origin, the SB give rise to PR. In cross-section, the PR are rather hexagonal in shape (Figs. 50 and 51). In the lungs of some species, e.g. the domestic fowl, the PR are separated by a band of connective tissue, the interparabronchial septa (IPRS; Figs. 50B, 51B and 52A,B) while in

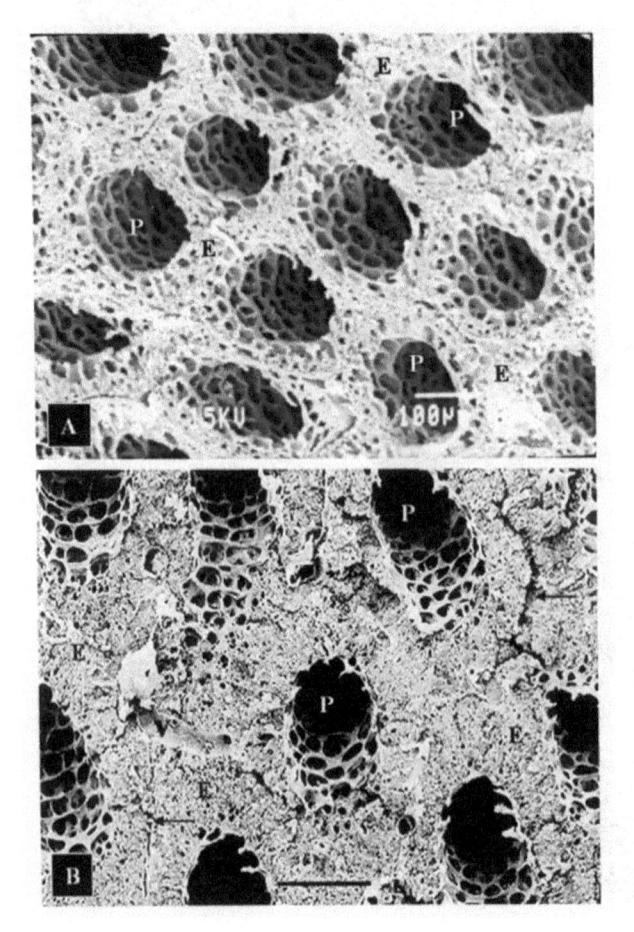

Fig. 50A,B. Parabronchi (*P*) of the lung of the domestic fowl, *Gallus gallus* variant *domesticus*, surrounded by exchange tissue (*E*). *Arrows* Interparabronchial septa; *V* blood vessel. *Scale bars* **B** 100 μm. (**A** from Maina 2003b; **B** from Maina 1994)

3.3 Airway (Bronchial) System

others, e.g. the house sparrow, *Passer domesticus* (Fig. 51A), and the ostrich, *Struthio camelus* (Fig. 53), IPRS are lacking. The extents of development of the IPRS vary greatly between different species of birds (Duncker 1971). In the lungs of seven orders of birds examined by Maina et al. (1982a), IPRS were completely lacking in the passeriform, columbiform, cuculiform, and psittaciform species; absent or very thin in anseriform; and well developed in charadriiform and galliform species. In the lungs of weak flyers and flightless species, e.g. penguins and galliform species, the PR are typically wide in diameter, have prominent atria, thin gas-exchange mantle relative to the PRL, and

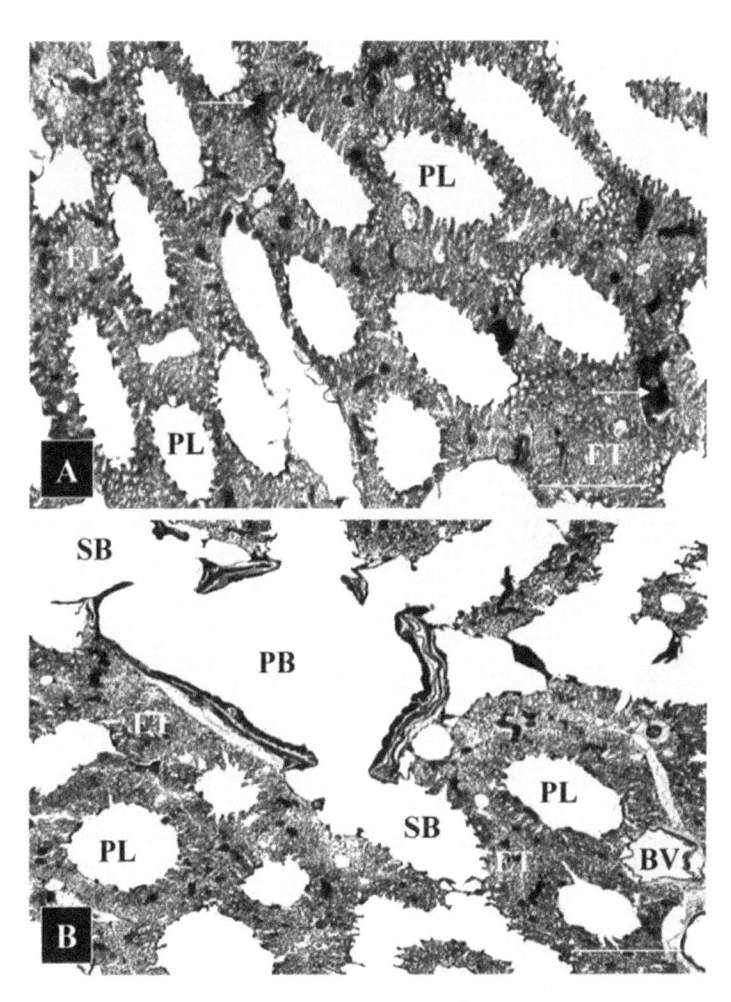

Fig. 51A,B. Lungs of the house sparrow, *Passer domesticus*, and the domestic fowl, *Gallus gallus* variant *domesticus*, showing parabronchial lumina surrounded by exchange tissue (*ET*). *Arrows* Interparabronchial blood vessels; *PB* primary bronchus; *SB* secondary bronchus; *BV*, blood vessel. Note that interparabronchial septa are lacking in **A**. *Scale bars* 100 μm

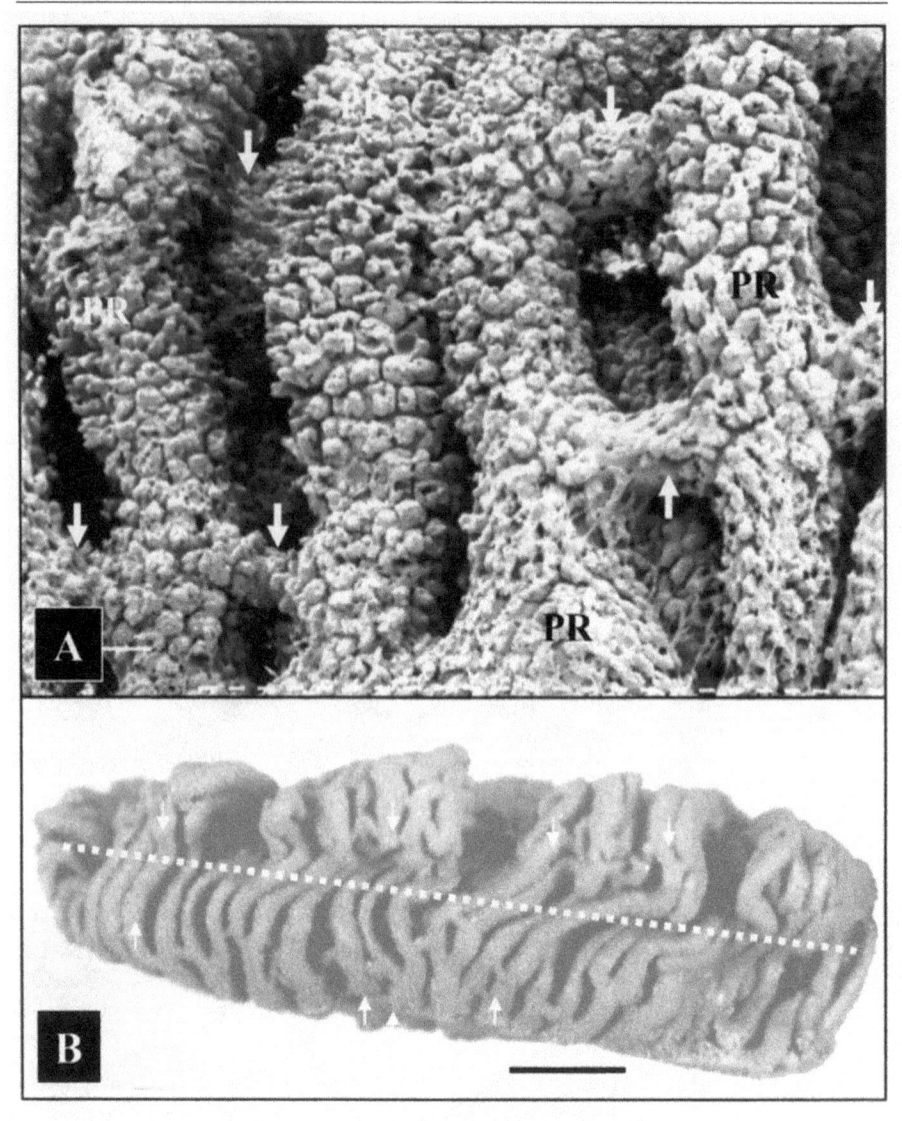

Fig. 52. A Lung of the domestic fowl, *Gallus gallus* variant *domesticus*, showing stacks of parabronchi (*PR*) anastomosing along their lengths (*arrows*) (Maina 1988). **B** Mediodorsal and medioventral secondary bronchi (*arrows*) anastomosing on the dorsal vertebral surface of the lung through hoop-like arranged parabronchi (*arrows*). *Scale bars* **A**, 2 mm; **B**, 1 mm

thin-to-thick IPRS. In contrast, in the lungs of small, strong fliers, the PR have a small diameter, the proportion of ET is greater, and the IPRS are thin or absent. The smallest PR and the greatest proportion of ET occur in the lungs of the psittaciform and passeriform species (Duncker 1971; Maina et al. 1982a).

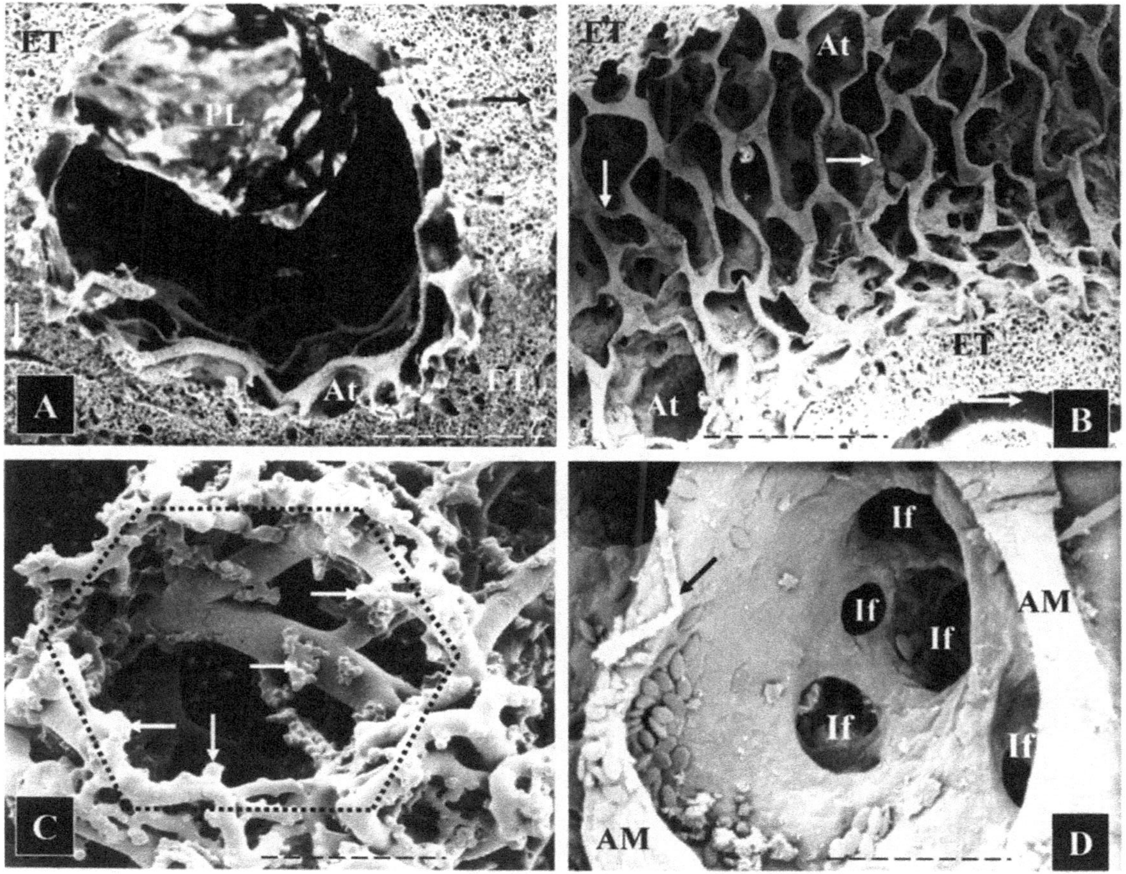

3.3 Airway (Bronchial) System 81

Fig. 53. A,B Lung of the domestic fowl, *Gallus gallus* variant *domesticus*, showing parabronchial lumina (*PL*) surrounded by exchange tissue (*ET*). *Arrows* (**A**) Interparabronchial septum; *arrows* (**B**) interatrial septa; *At* atria. **C** Cast of a parabronchus showing interparabronchial arteries (*dotted hexagon*) giving rise to intraparabronchial arteries (*arrows*). **D** Close-up of an atrium surrounded by interatrial muscles (*AM*) giving rise to infundibulae (*If*). *Scale bars* **A–C**, 0.5 mm; **D**, 0.2 mm. (**A** from Maina 2002c; **C** from Maina and Cowley 1998)

In the lungs of the avian species where IPRS occur, IPRBVs are located at the middle of adjacent PR (Figs. 50B and 51B). In those lungs where IPRS are lacking, the IPRBVs mark the extents of the gas ET mantles of adjacent PR (Figs. 50B and 51). The PR anatomose profusely (Fig. 52). The MVSB and the MDSB are coupled by hoop-like oriented PR that on the whole form the paleopulmonary part of the lung (Fig. 52B): the neopulmonary section is formed by the PR that connect the LDSB, LVSB, and the caudal ASs (Duncker 1971; Sect. 3.6). Visible on the vertebral surface of the lung, the planum anastomotica marks the site where PPPR connect the MVSB and MDSB (Fig. 52B).

3.3.4
Atria, Infundubulae, and Air Capillaries (ACs)

The walls of the PR and those of sizeable lengths of the SB are formed by a gas-exchange mantle (Figs. 50, 51 and 53A,B). Blood vessels mark the periphery of individual PR (Fig. 53C). The thickness of the mantle ranges from 200–500 μm in different species (Duncker 1974, 1979a). Circumferentially, the PRL are perforated by many openings that lead into fairly spherical compartments, the atria (Fig. 53A,B,D). The atria are delineated by bundles of connective tissue, the interatrial septa (Figs. 53B,D and 54). In the pigeon, *Columba livia*, and the gull, *Anas platyrhynchus*, the atria measure from 60–100×120–130 μm in diameter (West et al. 1977). In small species of birds that are characteristically good fliers, atria are shallower (e.g. Duncker 1974; Maina et al. 1982a). Three to eight funnel-shaped ducts, the infindibulae (Figs. 53D, 55, 56 and 57A), project from the atrium. The infundibulae open into ACs (Figs. 57B, 58 and 59). In *Columba* and *Anas*, the infundibulae have a diameter of 25–40 μm and extend into the ET for a depth of about 100–150 μm (West et al. 1977). The ACs anatomose with each other profusely (Figs. 57B, 58 and 59). In contrast to the dead-ending alveoli that form the terminal parts of the airway system of the mammalian lung, anatomosing profusely, strictly, the ACs do not form cul-de-sacs (Figs. 58 and 59). The ACs and BCs constitute about 90 % of the volume of the ET mantle (e.g. Maina et al. 1982a) and intertwine intimately.

Historically, the terms 'air capillary' and 'blood capillary' were derived from observation that the ET of the avian lung mainly consisted of a network of minute air and vascular units (Huxley 1882; Schulze 1908). Our recent study by three-dimensional computer reconstruction of serial sections (Woodward and Maina 2005; Fig. 60) has shown that the entrenched notion that the ACs are straight (nonbranching), blind-ending tubules that project outwards from the PRL and the BCs direct tubules that run inwards parallel to and in contact with the ACs to be is overly simplistic, misleading, and totally incorrect. The ACs are rather rotund structures that interconnect directly or via short, narrow passageways (Figs. 60F and 61A) while the blood capillaries (BCs) comprise of diffusely coupled segments of rather constant dimensions (Fig. 61B).

3.3 Airway (Bronchial) System

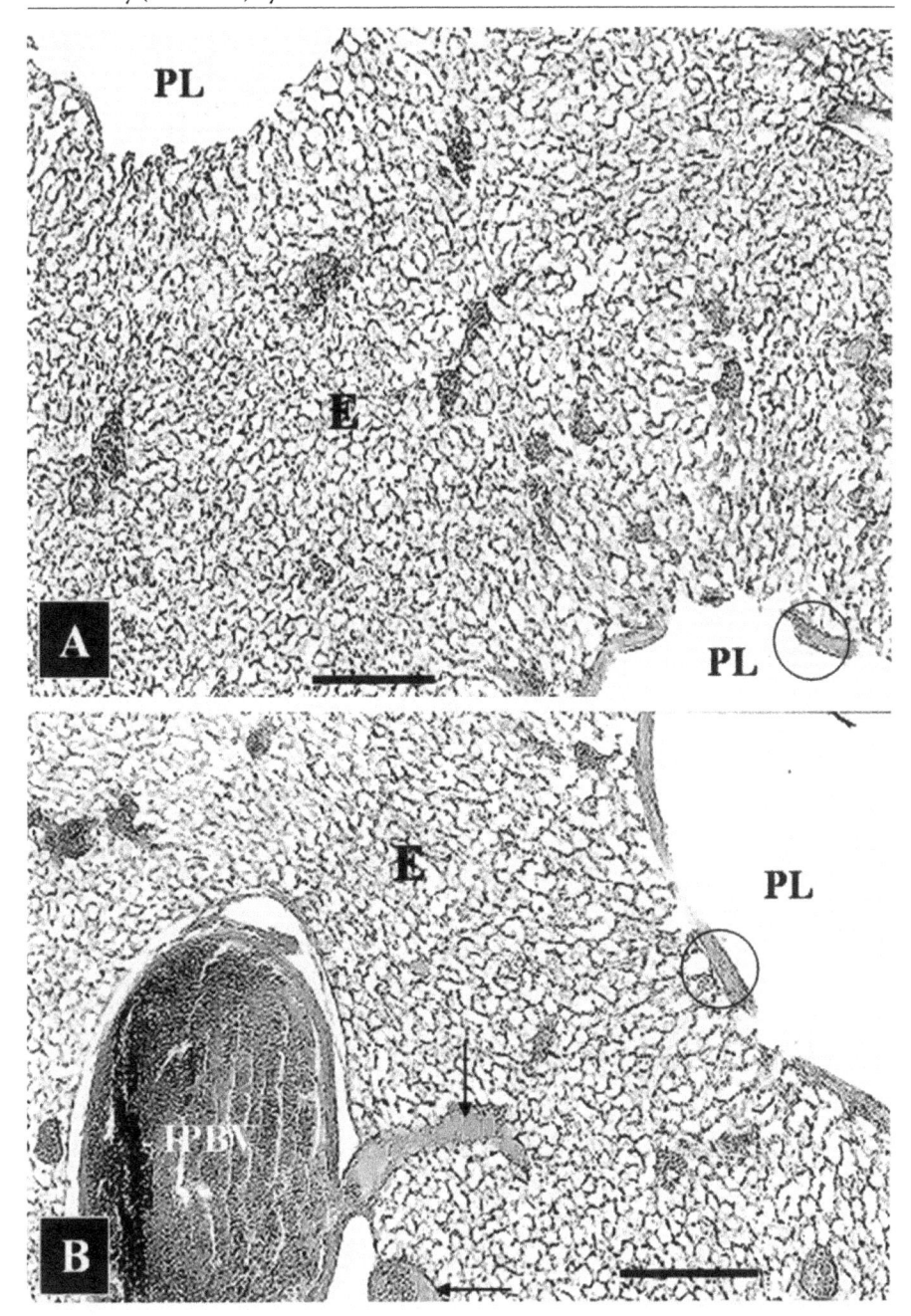

Fig. 54A,B. Lung of the ostrich, *Struthio camelus*, showing parabronchial lumina (*PL*), exchange tissue (*E*) and an interparabronchial blood vessel (*IPBV*) giving rise to an intraparabronchial blood vessel (*arrows*). *Circle* Atrial muscle. *Scale bars* 100 µm. (Maina and Nathaniel 2001)

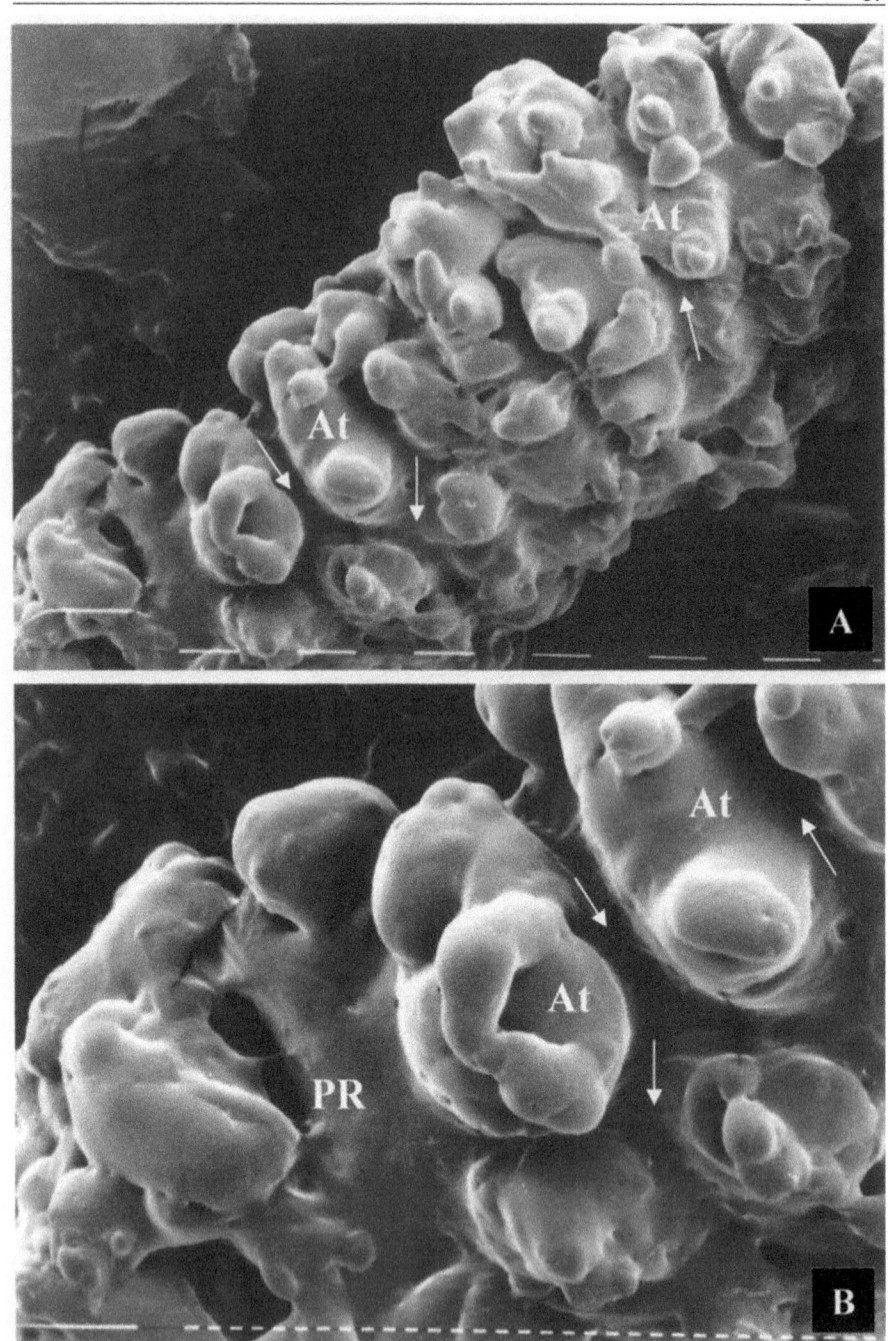

Fig. 55A,B. Casts of the lung of the domestic fowl, *Gallus gallus* variant *domesticus*, showing parabronchi (*PR*) with the atria (*At*) separated by interatrial septa (*arrows*). *Scale bars* **A** 100 µm; **B** 10 µm

3.3 Airway (Bronchial) System 85

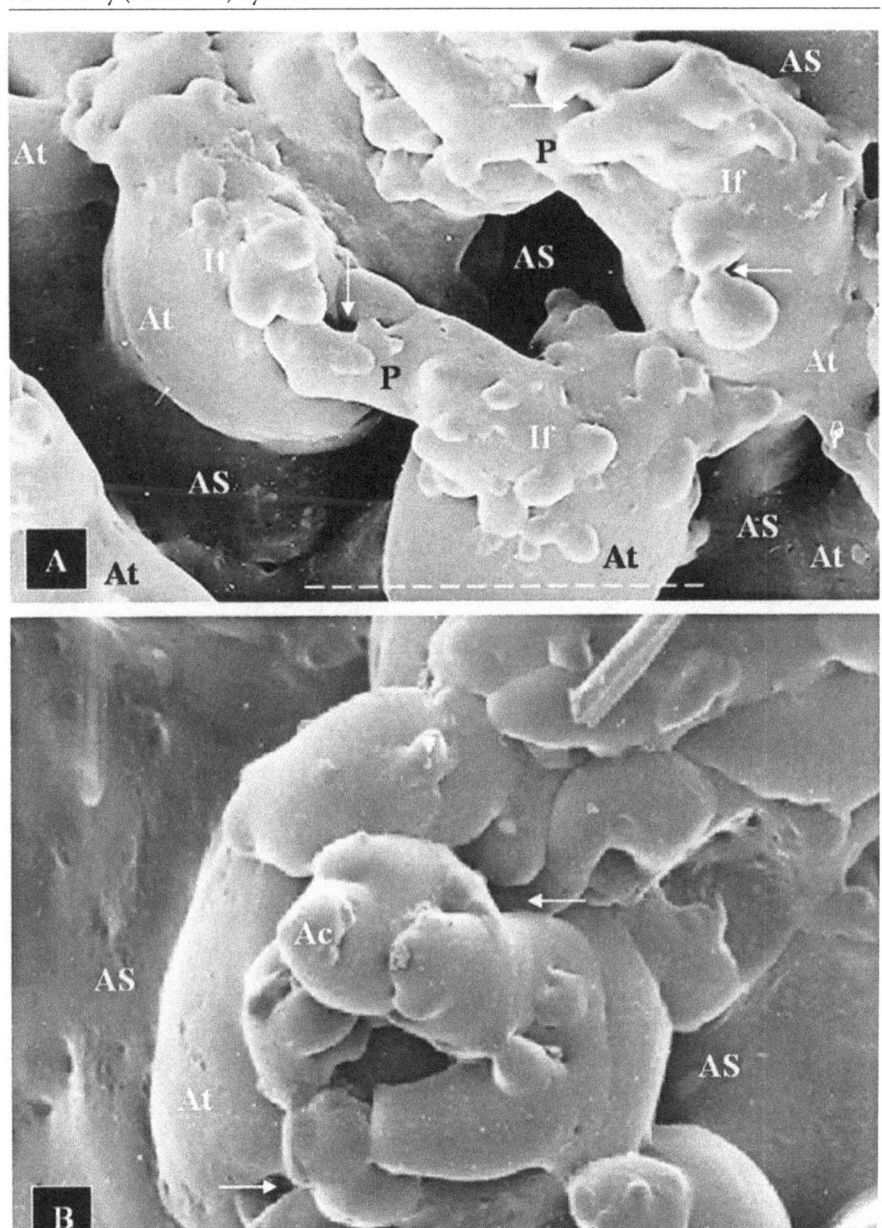

Fig. 56A,B. Casts of the lung of the domestic fowl, *Gallus gallus* variant *domesticus*, showing atria separated by interarial septa (*arrows*). Infundibula and air capillaries of different atria anastomose profusely. *AC* Air capillaries. *Scale bars* **A** 100 µm; **B** 2 µm. (**A** from Maina 1994; **B** from Maina 1998)

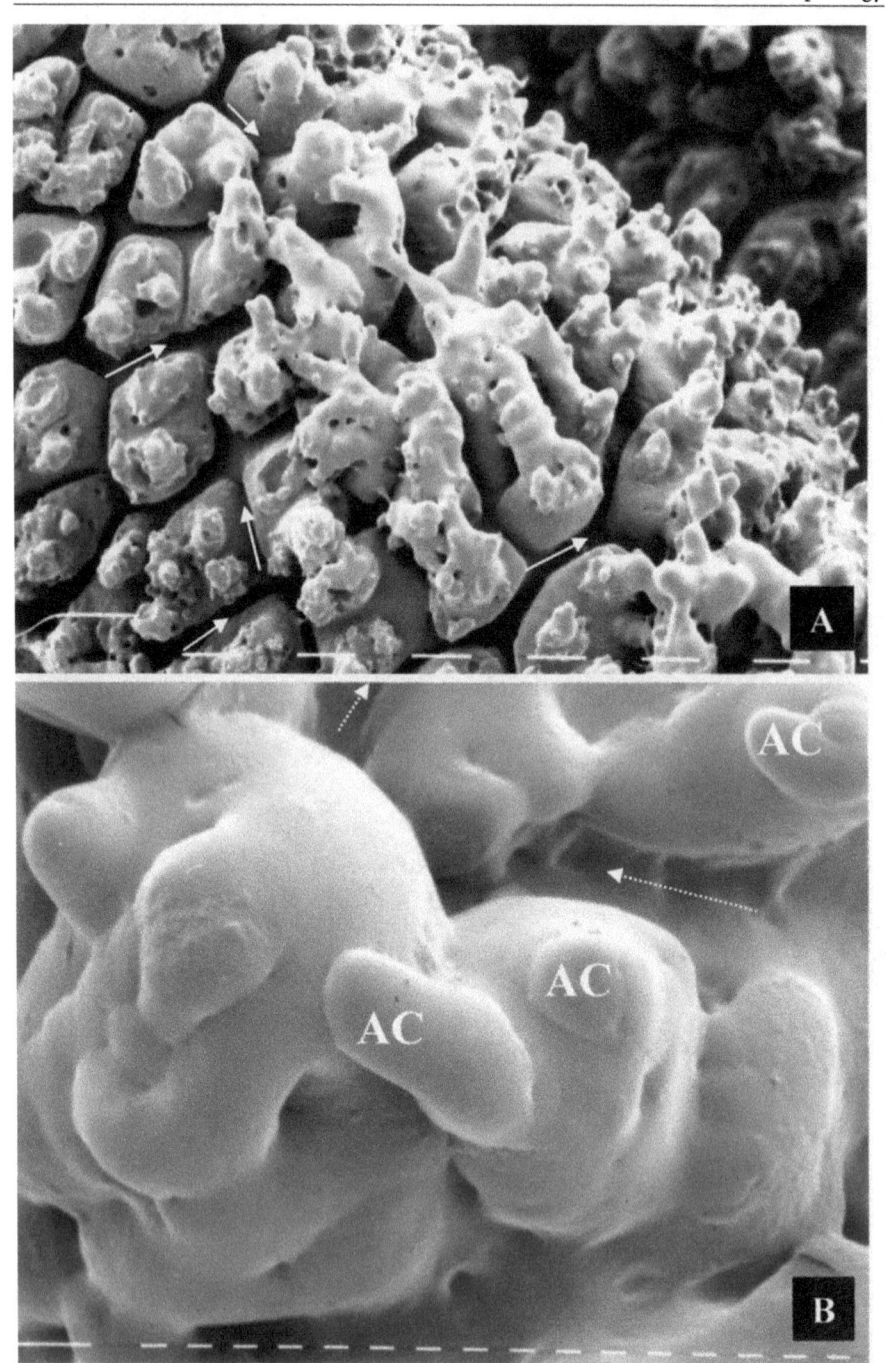

Fig. 57A,B. Casts of the lung of the domestic fowl, *Gallus gallus* variant *domesticus*, showing close-ups of anastomosing air capillaries (*AC*) of different atria (*P*). Separated by interatrial septa (*AS*), the atria (*AT*) give rise to infundibula (*If*) that in turn generate air capillaries. *Arrows* Areas occupied by the blood capillaries. *Scale bars* 5 μm. (**A** from Maina 1988)

3.3 Airway (Bronchial) System 87

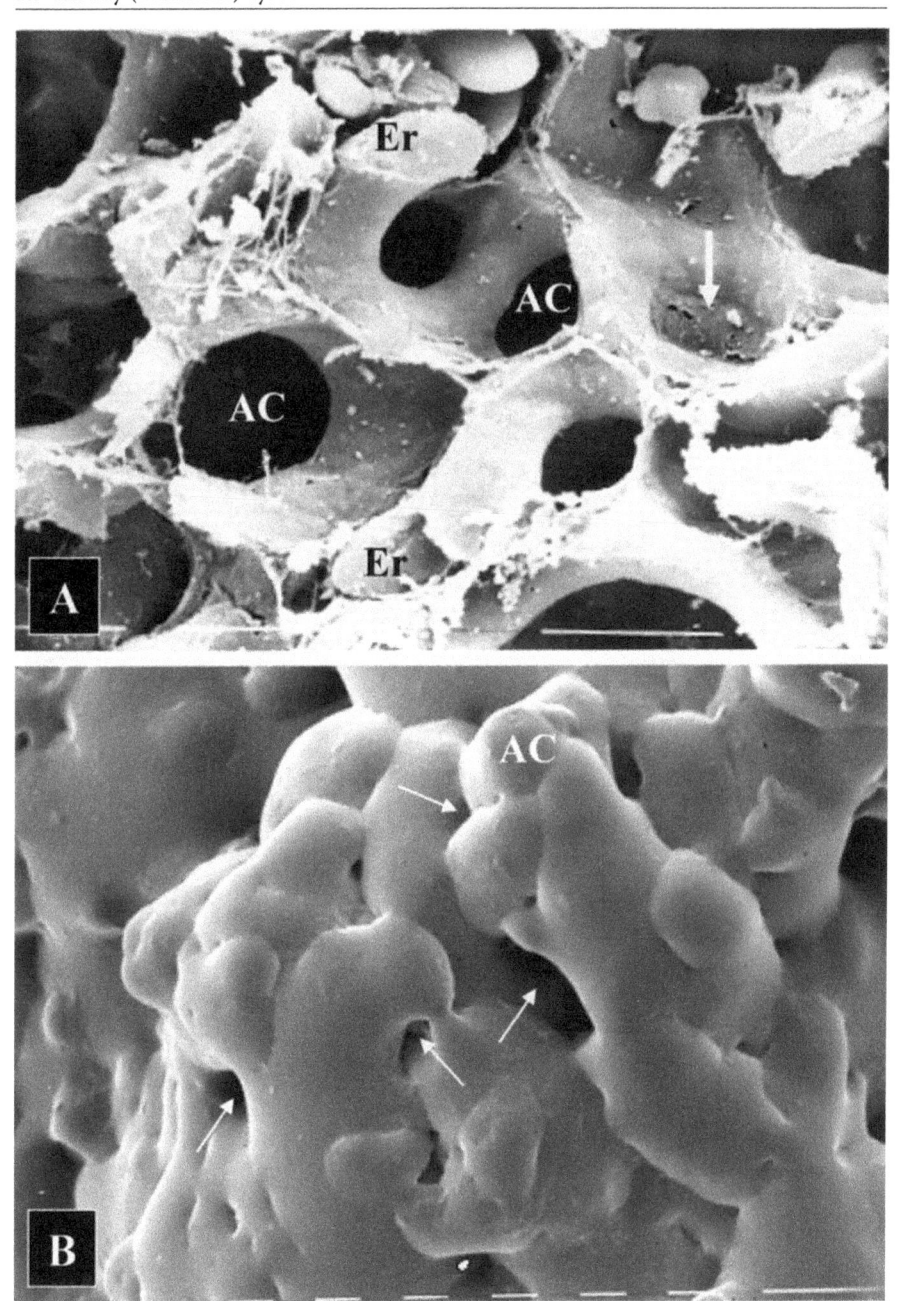

Fig. 58. A Lung of the domestic fowl, *Gallus gallus* variant *domesticus*, showing air capillaries (*AC*) and blood capillaries containing erythrocytes (*Er*). *Arrow* An area where air capillaries lie adjacent to each other. **B** Close-up of a cast of the lung of the domestic fowl, *Gallus gallus* variant *domesticus*, showing air capillaries (*AC*) anastomosing profusely. *Arrows* Blood capillaries. *Scale bars* **A** 20 µm; **B** 2 µm. (**A** from Maina 1982a; **B** Maina 2002b)

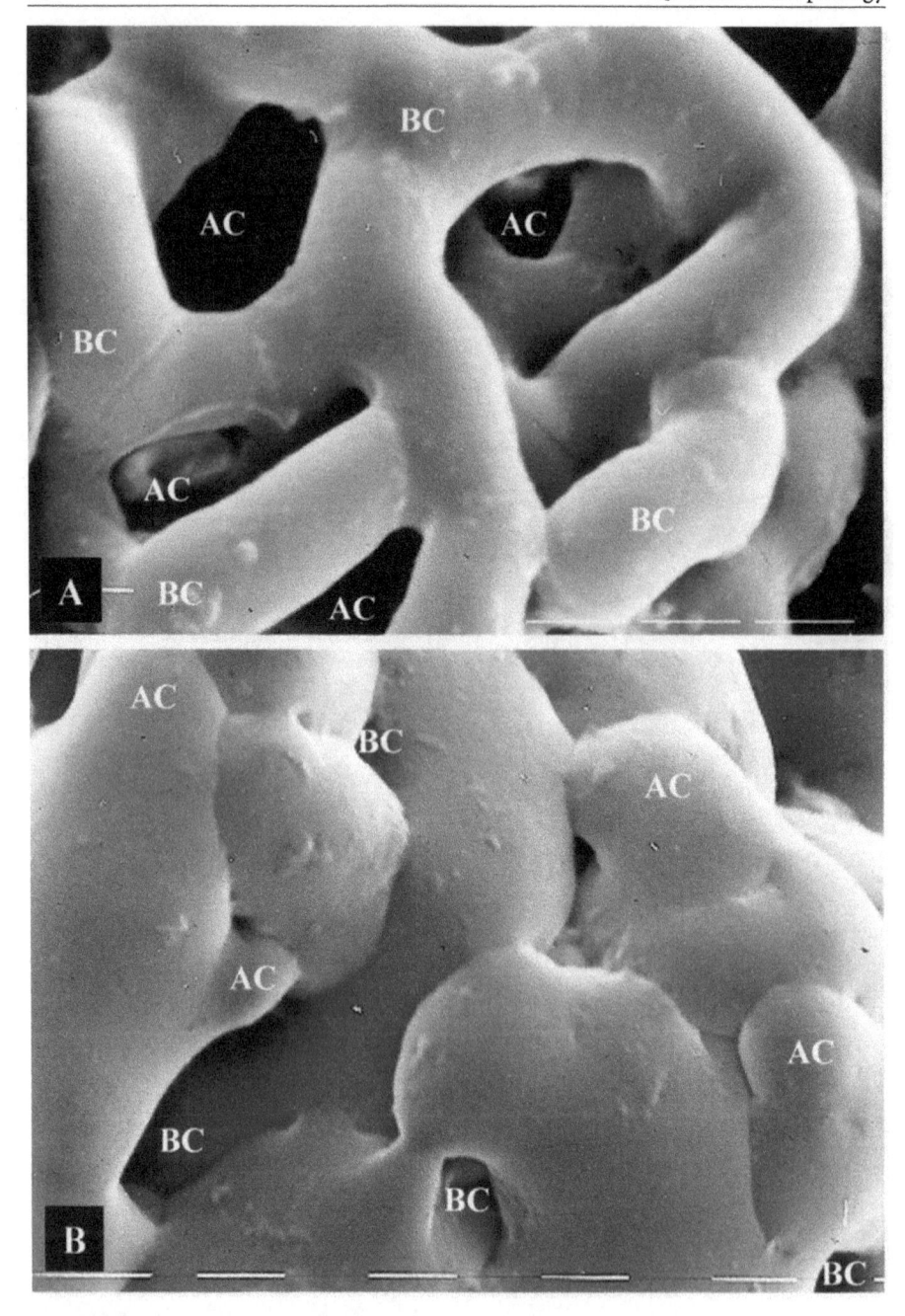

Fig. 59A,B. Casts of blood capillaries (*BC*) and air capillaries (*AC*) of the lung of the domestic fowl, *Gallus gallus* variant *domesticus*. *Scale bars* 5 µm. (Maina 1982a)

3.3 Airway (Bronchial) System

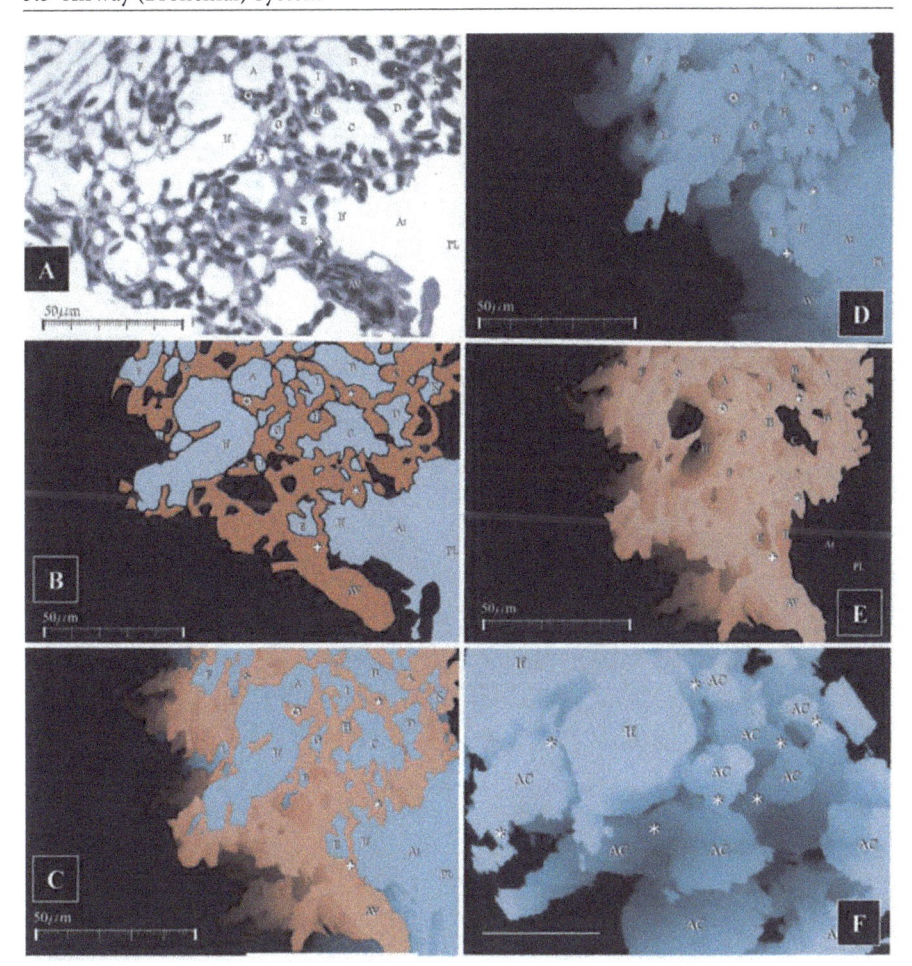

Fig. 60. A Toluidine blue stained section of the lung of the Muscovy duck, *Cairina moschata*, showing the area (to the *right of the dashed line*) on which three-dimensional reconstruction was performed. **B** Shaded airway components (*blue*) and vascular units (*red*). **C** Reconstruction of the airway (*blue*) and the vascular (*red*) systems. **D** Isolated airway system. **E** Isolated vascular system. In **A–E**, the individual units are labeled in *single letters* and *same symbols* to show the same units (for ease of comparison). *If* Infundibulum; *At* atrium; *PL* parabronchial lumen; *AV* atrial vein. **F** Close-up of three-dimensional reconstruction of the air capillaries. *AC* Air capillaries; *If* infundibulum; *stars* connections between air capillaries. *Scale bar* 10 μm. (Woodward and Maina 2005)

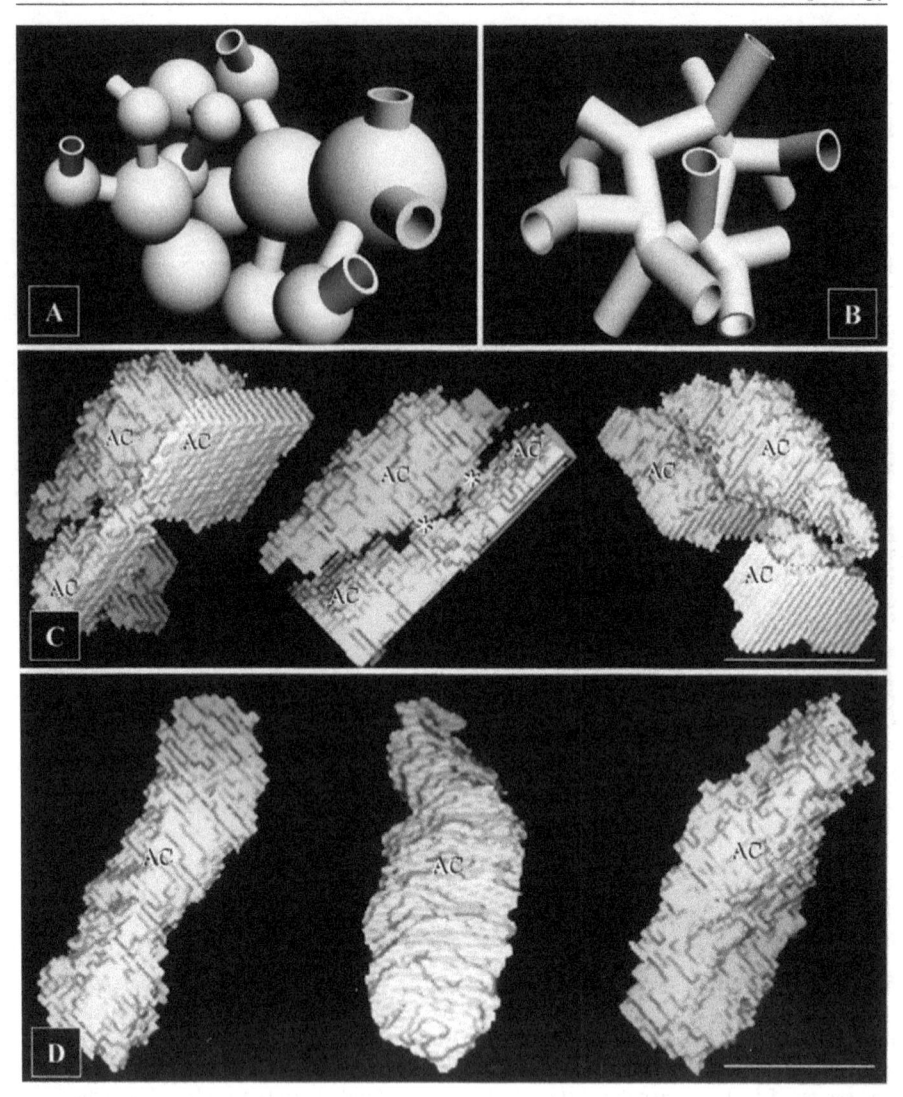

Fig. 61. A, B Schematic views of the air capillaries and blood capillaries. The air capillaries (**A**) are rotund structures that connect through narrow passages while the blood capillaries are formed by short segments of tubes that interconnect in three dimensions (**B**). **C, D** Most air capillaries connect (**C**) while few are discrete (**D**). *Scale bars* 10 µm. (Woodward and Maina 2005)

3.3 Airway (Bronchial) System

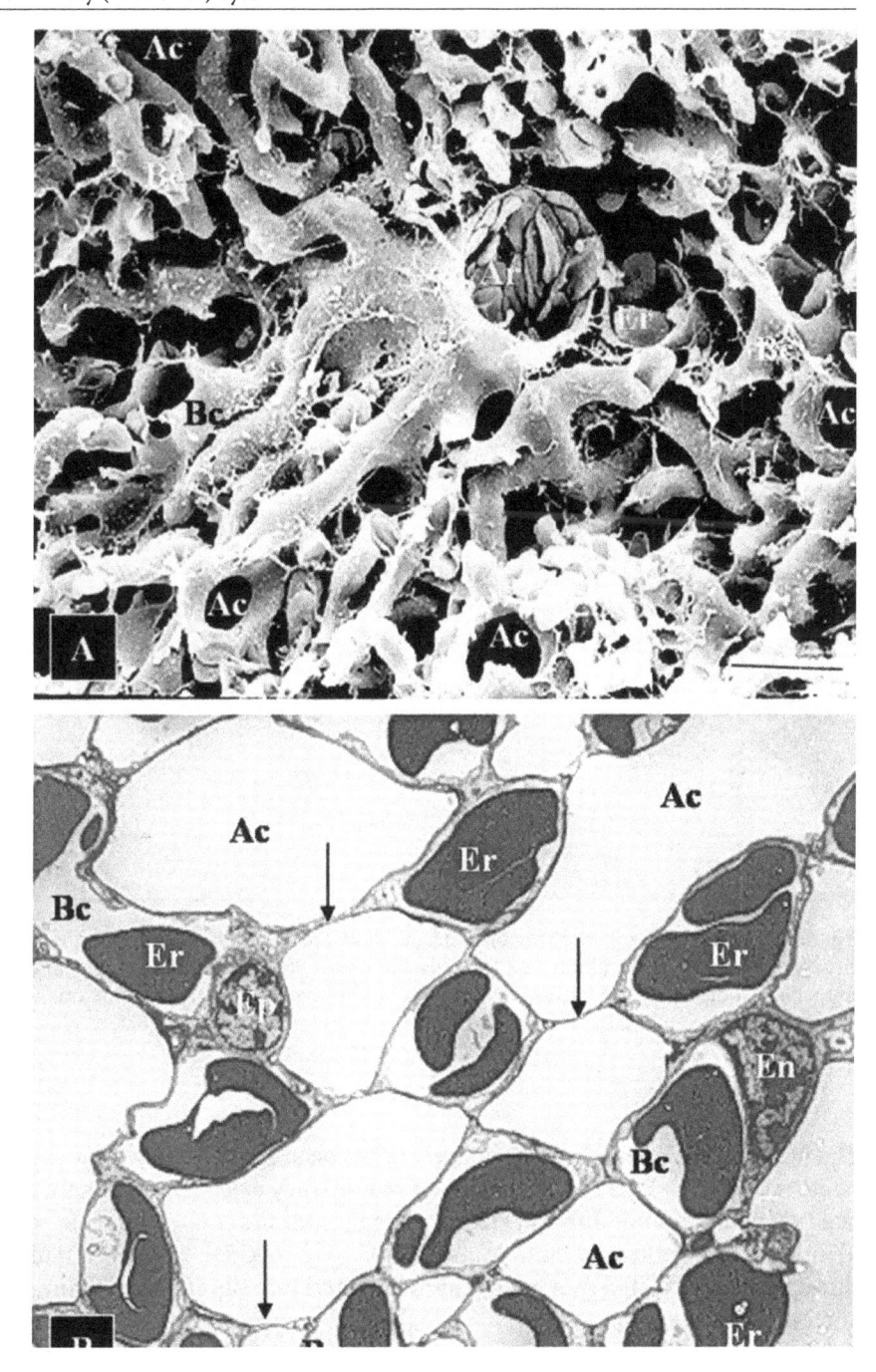

Fig. 62. A Lung of the domestic fowl, *Gallus gallus* variant *domesticus*, showing an arteriole (*Ar*) giving rise to blood capillaries (*Bc*). The spaces between the blood capillaries constitute the air capillaries (*Ac*). **B** Air capillaries (*Ac*) and blood capillaries (*Bc*) forming a network. *Er* Erythrocytes; *arrows* areas where air capillaries lie adjacent to each other; *En* endothelial cells. *Scale bars* **A** 50 µm; **B** 10 µm. (**A** from Maina 1994; **B** from Maina 2002b)

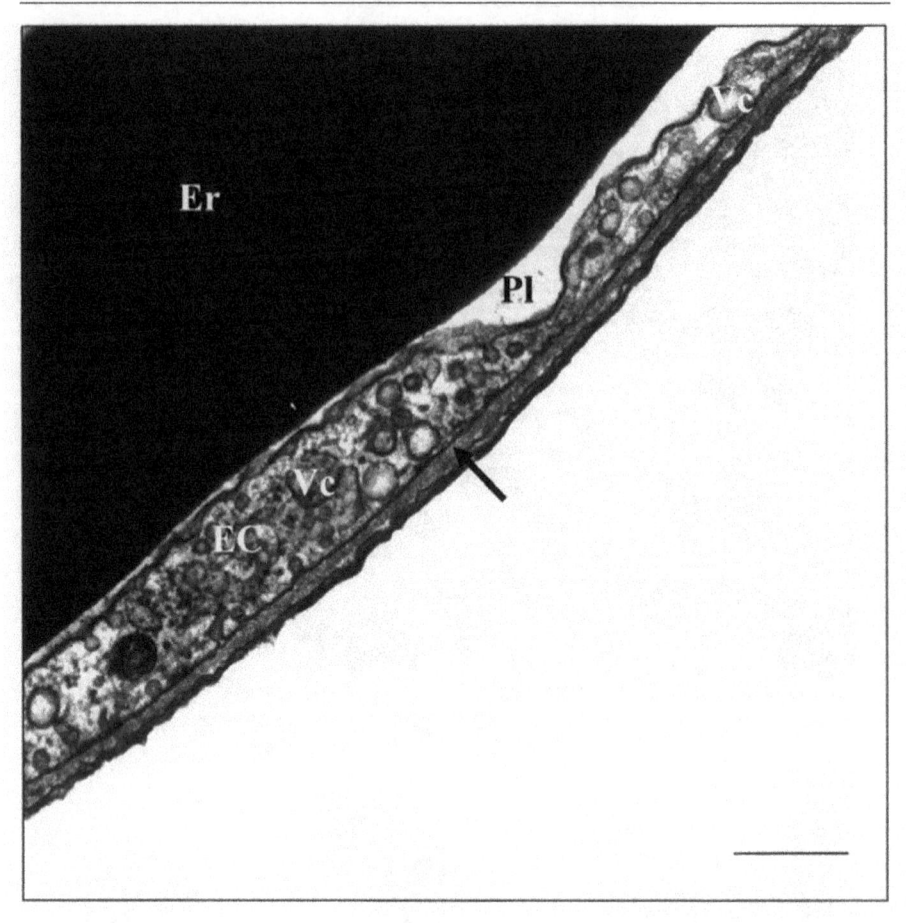

Fig. 63. Close-up of the BGB of the blackheaded gull, *Larus ridibundus*, showing an erythrocyte (*Er*) lying in the blood capillary; plasma layer (*Pl*), endothelial cell (*Ec*), and an epithelial cell overlying the extracellular matrix space (*arrow*). *Vc* Micropinocytotic vesicles. *Scale bar* 0.5 μm. (Maina 1998)

The ACs and BCs are not mirror images, as has been suggested by some investigators (Fig. 60C–E): a certain degree of redundancy exists. Interestingly, isolated ACs, that did not differ in size and shape from the connected ones, were identified by Woodward and Maina (2005; Fig. 60C,D). The mechanism through which isolated ACs form and their functional significance, if any, are currently unclear.

The diameters of the ACs range from 3 μm in the songbirds, through 10 μm in penguins (Spheniscidae), swans (Anatidae), the coot, *Fulica atra* (Duncker 1972), to 20 μm in the ostrich, *Struthio camelus* (Maina and Nathaniel 2001). Compared with the BCs that anastomose more regularly the ACs vary greatly in size, are larger in diameter, and anastomose more irregularly (Fig. 62). The

3.3 Airway (Bronchial) System

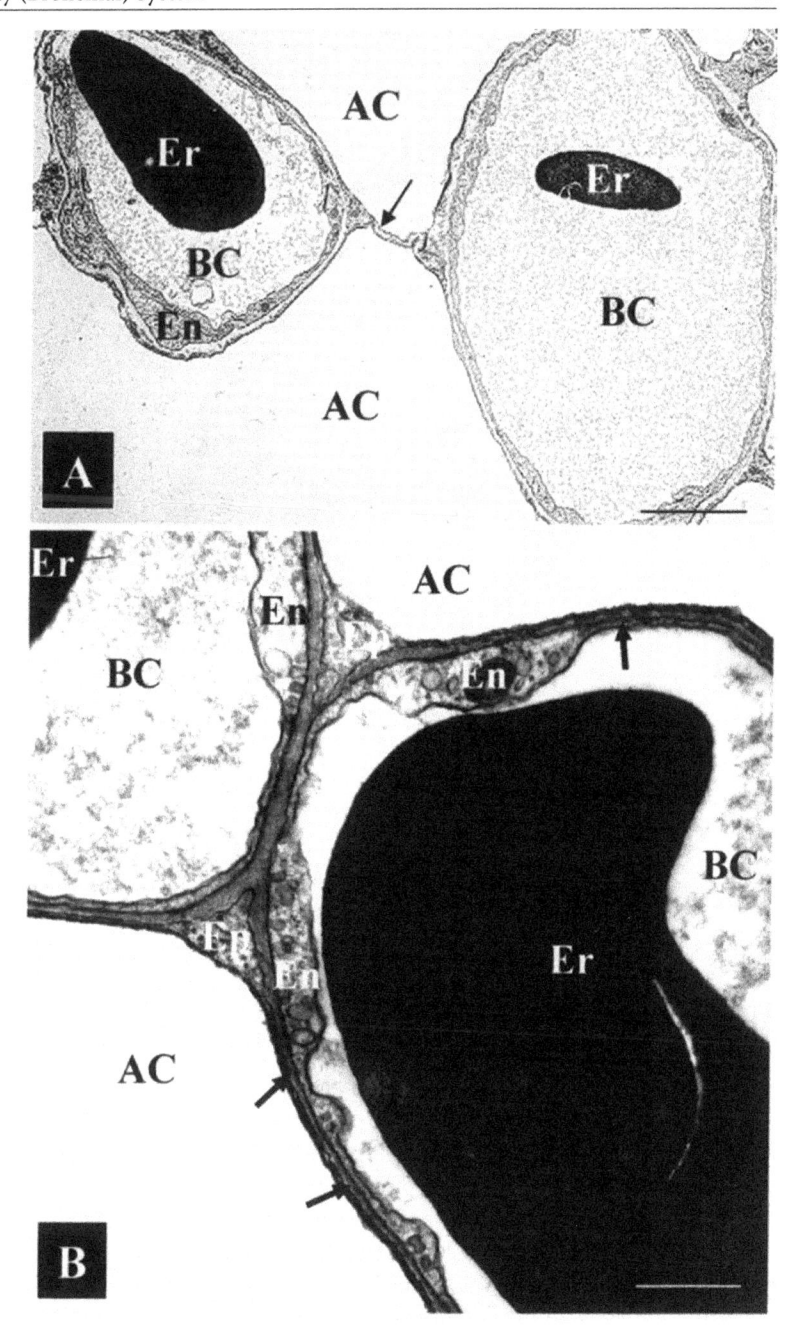

Fig. 64. A Lung of the domestic fowl, *Gallus gallus* variant *domesticus*, showing blood capillaries (*BC*) containing erythrocytes (*Er*) suspended in air capillaries (*AC*). *En* Endothelium; *arrow* site where AC lie next to each other. **B** Exposure of blood to air in the exchange tissue of the lung of the blackheaded gull, *Larus ridibundus*. *AC* Air capillaries; *BC* blood capillaries; *En* endothelium; *Ep* epithelium; *Er* erythrocyte; *arrow* thin parts of the BGB. *Scale bars* **A** 10 μm; **B** 5 μm. (**A** from Abdalla et al. 1982; **B** from Maina and King 1982a)

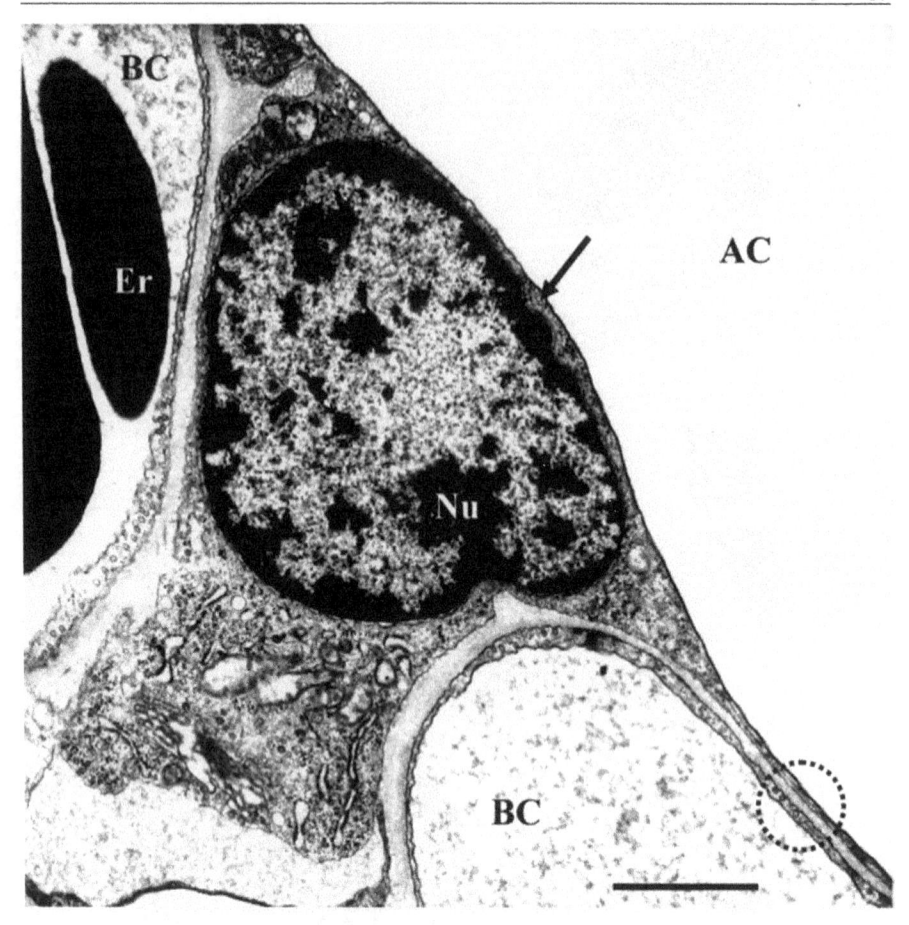

Fig. 65. Surface of the respiratory surface of the domestic fowl, *Gallus gallus* variant *domesticus*, showing the perikaryon of a type 1 epithelial cell (*arrow*) giving rise to peripherally extremely attenuated parts (*dashed circle*) that overlie the blood capillaries (*BC*). *Er* Erythrocytes; *Nu* nucleus; *AC* air capillaries. *Scale bar* 1 μm

BCs and ACs intertwine to form a network where blood is optimally exposed to air across a thin, sporadically attenuated blood–gas barrier (BGB; Figs. 63 and 64). The type II (granular) pneumocytes are lacking in the ET of the avian lung: the cells are confined to the atria and infundibulae. On the whole devoid of organelles, the type I (squamous) pneumocytes have extremely thin cytoplasmic flanges. The perikarya (of the type I cells) that occur in the ET are located at the corners of the BCs and ACs (Fig. 65), i.e. away from the BGB itself. This enhances gas exchange across a thin barrier.

3.4
Blood–Gas Barrier (BGB)

In its thinnest parts, the BGB is formed by an endothelial cell that fronts the pulmonary BCs, an intermediate extracellular matrix layer or BL, and an epithelial cell that faces air (Fig. 63). In the avian lung, the typically three-ply (laminated) design of the BGB is well formed at hatching (Chap. 2.5). The endothelial cell contains numerous micropinocytotic vesicles and particularly displays marked sporadic attenuation (Fig. 64), while the epithelial cell is extremely thin and has few discernable organelles. The BL is well developed and is always associated with the endothelial cell (Fig. 34). In areas where ACs lie adjacent to each other, a BL is lacking: in such cases, epithelial cells contact directly, back-to-back. The epithelium is covered by an osmiophilic surface layer, the surfactant (e.g. Maina and King 1982a).

The endothelium, the BL, and the epithelium comprise about 67, 21, and 12% of the volume of the BGB, respectively (Maina and King 1982a). The remarkable thinness of the BGB in the avian lung (e.g. Maina 1989b; Maina et al. 1989a) arises from lack of an interstitial space between the individual basal laminae of the epithelial and endothelial cells: connective tissue and cellular elements are lost early in development (Figs. 32–34). When and where they rarely occur, such components are shifted to nonrespiratory sites (Fig. 65). The sporadic thinning of the BGB (Figs. 63 and 64) should generate extreme thinness, while mechanical integrity is guaranteed by the thicker parts: a thin BGB offers less resistance to diffusion of respiratory gases.

3.5
Surfactant

Contrary to early assertions by, e.g. Miller and Bondurant (1961), it is now incontrovertible that surfactant occurs on the respiratory surface of the avian lung (e.g. Pattle 1978; del Corral 1995; Figs. 60–62). The composition and concentration of the surfactant per unit respiratory surface are comparable to that in the mammalian lung (e.g. Fujiwara et al. 1970). Two forms of surfactant, namely the lamellated osmiophilic bodies (LOBs) and trilaminar substance (TLS), occur on the respiratory surface of the avian lung. The LOBs resemble those of other vertebrates and are discharged from the type II epithelial cells of the atria by merocrine secretion (e.g. Akester 1970), while the TLS is unique to birds. In the developing lung, the LOBs appear before the TLS. Biochemically, compared with the LOB, the TLS has a low lipid-to-protein ratio, is more abundant, the laminae display wider spacing, and it does not dissolve during tissue processing for electron microscopy (e.g. Pattle 1978). TLS is found mainly in the lumina and cells of the atria and lasts on the surface of the ACs for only up to 2 weeks after hatching (e.g. Jones and Radnor

1972a,b; Pattle 1978). Interestingly, in some instances, LOBs and TLS occur in the same epithelial cell (e.g. Pattle 1978).

It is generally thought that, in the avian lung, LOBs represent the archetypal surfactant and may be involved in lowering surface tension at the gas–liquid interface and hence respiratory impedance. The role of the surfactant in the rigid, noncompliant avian lung, where stabilization of the terminal gas-exchange units (ACs) during ventilatory activity should be unwarranted, is rather paradoxical. Fedde (1980) suggested that the surfactant may be a feature that has been brought over during the evolution of the avian lung from the reptilian one and may be involved in prevention of transudation of plasma onto the respiratory surface rather than in reduction of surface tension. Additionally, the TLS may be associated with roles such as coagulation of blood that may leak through the BGB, hydration of the surfactant, and absorption of fluid that may collect on the surface of the ACs (Pattle 1978).

Compared with the rather spherical alveoli of the mammalian lung (Fig. 44B,C), earlier reported to be tubular in shape and blind-ended, the ACs are interconnected, rather spherical, respiratory units (Woodward and Maina 2005; Figs. 60F and 61A). The diameter of the ACs ranges from 3–20 µm, values that are on average about one-tenth those of the alveoli of the mammalian lungs (Tenney and Tenney 1970; Duncker 1971, 1974; Maina 1982a; Maina and Nathaniel 2001). The smallest alveoli of the mammalian lung of 35 µm diameter have been reported in the shrew (Tenney and Remmers 1963). Owing to their relatively small sizes, the surface tension in the ACs should be very high. Lacking connective tissue support (Maina and King 1982a), the ACs are inexplicably very stable: mechanical compression of the lung does not cause them to collapse significantly (Macklem et al. 1979); the exceptional stability of the ACs was associated with the presence of the TLS, though interestingly very little of it occurs on the respiratory surface of the mature lung. The stability of the ACs may, however, largely be a consequence of the general rigidity of the avian lung and its firm attachment to the ribs, rather than an intrinsic structural property of the ACs themselves.

3.6
Air Sacs (ASs)

3.6.1
Topographical and Structural Morphology

The ASs are capacious, transparent structures that connect to the avian lung (Fig. 38). In the chick embryo, six pairs of primordial ASs initially develop (e.g. Romanoff 1960; Chap. 2.3; Fig. 22). In most species, however, the final number of ASs ends up being less than this. During their development, except for the AASs that pass through the postpulmonary septum into the peritoneal cavity, all the other ASs invade the septum, dividing it into horizontal and oblique septa (Duncker 1978). The locations, sizes, connections, and diverticulae (expansions) of the ASs vary between species (e.g. King 1966; Duncker 1971). Commonly, the CeASs are small, paired structures that lie on the craniodorsal part of the thoracic cavity and the base of the neck, pneumatizing (aerating) the cervical vertebrae, the thoracic vertebrae, and the vertebral ribs (Duncker 1971): the ASs are missing in the loons (Gaviidae) and the grebes (Podicipedidae). In the turkey, *Meleagris gallopavo*, connection between the left and right CeASs occurs on day 17 of incubation (King and Atherton 1970). In some species, e.g. the gannet, *Morus bassanus*, and the ostrich, *Struthio camelus*, the CeASs form extensive subcutaneous diverticulae (e.g. Bezuidenhout et al. 2000).

In most birds, the ClAS is relatively large, unpaired, and is formed by fusion of right and left primordia. It occupies the cranioventral area of the thorax, the base of the neck, and most of the right and left axillary space. In birds like gulls (Laridae), the primordia remain separated (e.g. Locy and Larsell 1916a,b) while, in some species such as the pigeon, *Columba livia* (Müller 1908), and in the house sparrow, *Passer domesticus* (Wetherbee 1951), the ClAS and CeAS connect to form a single, large cervicoclavicular AS. In passeriform species and Trochilidae, the ClAS connects to the CrTAS (Wetherbee 1951; Duncker 1971). In such cases, the esophagus, trachea, syrinx, and the PB are located on the dorsal aspect of the AS. The topographical relationship between the ClAS (a compliant structure that fills with air and empties with ventilation; Fig. 41) and the syrinx, the organ of phonation in birds (e.g. King 1989), is thought to modulate vocalization (e.g. Brackenbury 1989). Several diverticulae extend from the ClAS to the heart, to the surface of the body, and to adjacent bones, e.g. sternum, coracoid, and ribs.

The CrTASs and the CaTASs are paired. They are located in the subpulmonary cavity, i.e. in the space ventral to the lung and the horizontal septum. In general, the CrTASs are smaller than the CaTASs. However, in some species, e.g. *Fulica atra*, the CrTAS is much larger (Duncker 1971), while, in others, e.g. *Gallinula chloropus*, the two sets of ASs are comparable in size (Groebbels 1932). The CrTASs are remarkably small in penguins (Spheniscidae) and passeriforms, while, in the hummingbirds (Trochlidae), the CaTAS is the largest AS (Stanislaus 1937). In the turkey, *Meleagris*, the CaTAS is lacking (King and Atherton 1970). In general, the CrTASs and the CaTAS have few, if any, diverticulae (King 1966, 1979).

The AAS is a paired structure that occupies the intestinal peritoneal cavity. It normally lies dorsal to the viscera. In the kiwi (Apterygidae), the AS is located in the subpulmonary cavity (Duncker 1979a). Owing to the asymmetry of the viscera, the left AAS is normally smaller than the right one. However, in some species such as the common loon, *Gavia immer*, and the herring gull, *Larus argentatus*, the left AS is larger (e.g. Gier 1952). To different extents, the left and right AASs connect and in some species communicate directly (Duncker 1971). While particularly large in some species, e.g. *Phoenicopterus* (Groebbels 1932), AASs are poorly developed in birds like penguins and the rhea (Duncker 1971), and are exceptionally small in the hummingbirds (Trochilidae; Groebbels 1932; Stanislaus 1937). The AASs pneumatize the synsacrum, pelvic girdle, femur, and caudal vertebrae (King 1975, 1979) and, in some species, e.g. the pigeon, *Columba*, lie subcutaneously above the pelvis (Müller 1908).

3.6.2
Ostia

The ASs connect to the lungs at sites called ostia (Figs. 38 and 66). Two kinds of connections, direct and indirect, occur. Direct connection exists when the ASs are joined to the lung by PB and/or SB while indirect ones are formed when PR connect the ASs to the lung. Most of the ASs have one or two direct connections and numerous indirect ones. Indirect connections are lacking in penguins (Spheniscidae; Vos 1937). With a single ostium that is located on the

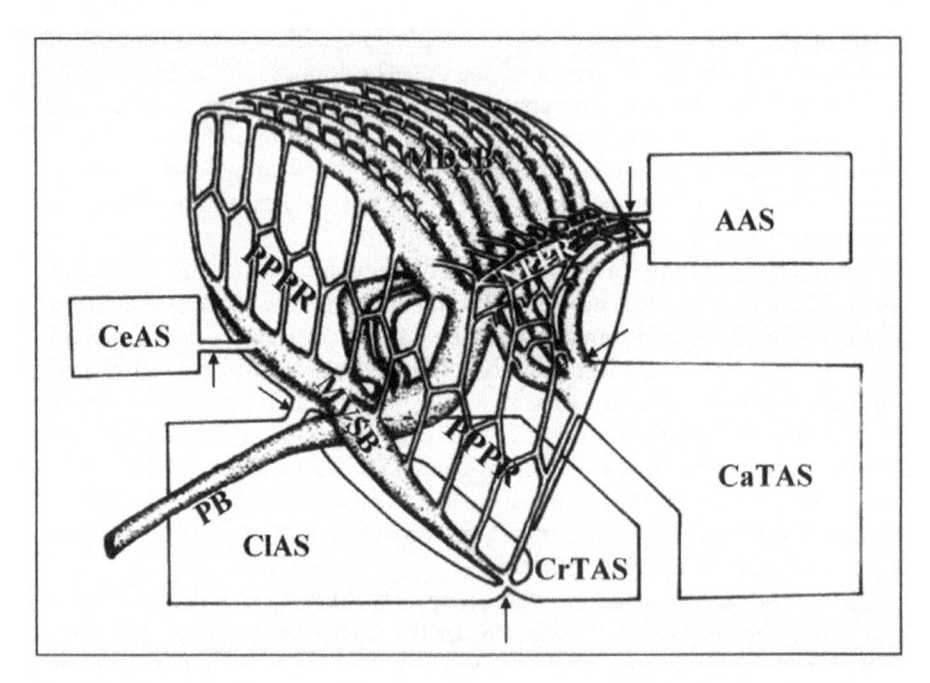

Fig. 66. Schematic diagram of the lung-air sac system of birds showing the primary bronchus (*PB*) running through the lung. In transit, it gives rise to medioventral secondary bronchi (*MVSB*), and mediodorsal secondary bronchi (*MDSB*). The secondary bronchi are connected by different groups of parabronchi, namely the paleopulmonic parabronchi (*PPPB*) and neopulmonic parabronchi (*NPPB*). *Arrows* Ostia; *AAS* abdominal air sac; *CaTAS* caudal thoracic air sac; *CrTAS* craniothoracic air sac; *ClAS* clavicular air sac; *CeAs* cervical air sac. (Redrawn from Duncker 1974, with permission from the publisher)

septal surface of the lung, in the domestic fowl the CeAS has a direct connection that arises from the first MDSB: indirect connections are lacking (Biggs and King 1957). The ClAS has two ostia: the medial ostium is connected directly to the third MDSB while the lateral one is directly connected to the first MDSB. Indirect connection occurs between the PR of the MVSB and the ClAS. A lateral ostium is lacking in the lungs of hummingbirds (trochlids; Stanislaus 1937). Inexplicably, in columbiform and psittaciform species, the ClAS is reported to have no connection with the lung (Juillet 1912). The CrTASs are commonly associated with two ostia: the medial ostium is connected directly to the third MDSB while the lateral ostium is indirectly connected through PR of one or more MDSB. In the hummingbirds, the CrTASs are indirectly connected to the lung (Stanislaus 1937). In many species, the CrTAS has a single ostium situated at or near the caudal part of the costal septal border of the lung: direct and indirect connections exist. In species such as penguins (Spheniscidae) where neopulmo (Sect. 3.6) is lacking, indirect connections to the CaTASs do not exist. The AASs have a single ostium in the lateral part of the caudal border of the lung (Fig. 41) that has both direct and indirect connections: the direct connection projects to the PB while the indirect ones comprise of the PR of the LDSB, LVSB, and the last MDSB. In the hummingbirds, only a single indirect connection to the ASSs occurs (Stanislaus 1937).

3.6.3
Cytoarchitecture of the Wall of Air Sacs

The wall of the ASs consists mainly of a simple epithelium supported on a thin layer of connective tissue (e.g. Walsh and McLelland 1974). The epithelium consists of squamous cells but, near the ostia, ciliated cuboidal and columnar cells occur (e.g. Fletcher 1980). In the domestic fowl, pseudostratified, ciliated columnar epithelium with goblet cells extends as a broad band from the PB into the AAS (Cook and King 1970). On the surface of the CaTAS, Cook et al. (1987) observed a pseudostratified, ciliated, cuboidal-to-columnar epithelium. In penguins, the epithelium of the ASs is generally tall, almost cuboidal. The epithelial cells are joined by junctional complexes at the luminal aspect and laterally by interdigitation. Microvilli project into the luminal space and electron-dense lysosome-like granules occur in the cytoplasm (e.g. Carlson and Beggs 1973; Walsh and McLelland 1974). Scanty muscle cells and clusters of fat cells have been reported in the walls of the ASs of some species of birds (e.g. Fletcher 1980). According to Trampel and Fletcher (1980), the smooth muscle tissue in the wall of the ASs is an extension of the layer that surrounds the PRL.

The ASs are characteristically avascular (e.g. Fletcher 1980) and play no significant role in gas exchange (e.g. Magnussen et al. 1976). Adrenergic and cholinergic nerve plexuses have been described in their walls (e.g. Rawal 1976; Cook et al. 1987). In the wall of the ASs of the domestic fowl, Cook et al. (1987)

showed VIP-, substance P-, somatostatin- and enkephalin-immunoreactive fibers.

3.7
Paleopulmo and Neopulmo

Hans Reiner Duncker (1971) was first person to point out that the structure of the avian lung was far from homogenous: it consisted of two distinguishable anatomical and functional parts. Based on what he envisaged to be their chronological appearance during the evolution of the avian lung, he named the parts paleopulmo (ancient=old lung) and neopulmo (modern=new lung). The paleopulmo occurs in the lungs of all birds while the neopulmo is totally lacking in some birds, especially the ancient (primitive) species, while in others it develops to varying extents. While the rationale of categorizing the avian lung into paleo- and neopulmonic sections is still tenuous (López 1995), the change in the perception of the structure of the avian lung meaningfully contributed to better understanding of certain aspects of its physiology. Previously, it was difficult to reconcile the lower concentration of CO_2 in the inspired air with the higher one in the caudal ASs. The explanation is that while much of the inspired air passes through the IPPB en route to the caudal ASs (direct connection; Sect. 3.6.2), in those birds that have it, some of the air flows through the ET of the NPPR (indirect connection; Fig. 66) collecting CO_2 (e.g. Piiper 1978). In birds where the neopulmo is lacking or is poorly developed, the MDSB occur superficially. In such cases, the SB are easily accessible from the costal surface of the lung for physiological studies, e.g. of air flow and gas composition.

The main structural and functional differences between the PPPR and the NPPR of the avian lung are: (1) the PPPR are located on the craniodorsal part of the lung while the NPPR are placed caudoventrally (Figs. 20C and 40A); (2) the PPPR occur in the lungs of all species while the NNPR are totally lacking in some species and where well developed may constitute as much as one-third of the volume of the lung; (3) the PPPR are arranged as hoop-like stacks that connect the MVSB and MDSB (Figs. 40 and 52B), while the NPPR anastomose intensely; (4) the air flow in the PPPR is caudocranial, i.e. from the MDSB to the MVSB and is continuous and unidirectional while that in the NPPR is tidal (=bidirectional), i.e. towards the caudal ASs during inspiration and in the reverse direction, i.e. into the MDSB during expiration; and (5) embryologically, the PPPR form earlier than the NPPR (Romanoff 1960; Maina 2003a,b). The term 'planum anastomoticum' describes the area where the PPPR connect the MVSB to the MDSB (Fig. 52B), while the dense network formed by the anastomoses of the NPPR is termed 'pulmo reteformis' (e.g. Duncker 1971; King 1979; López 1995).

The degree of development of the PPPR and NPPR differs between lungs of different species (Duncker 1971): NPPR are poorly developed in birds like stocks (Ciconiidae) and the emu, *Dromaius novaehollandiae*, where they are formed by only a few LDSB arising from the most caudal part of the PB, their laterally directed PR, and the connections of the PR with the CaTASs and the AASs. The increase in size, number, and the dorsocranial extension of the NPPR imparts the curvature that the PB shows as it transits the lung (Figs. 40B and 47B). In the cormorants (Phalacrocoracidae) and the auks (Alcidae), the neopulmo includes the LVSB and their laterally directed PR. In the cranes (Gruidae), the LDSB that arise from the caudal part of the PB are superimposed onto the NPPR, displacing the PB into the interior of the lung. In plovers (Charadriiformes), ducks (Anatidae), gulls (Laridae), owls (Strigidae), buzzards (Accipitridae), and parrots (Psittacidae), the NPPR contain laterally directed PR that originate from the proximal parts of the MDSB. In the pigeons (Columbidae), galliform species, and passerine birds, the NPPR and the PPPR interconnect. In the galliform species and passerine species, where the NPPR are very well developed, the diameter of the terminal part of the PB is smaller than that of a single PR.

Except for the topographical locations, sizes, and arrangements, there are no significant differences in the detailed morphology and morphometry of the structure of the ET in the PPPR and NPPR regions of the lung (Maina 1982b; Maina et al. 1983): the thickness of the BGB and the length densities of the BCs are similar. The bidirectionally (tidally) ventilated NPPR may serve as a reservoir, averting excessive flushing out of CO_2 from the ET. This may avert occurrence of respiratory alkalosis (e.g. Jones 1982), a physiological complication that soon befalls thermal stressed, panting birds with excessive CO_2 washout in the undirectionally ventilated PPPR. The ostrich, *Struthio camelus*, pants ceaselessly for as long as 8 h without experiencing acid–base imbalance (e.g. Schmidt-Nielsen et al. 1969). Interestingly, in the duck lung, the NPPR are well ventilated, even at rest (Holle et al. 1977). Without thorough experimental substantiation, Duncker (1972) suggested that, where well developed, the NPPR may form the main or exclusive site for gas exchange during rest while the PPPR may only become significantly involved during exercise. Inverse correlation was reported between the development of the PB and that of the NPPR (Duncker 1971): since NPPR provide an alternative pathway for air flow en route to the caudal AASs, the reduced volume of air flowing through the PB may allow it to become narrower.

3.8
Pulmonary Vasculature

The organization of the avian pulmonary vasculature (PV) has been investigated by various investigators (e.g. Radu and Radu 1971; Abdalla and King 1975, 1976a,b, 1977; West et al. 1977; Maina 1982a, 1988). Abdalla (1989) gave a detailed review of the distribution of the PA and PV in the lung of the domestic fowl. Unlike in the mammalian lung where the patterning of the arterial system follows that of the airways (Maina and van Gils 2001; Fig. 21), in the avian lung the PA and PV do not track each other closely, nor do they follow the bronchial system. The PA enters the lung at the hilus ventral to the first MVSB (Fig. 43A) and then divides into four main branches. The accessory branch of the PA (ABPA) is the smallest: the branch is reportedly missing in a number of species, e.g. the guinea fowl, *Numida melleagris*, the turkey, *Meleagris gallopavo*, and muskovy duck, *Cairina moschata* (Abdalla and King 1977). The ABPA supplies blood to a small area of the lung ventral to the hilus. The cranial branch of the PA (CBPA) runs both dorsally and caudally in a plane more or less parallel to and lateral to the origins of the MVSB supplying blood to the craniodorsal section of the lung, i.e. cranial to the third costal sulcus. The caudomedial branch of the PA (CMBPA) supplies blood to much of the substance of the lung. Given its direct origin, the branch is the most likely continuation of the PA. The caudolateral branch of the PA (CLBPA) supplies blood to the caudolateral, ventral, and caudoventral parts of the lung.

The blood supply to the avian lung can more-or-less be divided into two regions: the cranial area is supplied by the CBPA and the ABPA while the caudal one is supplied by the CMBPA and CLBPA. No anastomoses occur between the four branches of the PA and their terminal branches (Abdalla and King 1976a,b). The four main branches of the PA give rise to first order branches, the interparabronchial arteries (IPRBArs; Figs. 67 and 68) that in turn give

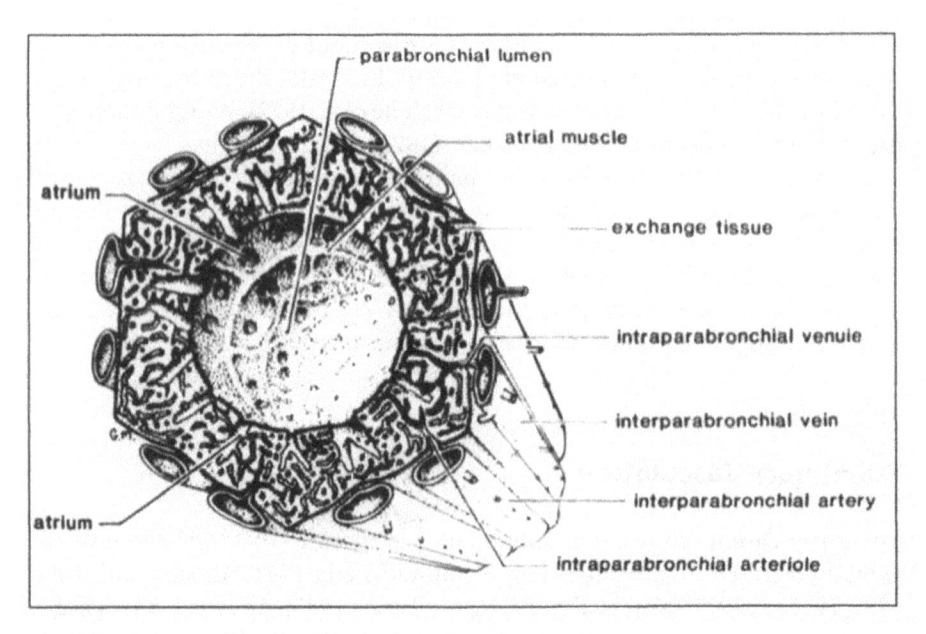

Fig. 67. Schematic diagram of a parabronchus showing its various component parts and the geometric relationships between the air flow and the blood flow. (Maina 2002c)

3.8 Pulomonary Vasculature

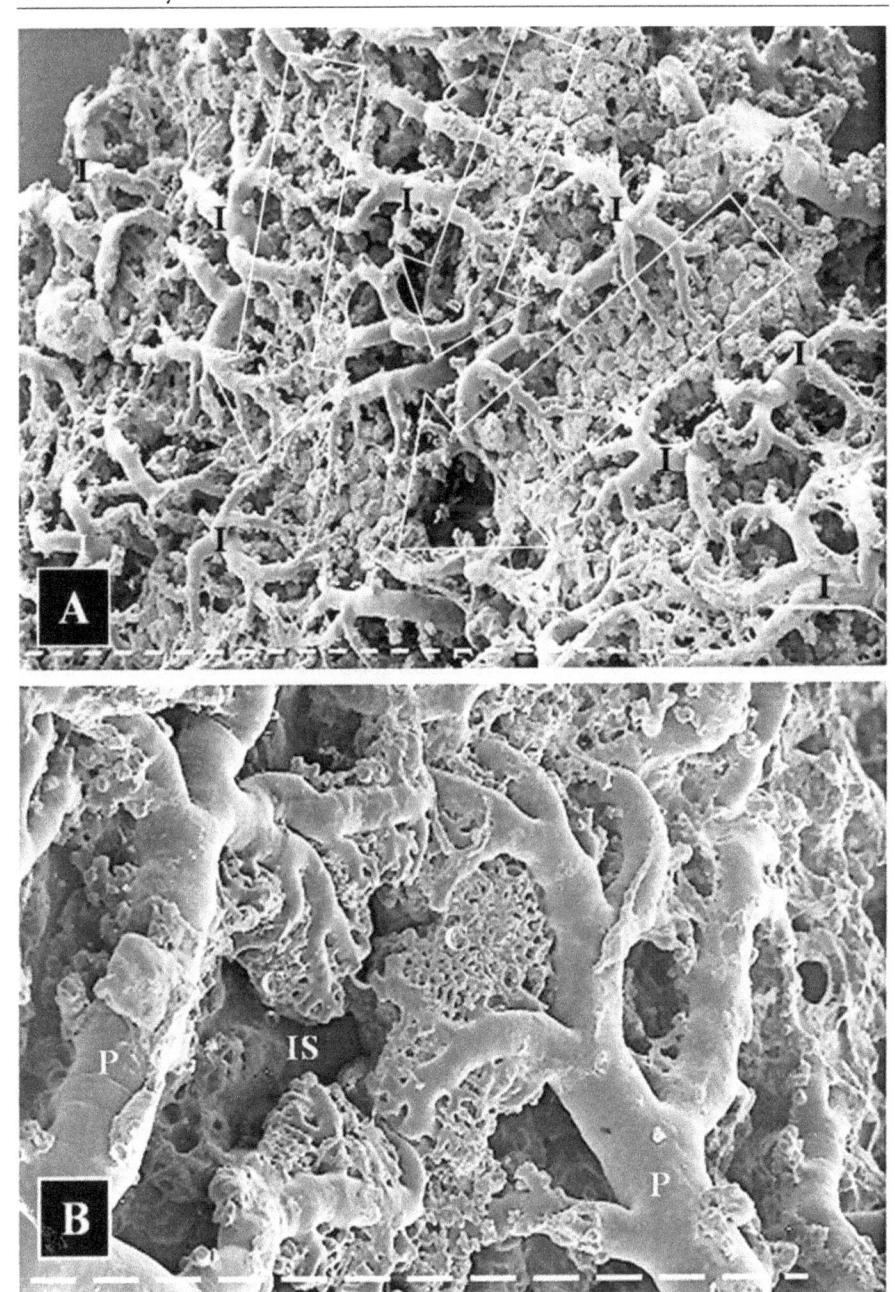

Fig. 68. Double latex cast preparation (i.e. injection of latex rubber into the airway and vascular systems) of the lung of the domestic fowl, *Gallus gallus* variant *domesticus*, showing the cross-current (**A**) and counter-current designs. **A** Interparabronchial arteries (*I*) giving rise to intraparabronchial arteries that interact with the air flow in the parabronchial lumina (*arrows*) in a perpendicular manner. *Scale bar* 0.5 mm. **B** Intraparabronchial arteries (*P*) give rise to blood capillaties (*C*) that run in an opposite direction to that of the air capillaries that emanate from infundibulae. *IS* Interatrial septum. *Scale bar* 50 µm. (Maina 1988)

rise to a series of smaller intraparabronchial arterioles (IPRBAos) that penetrate the ETs of the PR. The BCs arise from the IPRBAos and intertwine with the ACs that emanate from the infundibulae (Sects. 3.8 and 3.9).

3.9
Arrangement of the Structural Components for Gas Exchange

The arrangement of the IPRBArs and the IPRBAos relative to the orientation of the PR (Figs. 67 and 68) determines how deoxygenated blood is delivered and exposed to air in the ET. The trajectories between the bulk air flow in the PRL and that of the venous blood (from the IPRBArs) are essentially perpendicular. Moving centripetally (i.e. inwards), blood of uniform composition in O_2 content is delivered practically at the same time to all parts of the ET of the PR. Blood thus equilibrates with flowing air of varying composition along the length of the PR as O_2 is extracted and CO_2 added to the parabronchial gas. At the entrance of the PPPR, i.e. the caudal ends (=the sides facing the MDSB), blood is exposed to air with high PO_2 while the reverse is true at the opposite end. The concentration of O_2 in the arterial blood, i.e. the quantity in blood returning to the heart via the PV, arises from pooling of small amounts of O_2 extracted sequentially at infinitely many points along the length of the PR where ACs and BCs contact (Figs. 67 and 68). The design is termed 'multicapillary serial arterialization system' (MCSAS). The MCSAS extends the time over which the gas-exchange media (air and blood) are presented and exposed to each other. With the process occurring in unidirectionally and continuously ventilated parabronchial ET, transfer of O_2 and CO_2 is greatly enhanced. In the domestic fowl, the entire length of the PR, i.e. if all the PR in the lung are connected end-to-end, is about 300 m (King and Molony 1971). Under certain conditions, e.g. hypoxia and exercise, the PCO_2 in the arterial blood ($PaCO_2$) is lower than that in the end-expired air ($PECO_2$; e.g. Powell and Scheid 1989): the opposite condition may apply for O_2, i.e. PaO_2 exceeds PEO_2.

The perpendicular disposition between the direction of the flow of the deoxygenated blood from the IPRBArs into the IPRBAos and then into the BCs relative to that of the bulk air flow in the PRL (Figs. 67–69) is termed 'cross-current design'. Morphologically (e.g. Abdalla and King 1975; West et

Fig. 69. A Double latex cast injection of the lung of the domestic fowl, *Gallus gallus* variant *domesticus*, showing parabronchi studded with atria. Interparabronchial blood vessels (*I*) give rise to intraparabronchial vessels (that supply blood to the parabronchi, *P*). The *arrow* shows the flow of air in the parabronchial lumen. The disposition between the flow of air and that of the blood in the intraparabronchial arteries is perpendicular, i.e. cross-current. *Circles* Areas where blood capillaries run in opposite directions to form a counter-current

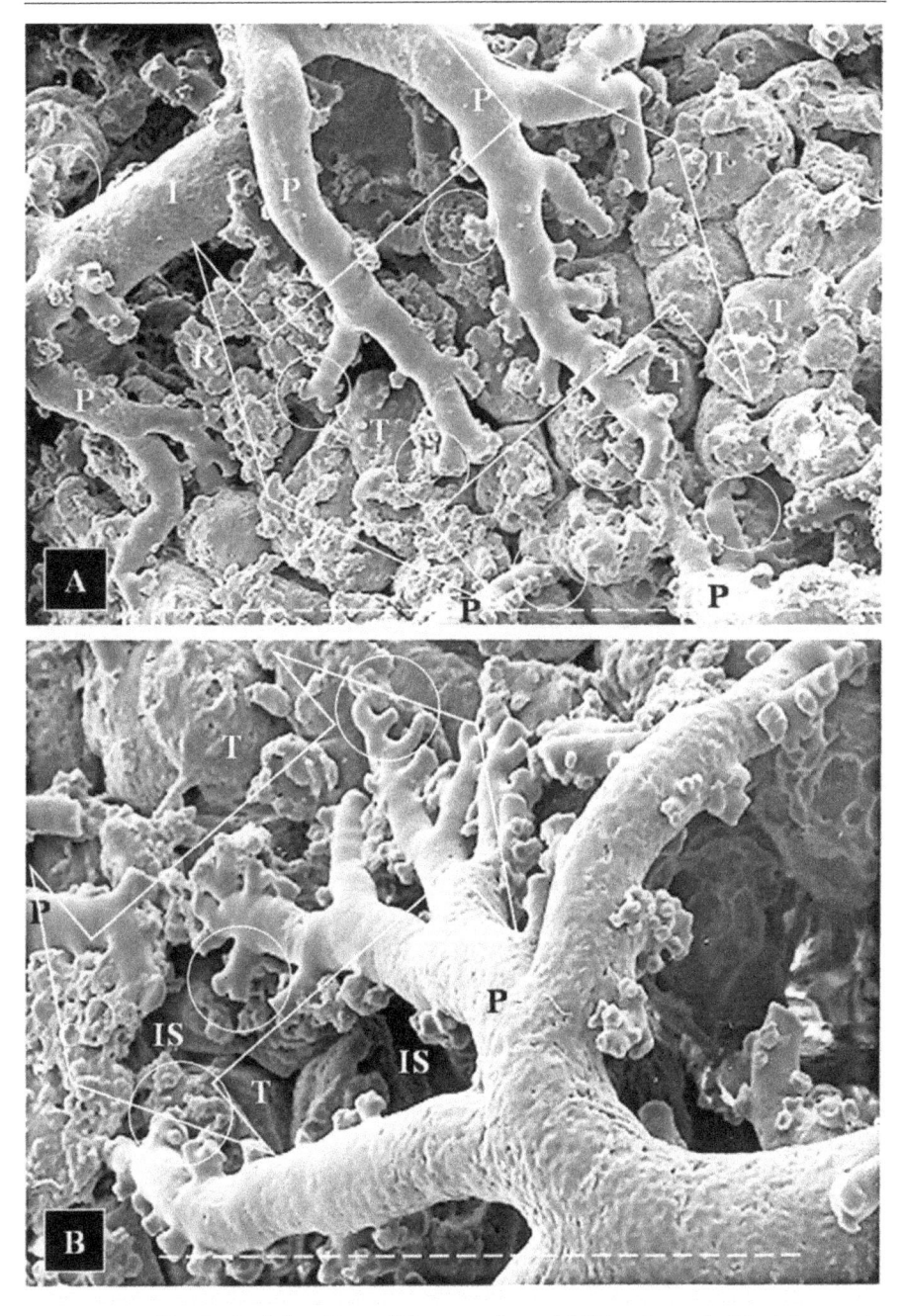

3.9 Arrangement of the Structural Components for Gas Exchange

arrangement; *T* atria. *Scale bar* 50 µm. **B** Intraparabronchial blood vessels (*P*) giving rise to a blood capillary network (*C*) that interfaces with the air capillaries arising from the parabronchial lumina through atria (*T*) and infundibulae. *IS* Interatrial septa; *arrow* air flow; *circles* interaction between air capillaries and blood capillaries. *Scale bars* **A** 50 µm; **B** 20 µm. (Maina 1988)

al. 1977; Maina 1988) and physiologically (e.g. Scheid et al. 1972), the arrangement has been unequivocally substantiated. Only of historical interest now, from the well-known respiratory efficiency of the avian lung, 'counter-current' arrangement of the structural components was assumed to occur in the avian lung (e.g. Schmidt-Nielsen 1971). Through an elegant experiment where the direction of the flow of the parabronchial air in a duck lung was reversed, Scheid and Piiper (1972) demonstrated that the PCO_2 and PO_2 in the exhaled air as well as in the arterial and mixed venous blood did not change significantly. If a 'counter-current' arrangement predominated in the avian lung, the procedure would have generated 'co-current' flow, i.e. at the gas-exchange level, air and blood would have flowed in the same general direction. In such a case, inadequate P would have been created and flux of respiratory gases would have been drastically diminished. In a 'cross-current' design with an inbuilt MCSAS, however, reversal of the direction of air flow (or for that matter that of blood – a technically more difficult experimental procedure) only alters the order in which capillary blood is arterialized: the overall gas-exchange efficiency, i.e. the quantity of O_2 and CO_2 exchanged, should not be significantly affected. Interestingly, in the avian lung, superimposed on the 'cross-current' design is a 'counter-current' one. The latter is formed by the centripetal (inward) flow of deoxygenated blood from the IPRBArs, IPRBAos and BCs and the centrifugal (=radial=outward) one of air from the PRL into the atria, the infundibula, and the ACs (Fig. 68 and 69). The role of the 'counter-current' arrangement to the overall gas-exchange process in the avian lung is reportedly inconsequential. Disparaging its functional significance, Piiper and Scheid (1973) dismissed it a "*counter-current-like mechanism*" while Scheid (1979) considered it as "*an auxiliary mechanism superposed on, and independent of, the basic cross-current arrangement between the capillary blood flow and the bulk parabronchial gas flow*". Based on the fundamental fact that convective flow occurs in the BCs and diffusion in the ACs, the arrangement cannot by strict definition constitute a true counter-current design. Structural and geometric properties of the ACs and BCs may further explain why the 'counter-current' arrangement in the avian lung may be of little functional impact: the ACs and the BCs intertwine closely in three dimensions (Figs. 58 and 59) and, as they network, they truly contact over very short distances. Such sites may fall well below the critical distances that are necessary for efficient gas exchange in a 'counter-current' system.

After blood is oxygenated in the BCs of the ET of the avian lung, it ultimately drains into the PV through an intricate venous pathway that includes atrial veins, intraparabronchial venules, and interparabronchial veins (IPRVs). The IPRVs and their branches do not follow the IPRArs nor do they pattern the bifurcation of the airways (Abdalla 1989): ETs of adjacent PR may drain into a single IPRV. Arteriovenous anastomoses do not occur between the PA and PV and their terminal branches (e.g. Abdalla and King 1975; 1976a,b; James et al. 1976; West et al. 1977; Holle et al. 1978; Parry and Yates 1979).

3.10
Cellular Defenses of the Lung

While the biology of the alveolar macrophages (AMs) of the mammalian lung has been well studied (e.g. Bowden 1987), relatively little is known about the surface (free) avian pulmonary macrophanges (SAPMs). Utterly divergent observations exist regarding the cellular defense strategies, mechanisms, and efficacies in the avian lung. Toth et al. (1988) asserted that *"the paucity of the free macrophages in the avian respiratory system suggests a deficiency in the defense system of the respiratory tract of poultry against bacteria, mycoplasma, fungi, and viruses, all of which are major causes of pneumonitis and air sacculitis"*. Stearns et al. (1987) remarked that *"one striking aspect of the avian gas exchange tissue region, in contrast to mammals, was our inability to find any surface macrophages"*. Klika et al. (1996) declared that *"in the most peripheral parts of the lung, i.e., the atria, infundibula and ACs, no free macrophages can be observed at either the light or electron microscopic level"*. Lorz and López (1997) stated that *"the avian respiratory macrophages are always located in the subepithelial connective tissue and they have never been observed on the luminal face of atrial or parabronchial epithelia, as occurs in the mammalian alveoli"*. Klika et al. (1996) maintained that the interstitial macrophages *"do not migrate from the subepithelial tissue compartment to the airway surfaces"*. Comparing the lungs of city and country pigeons, Lorz and López (1997) noted that *"subepithelial macrophages were rare or absent in the country birds"*.

Practical and technical reasons may partly explain the above perplexing conflicting reports. Some of these are: (1) since very small pieces of tissue are routinely sampled, processed, and viewed for electron microscopy and few cells are scattered over a vast surface, chances of the cells being missed are high, and (2) the cells may detach from the respiratory surface during the lengthy, caustic process of tissue processing for electron microscopy. With a scanning electron microscope, Maina and Cowley (1998) observed SAPMs on the atrial and infundibular regions of the lung of the pigeon, *Columba livia* (Figs. 70 and 71A). The SAPMs described by Ficken et al. (1986) in the domestic fowl are ultrastructurally similar to those in the pigeon, *Columba* (Maina and Cowley 1998; Fig. 71B,C), and do not significantly differ from the mammalian ones (e.g. Bowden 1987).

It is now widely recognized that the cellular elements of the mononuclear phagocytic system originate from the bone marrow (e.g. van Furth 1982). After necessary structural, pharmacological, and biochemical modifications (e.g. Chandler and Brannen 1990), the cells leave the circulatory system and ultimately settle in the body tissues where they form the so-called resident tissue macrophages (e.g. Lasser 1983). In a healthy mammalian lung, it is believed that the AM is the final stage of a complex developmental process in the maturation process of the blood monocyte, with the interstitial macrophage possibly being a transitory stage (e.g. Chandler et al. 1988). Inter-

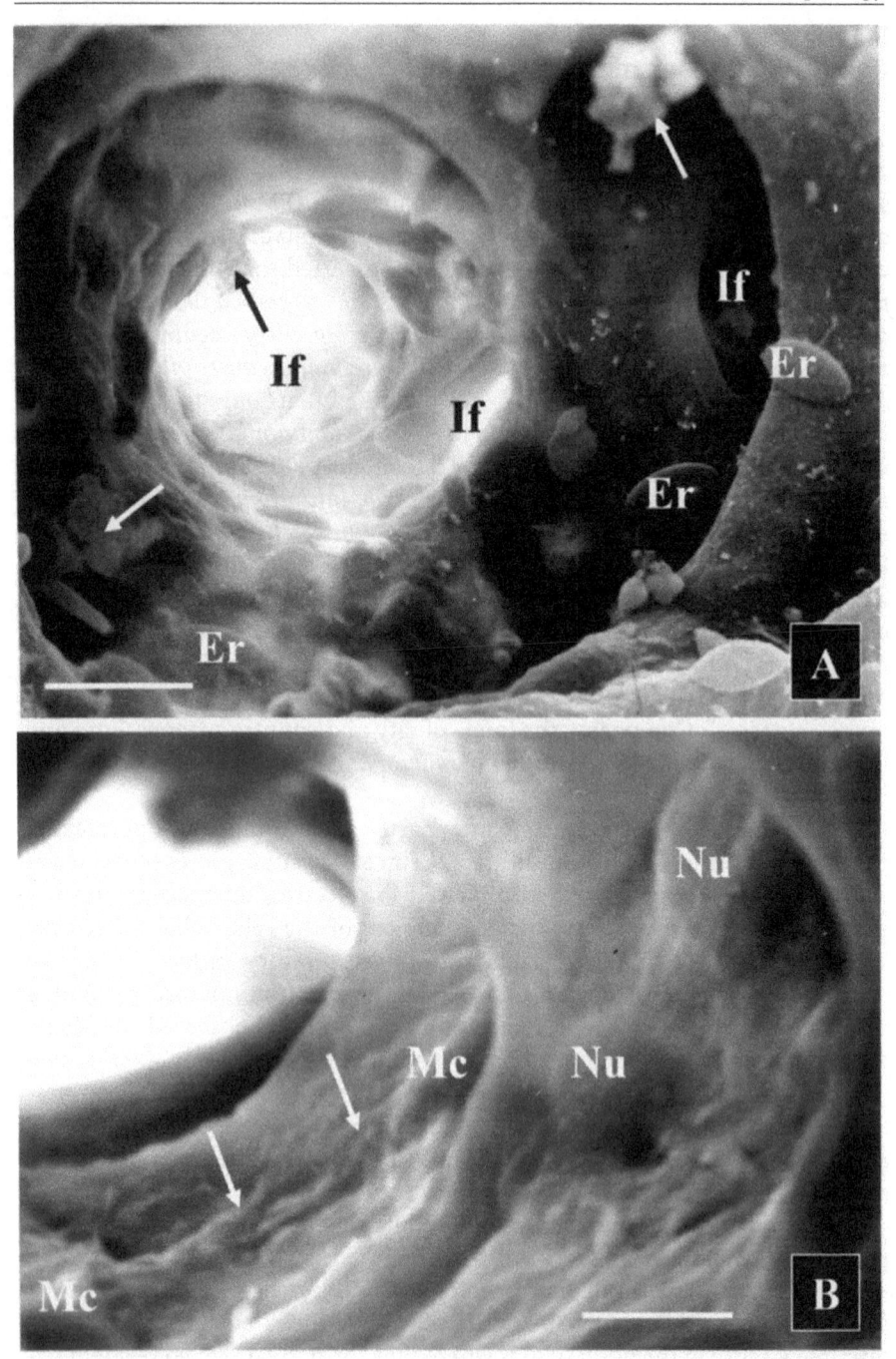

Fig. 70. A Lung of the domestic fowl, *Gallus gallus* variant *domesticus*, showing an infundibulum (*If*) and particulate matter (*arrows*). *Er* Erythrocytes. **B** Macrophages (*Mc*) on the surface of an infundibulum. *Arrows* Filopodia extending from the perikarya (*Nu*). *Scale bars* **A** 30 μm; **B** 50 μm. (Maina and Cowley 1998)

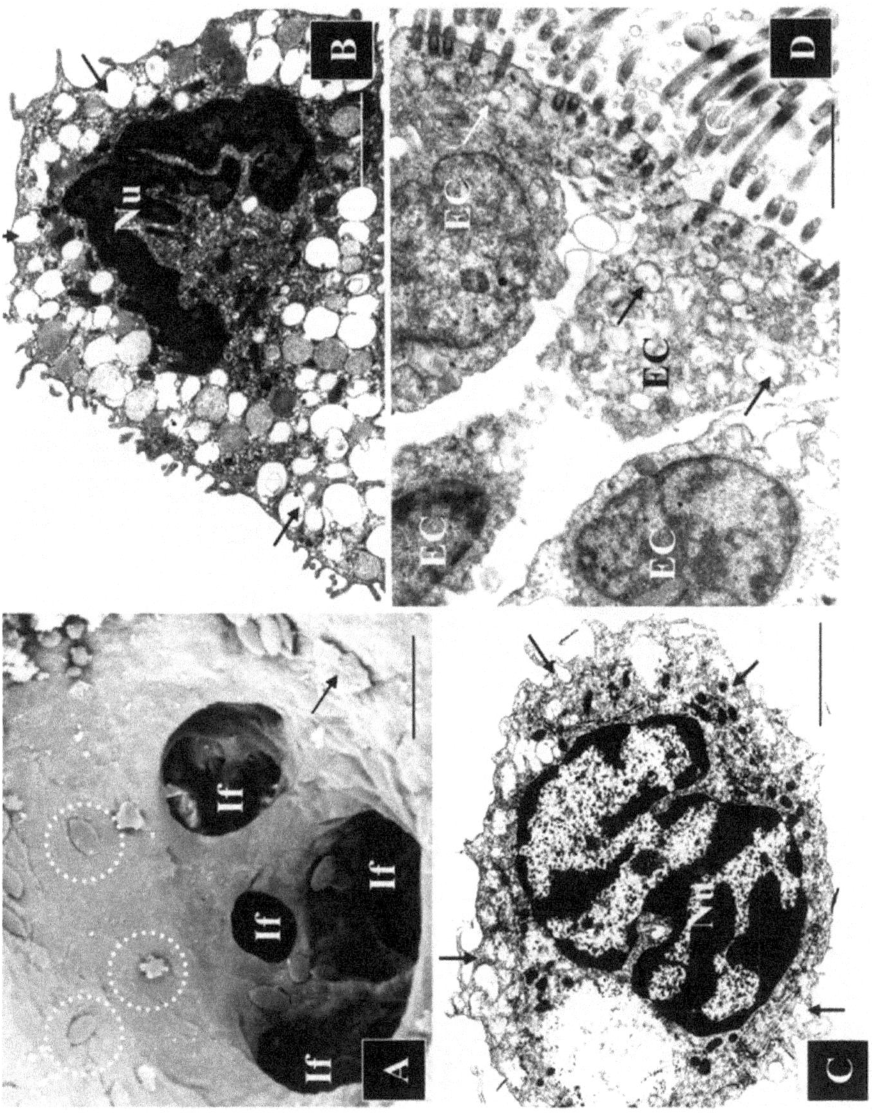

3.10 Cellular Defenses of the Lung 109

Fig. 71. Lung of the rock dove, *Columba livia*, showing **A** infundibulae (*If*) at the floor of an atrium. *Dashed circles* Particulate matter being phagocytosed by the epithelium; *arrow* a putative macrophage. **B** A macrophage with numerous vesicular bodies (*arrows*). *Nu* Nucleus. **C** Macrophage with filopodial extensions (*arrows*). *Nu* Nucleus. **D** Bronchial epithelial cells (*EC*) containing numerous vesicular bodies (*arrows*). *Ci* Cilia. *Scale bars* **A** 50 μm; **B** 3 μm; **C** 3 μm; **D** 30 μm. (**A, C, D** from Maina and Cowley 1998)

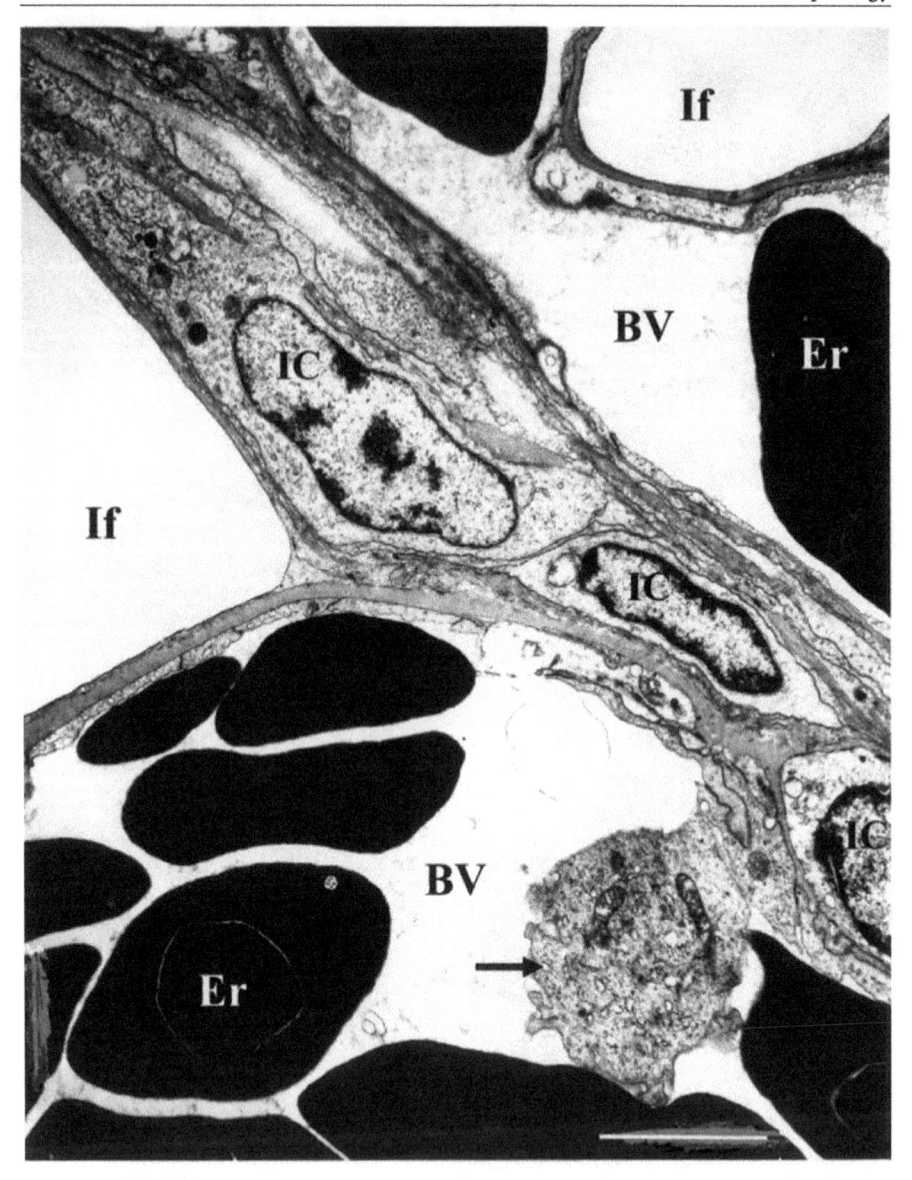

Fig. 72. Lung of the rock dove, *Columba livia*, showing blood vessels (*BV*) exposed to infundibulae (*If*). *Er* Erythrocytes; *IC* interstitial cells; *arrow* pulmonary intravascular macrophage. *Scale bar* 15 µm. (Maina and Cowley 1998)

estingly, some pulmonary phagocytes do not leave the pulmonary vasculature, i.e. they do not pass through the BGB to lie on the respiratory surface where they express full inventory of the functional properties of a completely differentiated macrophage (e.g. Betram et al. 1989). The unique subpopulation of pulmonary vascular resident macrophages that adhere to the endothelial cell is called 'pulmonary intravascular macrophages' (PIVMs). The cells widely occur in the mammalian lung (e.g. Dehring and Wismar 1989; Atwal et al. 1992) and have been reported in the avian lung (Maina and Cowley 1998; Fig. 72). Another kind of macrophage, the 'pulmonary subepithelial (interstitial) macrophage' (PSEM), leaves the vasculature to reside in the subepithelial space. Chandler et al. (1988) observed remarkable morphological and functional heterogeneity of the PSEMs, properties that they attributed to differences in stages of development and sites of origin. Quantitatively, little is known about the PIVMs and PSEMs.

Whereas in mammals AMs are easily and abundantly harvested from the lung by pulmonary lavage, the method is not as effective in the avian lung. This is fundamentally due to the greater architectural complexity of the avian respiratory system. From difficulties of harvesting cells for experimentation, many investigators (e.g. Kodama et al. 1976) extrapolated observations made on avian blood monocytes, splenic macrophages, and peritoneal exudate macrophages to respiratory macrophages. Microscopic studies of lungs exposed to respirable insoluble particulate matter (e.g. Klika et al. 1996; Scheuermann et al. 1997) have yielded important data on the mechanisms of avian pulmonary defense. In fact, investigators like Hazelhoff (1951) meaningfully applied deposition of particulate matter (carbon particles) to investigate the pattern of air flow in the avian lung. The technique has more recently been used to understand disease distribution patterns in the lung (e.g. Fletcher 1980).

Different investigators, e.g. Klika et al. (1996) and Spira (1996), have remarked on a very high predisposition of birds to pulmonary diseases. In the poultry industry, huge economical losses have been ascribed to mortalities arising from respiratory infections (e.g. Mensah and Brain 1982; Toth et al. 1988). Small numbers of SAPMs on the respiratory surface of the avian lung (e.g. Stearns et al. 1986; Maina and Cowley 1998; Nganpiep and Maina 2002) and enzymatic deficiencies in their oxidative metabolism (e.g. Penniall and Spitznagel 1975; Bellavite et al. 1977) have been reported. A number of morphological and physiological reasons may predispose the PRL to infections and afflictions. The most important ones are: (1) compared with mammals, for animals of similar body mass, birds have a BGB that is 56–67% thinner and a RSA that is 15% more extensive (Maina 1989b; Maina et al. 1989a; Chaps. 4.3 and 4.4), parameters that make the lung highly vulnerable to pathogens; (2) having a large tidal volume and the PPPR being ventilated continuously and unidirectionally (e.g. Fedde 1980), the propensity of entrance and deposition of harmful particles and microorganisms onto the respiratory surface is high; (3) in some species of birds, e.g. the ostrich, *Struthio camelus*, the ASs extend out to lie subcutaneously (e.g. Bezuidenhout et al. 2000) where, arising from trauma and infection, air sacculitis (infection of the ASs) can easily diffuse and affect the lungs.

In small mammals, e.g. mouse, rat, and guinea pig, yields of $0.55-1.55 \times 10^6$, $2.86-4.43 \times 10^6$, and $1.08-1.77 \times 10^7$ of AMs, respectively, have been estimated after pulmonary lavage. Some of the values are 20 times greater than those encountered on the respiratory systems of much larger birds (e.g. Holt 1979; Toth and Siegel 1986). The average number of SAPMs in the pigeon, *Columba livia*, lung is 1.6×10^5 (Maina and Cowley 1998), a value lower than that of 2.5×10^5 in the domestic fowl (Toth and Siegel 1986; Toth et al. 1987) and in the turkey (1.15×10^6; Ficken et al. 1986). The number of SAPMs per unit body mass in the rat was significantly greater than that of the domestic fowl and the duck, *Cairina moschata* (Nganpiep and Maina 2002; Fig. 73). In 30-year-old humans, 1.5×10^7 and 5.2×10^7 AMs were harvested by bronchopulmonary lavage from the lungs of a nonsmoker and a smoker, respectively (Hof et al. 1990). The recovery rate of the lavaged fluid after aspiration was 80–90 % in the domestic fowl (Toth and Siegel 1986), that in the rock dove was 72 % (Maina and Cowley 1998), and in the duck and the domestic fowl the values were 89 and 91 %, respectively (Nganpiep and Maina 2002). The recovery value of 74 % achieved for the human nonsmoker's lung by gravity (Hof et al. 1990) compares with the 72 % value reported by aspiration in the pigeon lung (Maina and Cowley 1998). Owing to the complexity of the respiratory system,

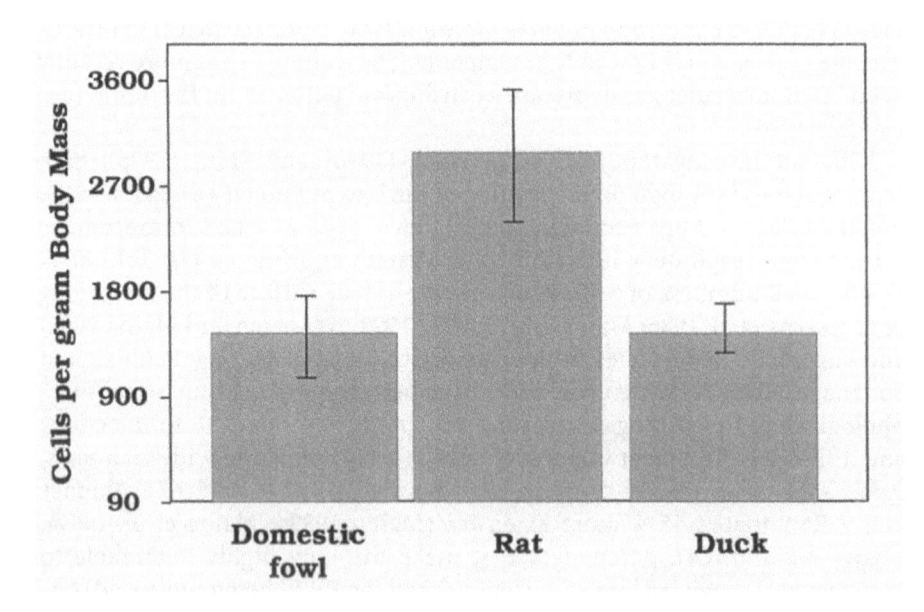

Fig. 73. Comparison of numbers of surface pulmonary macrophages per unit body mass in the domestic fowl, *Gallus gallus* variant *domesticus*, rat, *Rattus rattus*, and duck, *Cairina moschata*. (Nganpiep and Maina 2002)

in birds, very little of the lavage fluid can be recovered by gravity and total recovery of the fluid by aspiration is practically impossible.

For comparative interest, no surface resident macrophages occur in a non-challenged lung of the snake, *Boa constrictor* (Grant et al. 1981). The animal's low MR and ventilatory rate may explain the lack of macrophages: unphagocytosed materials persisted on the respiratory surface for up to 4 days. Challenge of the snake lung with inspirable particles increases surfactant secretion, elicits surfacing of nonphagocytic eosinophilic granulocytes, but interestingly does not set off release of mononuclear phagocytic macrophages (Grant et al. 1981). Surface macrophages occur in the lung of the tree frog, *Chiromatis petersi* (Maina 1989 c). Welsch (1983) activated a macrophagic response in the lungs of *Xenopus laevis* after aspiration of carbon particles.

Stearns et al. (1987) pointed out that two factors may explain the dearth of SAPMs in the avian lung. These include: (1) the extremely 'long' distances the cells would have to travel to convey the ingested particles to the ciliated parts of the lung for onward transport by mucociliary escalator system to the SB, PB, trachea, and larynx for final clearance: in the avian lung, only the trachea, the PB, and the initial sections of the SB are ciliated (e.g. López 1995; Pastor and Calvo 1995), and (2) the particularly narrow diameters of the ACs (3–20 µm; e.g. Duncker 1974; Maina and Nathaniel 2001) may not be sizeable enough to accommodate large, motile, phagocytic cells. Mensah and Brain (1982) reported removal of inhaled insoluble technetium from the lungs of birds but the actual mode of transfer at the PR level was not clarified. The entrapped particles may not necessarily be eliminated through the airway mucocilliary escalator system: in the domestic fowl, the ingested particles are first trapped by the TLS from where they are moved into the epithelial cells and subsequently to the basal aspects of the underlying PSEMs (Stearns et al. 1987). There, particles may be solubilized in situ or phagocytosed by the PIVMs (e.g. Lippmann and Schlesinger 1984).

Regarding the alleged defense incapability of the avian lung and the paucity of SAPMs, a number of scenarios are conceivable. These are: (1) a weak front-line defense may exist on the respiratory surface; (2) in addition to SAPMs, other defense line(s) may exist; and (3) SAPMs may be so efficient that few cells are needed to grant adequate protection. Based on the existence of a diversified defense armamentarium that includes PIVMs, PSEMs, phagocytic bronchial epithelial cells (Maina and Cowley 1998; Nganpiep and Maina 2002; Figs. 71D and 74A-F), and SAPMs abundantly endowed with lysosomes (Nganpiep and Maina 2002), the first scenario appears untenable. Regarding the second circumstance, Toth et al. (1988) noted substantial increases of SAPMs of polymorphonuclear leukocyte type (by three orders of magnitude) within 24 h after intratracheal administration of live, apathogenic *Pasteurella multocida* vaccine to chickens. Similarly, in the domestic fowl and duck, Nganpiep and Maina (2002) observed a substantial flux of macrophages onto the respiratory surface after only 2.5 min of lavage (Fig. 75). Capacity to mobilize and quickly transfer phagocytic cells to the

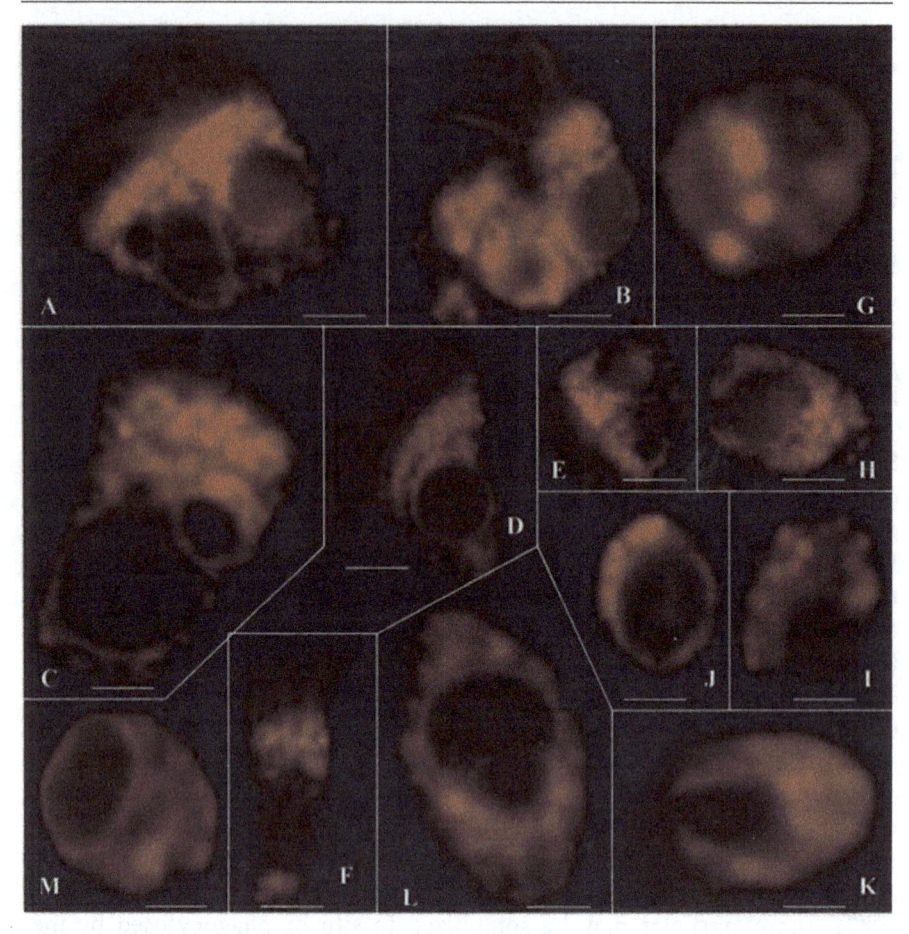

Fig. 74. A–F Fluorescence microscopy of the bronchial epithelial cells and surface macrophages (stained with lysotracker) of the lung of the domestic fowl, *Gallus gallus* variant *domesticus*. Lysozymal bodies are highly concentrated at the apical aspects of the epithelial cells (**A–F**) and the surface macrophages (**G–M**) are well endowed with lytic enzymes. *Scale bars* 5 μm. (Nganpiep and Maina 2002)

respiratory surface may explain why a large resident population of SAPMS may be lacking.

The deduction that birds have a weak pulmonary defense capacity has been largely based on observations made on domestic and captive species, especially the domestic fowl. Domesticated from the wild jungle fowl, *Gallus gallus*, of Southeast Asia some 8000 years ago (e.g. West and Zhou 1988), through intense genetic breeding, some 40 different breeds of birds of commercial value have been produced. While in the late 1940s broilers took about 90 days to grow to a slaughter body mass of 1.8 kg, in 1960, it took 70 days for a table bird to reach a similar live weight, and, in the 1980s, it took only 40 days

3.10 Cellular Defenses of the Lung

(Smith 1985; Gyles 1989). Presently, broilers reach a body mass of 2.5 kg in less than 40 days (e.g. Ross Breeders 1999). Arising from genetic breeding and better husbandry, much of the weight gain occurs during the first 2 weeks posthatching (Ricklefs 1985). Directed (forced) growth and productivity without giving the supporting structures time to 'catch up' (i.e. adjust) has inescapably precipitated structural-functional disequilibria. The chickens (particularly males), e.g., are incapable of reaching $VO_{2\,max}$ on treadmill exercise (e.g. Brackenbury 1984). Growth rate, productivity, and functional performance appear to have reached a limit in the domestic fowl (e.g. Konarzewski et al. 2000). Similar conclusions were independently reached by, e.g. Mason et al. (1983) and Jones (1998) on the performance of the horse, one more animal that has been phenotypically intensely bred for running speed and high aerobic capacity over several thousand years. Data collected by Jones (1998) indicate that the performance of the respiratory system of the thoroughbreds has been optimized: the winning times of the most elite British and American thoroughbreds over the past 70–150 years leveled out over the last two decades. More than 40–80% of thoroughbred horses are reported to come down with exercise-induced pulmonary hemorrhage during high intensity exercise (e.g. West et al. 1993). In the domestic birds, mortality of up to 10% of

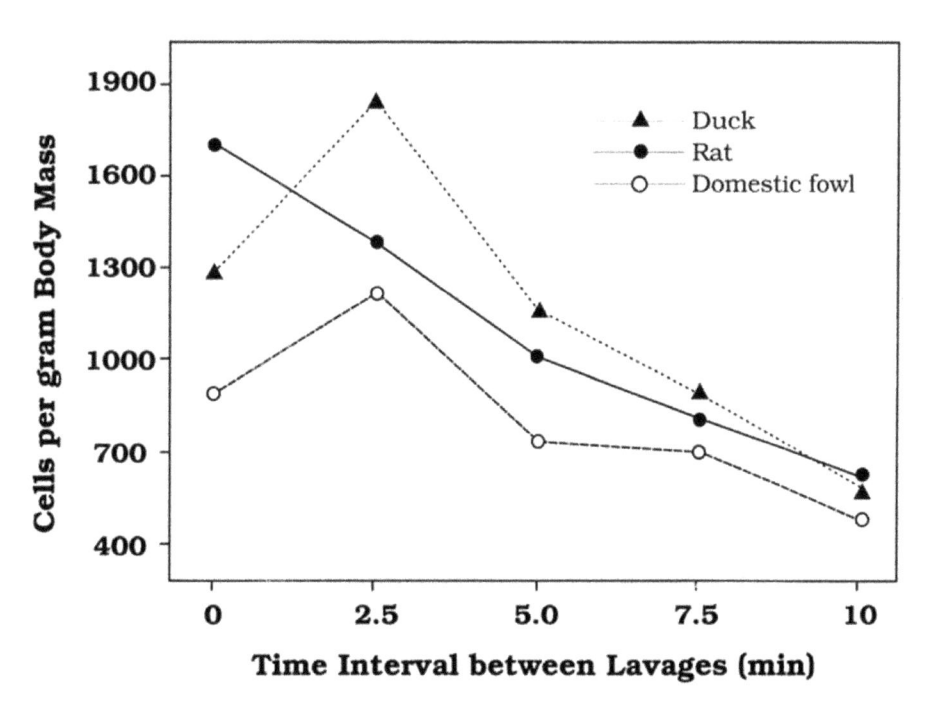

Fig. 75. Flux of macrophages to the surface of the lungs in the domestic fowl, *Gallus gallus* variant *domesticus*, rat, *Rattus rattus*, and duck, *Cairina moschata*, after lavages. (Nganpiep and Maina 2002)

flocks arises from metabolic diseases such as heart failure syndrome (e.g. Julian et al. 1984; Schlosberg et al. 1996; Silversides et al. 1997) and aortic rupture (e.g. Carlson 1960). Worldwide increase of incidence in ascites has been reported in young broilers (Julian and Wilson 1986; Maxwell et al. 1986a,b; Julian 1987). The syndrome has been associated with right ventricular hypertrophy (e.g. Julian et al. 1984; Julian and Wilson 1986; Huchzermeyer and de Ruyk 1986). Village free-ranging chickens that are not exposed to intense husbandry are not prone to ascites (e.g. Pizarro et al. 1970). In battery chicken production, the birds are kept in crowded spaces and placed on a strict feeding regimen that may include force-feeding. Under such conditions, stress may alone predispose birds to diseases, particularly those transmitted through aerosol.

In summary, contrary to observations made by some investigators, the lung–air sac system of birds seems to possess adequate defense capacity. Experimentally introduced bacteria are cleared within 24–48 h (Nagaraja et al. 1984) and injection of a suspension of incomplete Freund's adjuvant into the AASs (Ficken et al. 1986), intratracheal inoculation of heat-killed *Escherichia coli* (Toth et al. 1987), and exposure to live, apathogenic *Pasteurella multocida* vaccine (Toth et al. 1988) cause large production of SAPMs. Radioactive technetium particles exposed to conscious chickens are cleared from the lungs to the intestines within 1 h of exposure (Mensah and Brain 1982). Chickens with a high number of SPMs do not show any signs of respiratory disease (Toth et al. 1988). The respiratory surface of the avian lung has evolved various defense properties and strategies that include: (1) phagocytic epithelial cell lining of the atria and infundibula (Figs. 70 and 53D), and (2) epithelial cells abundantly endowed with lysosomes (Nganpiep and Maina 2002; Figs. 71D and 74A-F).

3.11
Control of Air Flow

Gas exchangers develop either by means of evagination or invagination (e.g. Maina 1998; 2002a,b): the former process entails growth away from a particular site of the surface of the body and the latter development into it. Normally called gills, those respiratory organs that form by evagination are the more ancient ones and are designed for water-breathing. The invaginated respiratory organs are the more derived category of gas exchangers: termed lungs, they evolved for air-breathing. Well adapted for water conservation, invaginated respiratory organs were a prerequisite for transition from water to land (e.g. Maina 1998). With a 'dead-ended' airway system, invaginated gas exchangers can only be ventilated tidally while the evaginated ones can be ventilated continuously unidirectionally. The avian lung is exceptional: itself an invaginated respiratory organ, tidal ventilation and through-flow (i.e. continuous unidirectional) ventilation occur. The first process entails mass

3.11 Control of Air Flow

movement of air in-and-out of the respiratory system and the second one back-to-front flow of air through the PPPR. The 'dual' ventilation of the avian respiratory system has been allowed by its complex morphology: the gas exchanger (the lung) has been totally uncoupled from the ventilator (the ASs); the air conduits, i.e. the PB, SB, and PR, are continuous and the PPPR are arranged in parallel to the PB (Figs. 40B and 66). The ASs are connected to the lung at various points called ostia (Fig. 38; Sect. 3.5.2). Functionally divided into a cranial group that comprises of the CeAS, ClAS and CrTAS, and a caudal group that consists of CaTAS and AAS (Figs. 38 and 66), the lung is ventilated in a bellow-like manner, back-to-front (e.g. Fedde 1980).

Owing to the complexity of the avian lung, many pathways for intrapulmonary air flow are theoretically possible: the actual route followed by the inspired air cannot be discerned by simple physical examination of the respiratory system. It takes two inspiratory and two expiratory cycles for a given volume of inspired air to move across the lung-air sac system. Inspired air flows through the IPPB (mesobronchus) into the caudal ASs, completely bypassing the openings of the MVSB. Banzett et al. (1987, 1991) and Wang et al. (1988) termed the mechanism by which the air is shunted past the openings 'inspiratory aerodynamic valving' (IAV). In fluid flow mechanics, the geometry and size of a conduit determines the dynamics of flow. To ascertain patency, the EPPB is supported by cartilages. In the majority of birds, the cartilages are C-shaped, with the body of each cartilage lying in the medial wall. However, in some species, e.g. in many ducks (Anatidae; Duncker 1971), the cartilages may almost encircle the EPPB while, in hummingbirds (Hirundinidae), they completely surround it (Warner 1972). As the EPPB enters the lung (at the hilus) to form the IPPB, the cartilages become shorter and are initially located more dorsally and subsequently medially (King 1966; Duncker 1971).

Early investigators, e.g. Dotterweich (1934) and Vos (1934), envisaged that mechanical/anatomical sphincters (valves) opened and closed in phase with the respiratory cycle, directing air flow in the lung-air sac system. While theoretically appealing, sphincters have not been found, even in sites where they would be expected to occur, e.g. at the openings of the MVSB (where IAV occurs) and at the junction of the IPPB and the MDSB where 'expiratory aerodynamic valving' (Brown et al. 1995) occurs. Given that unidirectional air flow continues in pump-ventilated paralyzed and fixed (dead) avian lungs, structure-specific aerodynamic properties should sustain the air-flow pattern in the intrapulmonary pathways of the avian lung. Finding no anatomical valves, Dotterweich (1936) proposed the concept of 'fluid valve' in which he argued that the shunting of the inspired air past the openings of the MVSB in the avian lung could be explained by aerodynamic forces caused by the sizes and geometries of the EPPB, IPPB, and the MVSB. For some considerable time (e.g. Banzett et al. 1987, 1991; Butler et al. 1988; Kuethe 1988; Wang et al. 1988, 1992; Maina and Africa 2000; Maina and Nathaniel 2001), this supposition was accepted as absolute truth, without thorough theoretical analysis, experimental testing, and morphological verification.

Explication of the mechanism(s) by which IAV is produced has been a matter of lingering debate and controversy. From gas density, flow velocity, and pressure differential studies on structural models, Banzett et al. (1987, 1991), Butler et al. (1988) and Wang et al. (1988) theorized that a narrowing of the EPPB occurred ahead of the origin of the MVSB. Using a radio opaque gas, Wang et al. (1992) recognized a constriction of the EPPB close to the origin of the first MVSB in the goose (*Anser anser*) lung. It was envisaged that the narrowing accelerated the flow of the inspired air, generating a forward convective momentum that propelled the air past the openings of the MVSB (Fig. 76). The constriction was termed 'segmentum accelerans' (SA) and was seen to change in size with respiratory rate (Wang et al. 1992): during ventilatory hyperpnia, the passage was wide and during resting breathing it was narrow.

A swelling that narrowed the lumen at the terminal section of the EPPB (close to the origin of the first MVSB) was morphologically demonstrated in the domestic fowl by Maina and Africa (2001; Fig. 77A,B). From the topographical location, the 'swelling' demonstrated functionally by Wang et al. (1992) and the one observed morphologically by Maina and Africa (2001) are undoubtedly one and the same structure. The SA is aerodynamically shaped

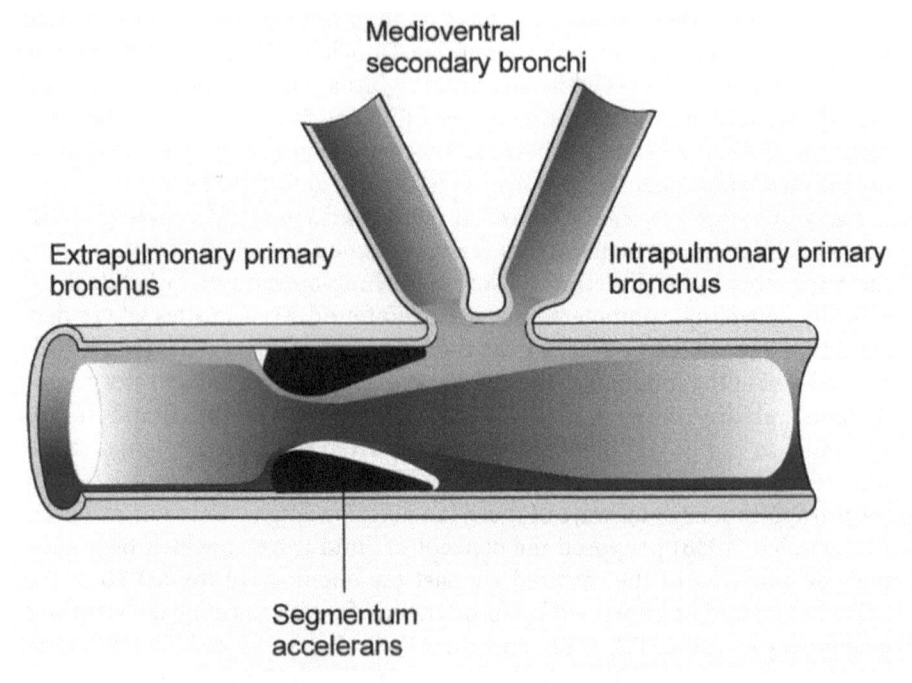

Fig. 76. Schematic diagram showing acceleration of air flow past the constriction of the extrapulmonary primary bronchus at the segmentum accelerans, resulting in the shunting of air past the medioventral secondary bronchi, i.e. inspiratory aerodynamic valving. (Maina 2002 c)

3.11 Control of Air Flow

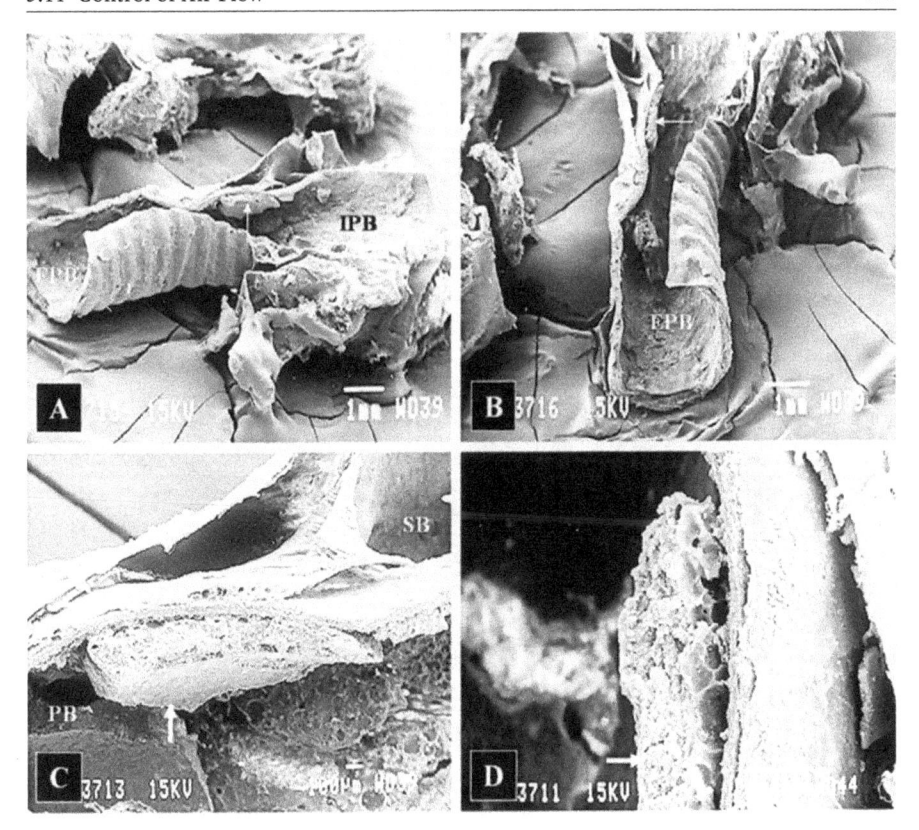

Fig. 77. A, B Extrapulmonary primary bronchus (*EPB*) of the domestic fowl, *Gallus gallus* variant *domesticus*, showing location of the segmentum accerelans (*arrow*). *IPB* Intrapulmonary primary bronchus. **C, D** Close-ups of the segmentum accerelans (*arrow*) showing its aerodynamic shape (**C**) and high degree of vascularization (**D**). (Maina and Africa 2000)

(Fig. 77C) and is highly vascularized (Figs. 77D and 78). It was envisaged by Maina and Africa (2001) that the SA may function like an erectile (cavenous) tissue, where the extent of enlargement may be controlled by the dynamics of influx and efflux of blood: thickening of the SA should narrow the lumen of the EPPB while 'slimming down' should widen the passageway. All other factors constant, narrowing of the opening should accelerate the flow of the inspired air, thrusting it past the openings of the MVSB (Fig. 76). Banzett et al. (1991) and Wang et al. (1992) theorized that during exercise and other states that invoke hyperpnea, the SA flattens to the diameter of the EPPB and speculated that, in such a state, the velocity of air flow is adequate to produce convective momentum that should push the air past the openings of the MVSB. However, during resting breathing, when the velocity of the air is slow, a constriction may be critical for generation of IAV.

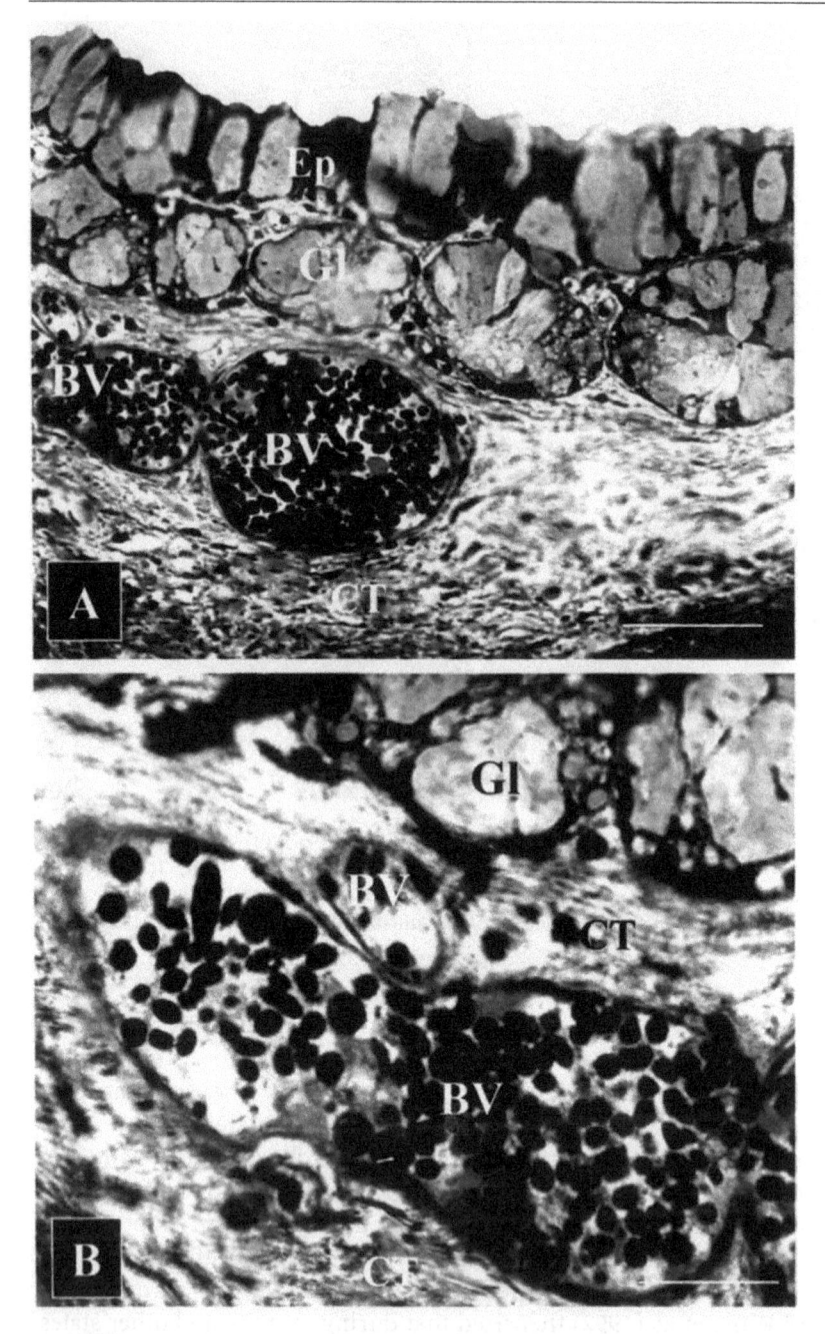

Fig. 78A,B. Area close to the segmentum accelerans showing intense vascularization. *BV* Blood vessels; *Gl* goblet cells; *Ep* epithelial cells; *CT* connective tissue. *Scale bars* **A** 15 µm; **B** 10 µm. (**B** from Maina and Africa 2000)

3.11 Control of Air Flow

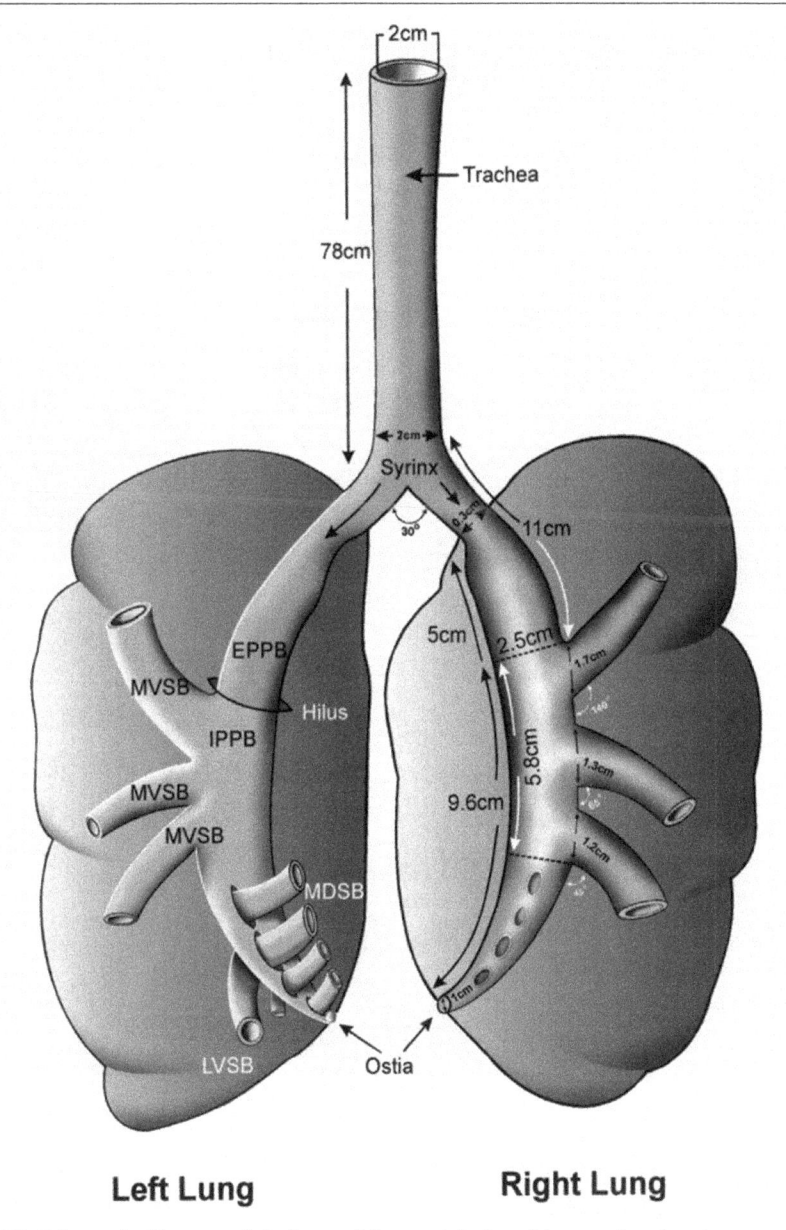

Fig. 79. Schematic diagram of the lung of the ostrich, *Struthio camelus*, showing measurements and geometries of some of the air-conducting airways. (Maina and Nathaniel 2001)

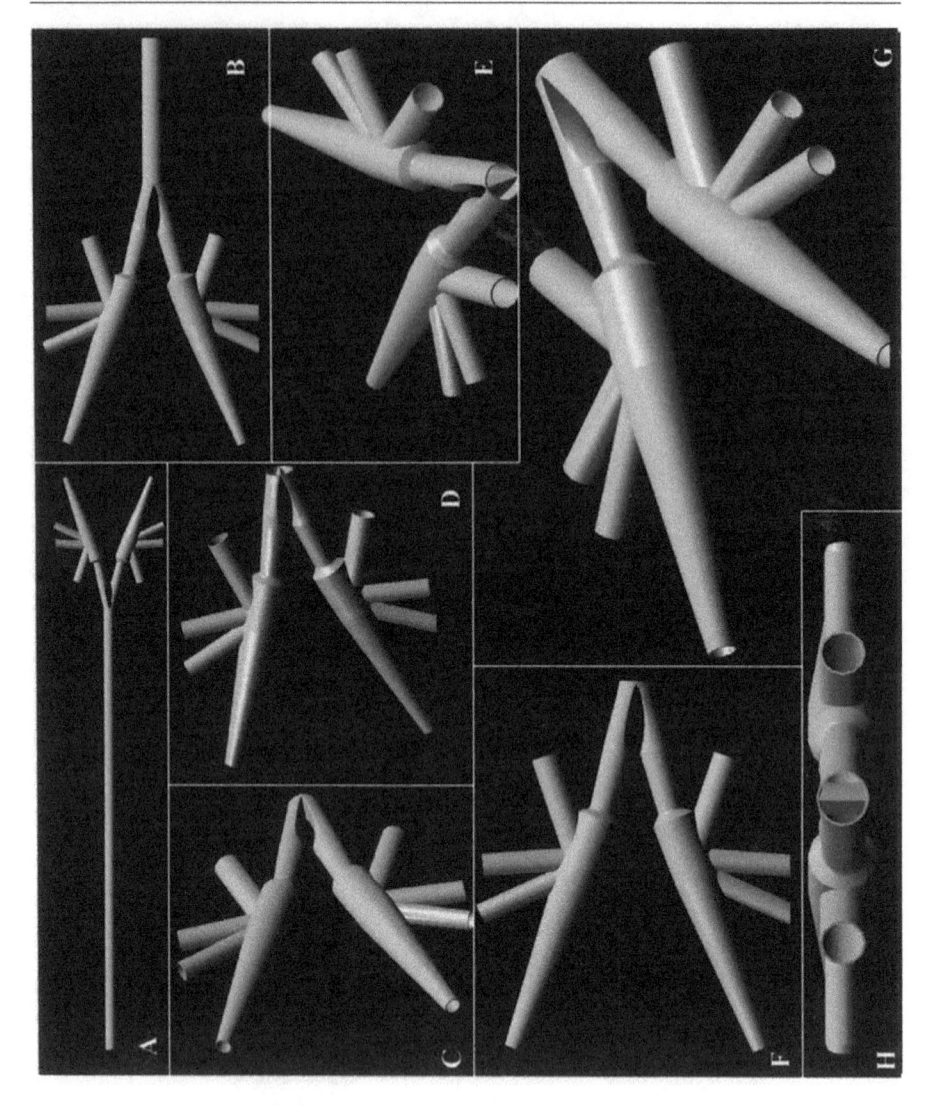

Fig. 80. A Computer-generated reconstructions from actual measurements of sizes and geometries of the trachea, primary bronchus, and medioventral secondary bronchi of the lung of the ostrich, *Struthio camelus* (see Fig. 78). **B–H** Close-ups of the primary bronchus and medioventral secondary bronchi. **H** Frontal view of the syringeal constriction, extrapulmonary primary bronchus, and medioventral secondary bronchi

3.11 Control of Air Flow

The specific mechanism by which the thickening of the SA is regulated is presently unknown. Chemoreceptors sensitive to CO_2 in the inspired air or mechanoreceptors that may detect velocity of air flow or diametric changes of the EPPB could be involved. Molony et al. (1976) observed that air-flow resistance across the openings of the MVSB was dependent on the PCO_2 in the inspired air: resistance was high at low PCO_2, and vice versa. Barnas et al. (1978) observed that the intrapulmonary smooth muscle was sensitive to changes in the concentration of CO_2. Interestingly, a SA does not appear to be ubiquitous to the respiratory systems of birds: it is lacking in the ostrich (Maina and Nathaniel 2001). Since the general air-flow pattern in the lung-air sac system of birds is basically similar, i.e. the ET is unidirectionally and continuously ventilated (e.g. Fedde 1980), an SA may not be the only factor involved in production of IAV. Together with an SA (where it occurs), structural factors/properties like angulation of the MVSB relative to the long axis of the IPPB (Figs. 47–49), sizes and shapes of the EPPB, IPPB, and SB (Figs. 79 and 80), and syringeal narrowing (Fig. 41) may to varying extents influence IAV. Model-based studies (e.g. Butler et al. 1988; Wang et al. 1988) have, however, demonstrated that the geometry, particularly the angulation and narrowing of orifices of the MVSB, does not significantly affect the inspiratory valve performance. Further studies are urgently needed to explicate the mechanisms by which IAV is generated and maintained.

4
Quantitative Morphology (Morphometry)

Although much fundamental insight can be gained by purely descriptive analysis of certain phenomena, in the end, statements about correlation between structure and function will have to be supported by solid quantitative data. Quantitative morphology or morphometry is hence an essential part of such studies. Weibel (1984)

4.1
General Observations

At the ordinary range of temperature and pressure, of the three physically occurring states of matter, i.e. solid, liquid, and gas, only fluids (water and air) are atomically/molecularly appropriately configured to 'dissolve' and transfer molecular O_2 to the respiratory site. Only relatively few animals, the so-called bimodal (=transitional) breathers, to varying extents utilize both respiratory fluid media as source of O_2 (e.g. Graham 1994; Maina 1998). Animals have had little choice as to what fluid medium to use: the majority utilize either water (water-breathers) or air (air-breathers). Under such prescriptive circumstances, it is axiomatic that similarities of design should have evolved in the gas exchangers, particularly at the elemental level of design. A common mechanistic model of a gas exchanger is that of a biological construction in which an external respiratory fluid medium (air and/or water) and an internal one (hemolymph/blood) are presented to each other across a thin, expansive tissue barrier. Contrary to earlier suppositions that claimed active transfer of molecular O_2 across biological tissues (e.g. Haldane and Priestley 1935; Forster 1996), it is now unequivocal that passive diffusion along prevailing P is the sole means by which flux of O_2 occurs. The fundamental structural features that determine transfer of O_2 through spaces and across tissue barriers are: (1) extensive surface area, (2) optimal volumes of the respiratory media, and (3) thin tissue partitioning (Figs. 81 and 82). Until fairly recently, mor-

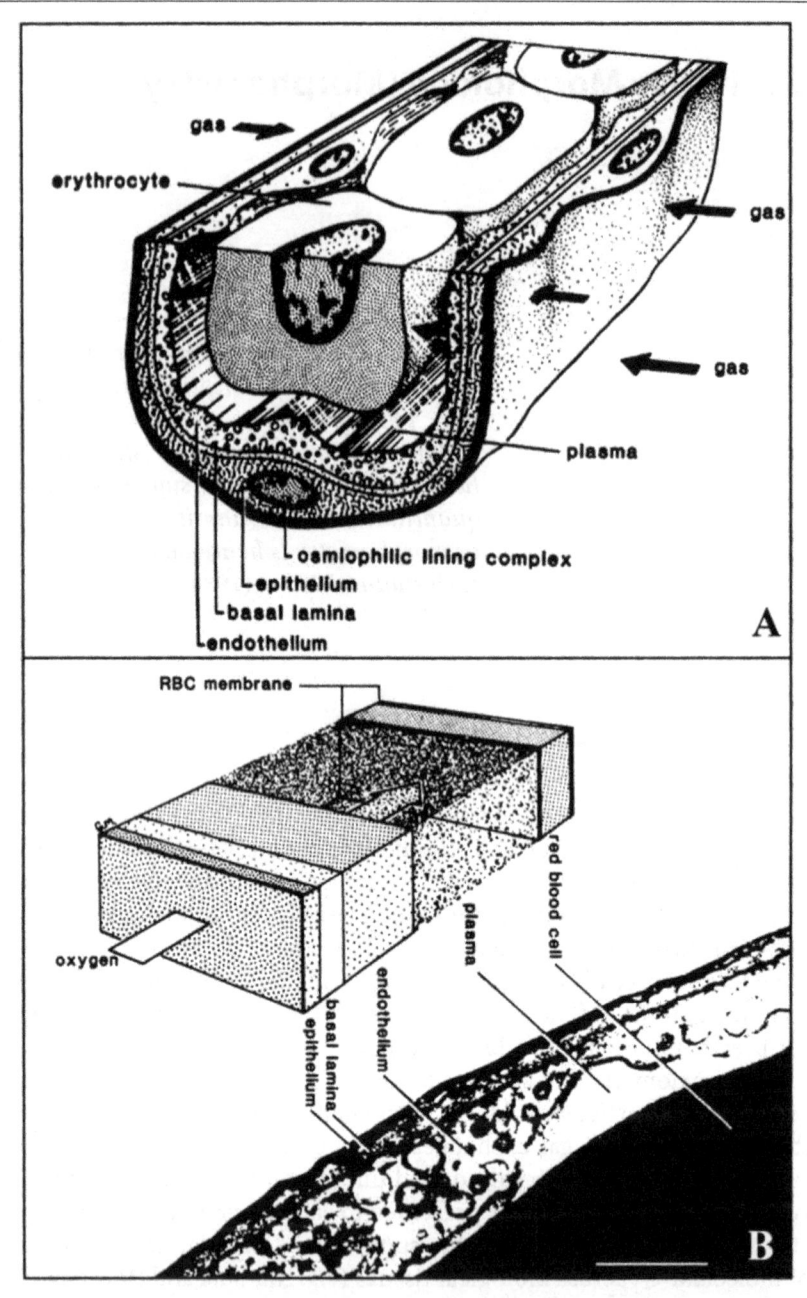

Fig. 81. A Cross section of a blood capillary in the lung showing the components through which oxygen diffuses, i.e. the BGB, the plasma layer, and the erythrocyte cytoplasm. **B** Schematic diagram and an electron micrograph of the lung of the black-headed gull, *Larus ridibundus*, showing the air-hemoglobin pathway, i.e. the distance that a diffusing oxygen molecule has to traverse. *Scale bar* **B** 0.2 μm. (Maina 1989b)

4.2 Volume of the Lung (VL)

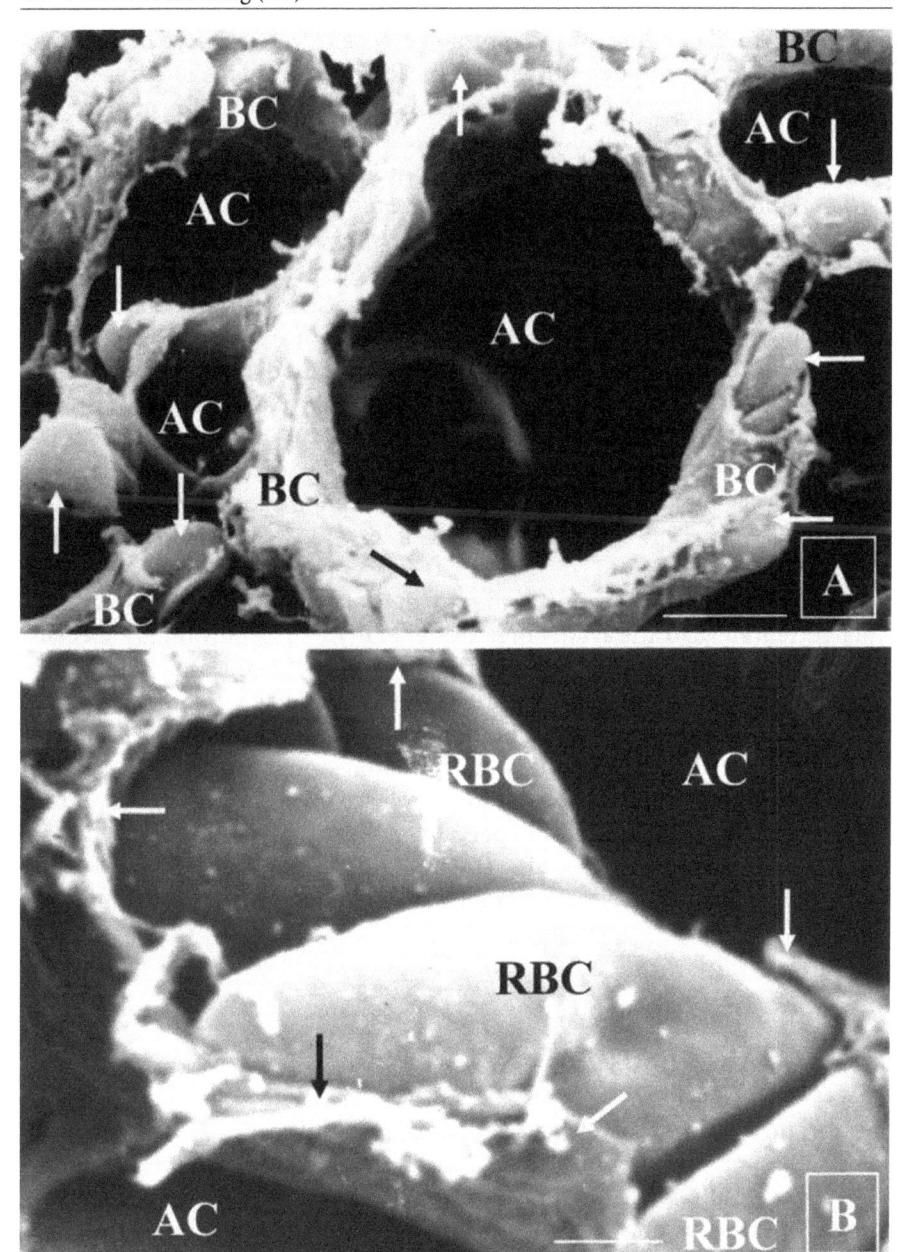

Fig. 82. A Air capillaries (*AC*) that are delineated by blood capillaries (BC). *Arrows* Erythrocytes. **B** Close-up of a blood capillary containing red blood cells (*RBC*). Oxygen diffuses from the air capillary (*AC*) across the BGB (*arrows*) into the red blood cells (*RBC*). *Scale bars* **A** 5 µm; **B** 3 µm. (Maina 1998)

phological studies on the avian respiratory system were entirely descriptive (qualitative). Although such studies significantly advanced the understanding of its essential design, the descriptive accounts fell far short of providing the exact details necessary for thorough explication of the essence of the functional design of the avian lung. Predictably, Fisher (1955) pointed out that *"without knowledge of the detailed structure, the manner in which the system (the avian respiratory apparatus) works will remain largely hypothetical"*.

The genesis of reliable quantitative (morphometric) methods in biology has been reviewed by Weibel (1979). Pioneered by geologists and material scientists who were interested in finding out the structure of composite materials like rocks (e.g. DelessÈ 1846) and metals, the methods were duly modified for application in biology through suitable tissue fixation, sampling, and analysis. The analytical techniques were based on reproducible, sound statistical and mathematical deductions (e.g. Buffon 1777; Snedecor 1956). Regarding pulmonary morphometry, the techniques were meaningfully applied in the seminal works of, e.g., Weibel and Gomez (1962), Dunnill (1962), and Weibel (1963).

The current pulmonary morphometric data show that: (1) compared with other vertebrate taxa, the avian lungs are generally structurally highly specialized, and (2) between different groups of birds, the degrees of pulmonary refinement reflect the different metabolic capacities that are in turn set by factors such as phylogenetic status, body mass, lifestyle, and habitat occupied. Comparisons of some pulmonary morphometric parameters between the lungs of nonflying mammals, bats, and birds are given in Tables 1–6 and in Figs. 83–88. Due to constraints of space, here, the discussion in this section is kept succinct. For more detailed accounts, original publications (e.g. Maina 1989b, 2002 c; Maina et al. 1989a) and others that are given here should be consulted.

4.2
Volume of the Lung (VL)

In birds, the total volume of the respiratory system (i.e. the VL, air sacs, and pneumatic spaces) constitutes about 20 % of the total body volume (Duncker 1971). In the mute swan, *Cygnus olor*, at 34 %, the value is much greater. For animals of the same body mass, the combined VL and that of the ASs is 3 to 5 times larger than the VL of a mammal and two times greater than that of a reptile (Tenney and Remmers 1963; Tenney and Tenney 1970). Displaced to dorsal aspect, i.e. the roof, of the coelomic cavity, and firmly attached to the ribs (Figs. 38, 39B and 43B), the avian lungs are relatively small (Fig. 38). Compared with a nonflying mammal of comparable body mass, the lungs are 27 % smaller (Maina 1989b; Maina et al. 1989a; Table 1; Fig. 83). Bats, the only volant mammals, however, have exceptionally large lungs (Maina et al 1982b, 1991; Maina and King 1984).

4.2 Volume of the Lung (VL)

Table 1. Volume densities (%) of the main components of the avian lung. RJF Red jungle fowl

Order, scientific and common name	n	Body mass (g)	Total lung volume (cm³)	Exchange tissue (%)	Lumina of parabronchi and secondary bronchi including atria (%)	Blood vessels larger than capillaries (%)	Primary bronchus (%)
STRUTHIOFORMES							
Struthio camelus							
Ostrich[a]	1	45000	1563	78.31	15.82	4.38	1.63
SPHENISCIFORMES							
Spheniscus humboldti							
Humboldt penguin[b]	1	4500	1368	51.29	37.71	7.44	3.56
ANSERIFORMES							
Anas platyrhynchos							
Mallard duck[c,d,e]	5	1038	30.6	40.55	50.37	6.70	2.38
Anser anser							
Greylag goose[c,d,e]	5	3838	95.3	40.37	50.99	6.91	1.73
Branta canadensis							
Canada goose[f,g]	–	–	–	36	–	7.5	30.2
Cygnus olor							
Mute swan[f,g]	–	–	–	31	–	4.9	32.5
Cairina moschata							
Muscovy duck (dom.)[h]	5	1627	48.1	49.24	37.47	8.24	5.05
FALCONIFORMES							
Falco tinnunculus							
Common kestrel[c,d]	2	66	3.1	51.9	9.03	7.33	1.79

Table 1. (Continued)

Order, scientific and common name	n	Body mass (g)	Total lung volume (cm³)	Exchange tissue (%)	Lumina of parabronchi and secondary bronchi including atria (%)	Blood vessels larger than capillaries (%)	Primary bronchus (%)
GALLIFORMES							
Gallus gallus							
Domestic fowl[i]	3	2141	27.0	46.35	30.56	13.65	9.34
Domestic fowl[f,g]	–	–	–	46.7	–	6.8	14.7
Domestic fowl[h]	5	1873	26.6	49.66	38.01	6.64	5.69
Domestic fowl[h] (RJF)	5	478	8.7	53.15	36.05	6.61	4.63
Numida meleagris							
Guinea fowl (dom.)[j,k]		1839	25.8	47.35	29.37	13.09	10.19
GRUIFORMES							
Amaurornis phoenicurus							
White-breasted water hen[h]	5	146	5.1	55.0	36.73	5.20	3.05
Fulica atra							
Coot[f,g]	–	–	–	50.0	–	7.3	10.8
CHARADRIIFORMES							
Alca torda							
Razorbill[c,d,i]	2	487	18.1	32.79	53.43	11.40	2.38
Cephus carbo							
Spectacled guillemot[c,d,i]	9	737	24.1	33.40	47.92	13.86	4.82
Larus argentatus							
Herring gull[c,d,i]	2	654	18.2	32.64	60.41	5.56	1.39
Larus canus							

4.2 Volume of the Lung (VL)

Species	n						
Common gull[c,d,i] *Larus ridibundus*	1	302	7.1	30.85	61.70	5.39	2.06
Black-headed gull[c,d,i]	6	253	7.5	35.83	54.88	7.17	2.12
COLUMBIFORMES							
Columba livia Rock dove[c,d]	1	216	7.4	43.57	32.26	14.85	9.32
Columba livia Domestic pigeon[f,g]	–	–	–	49.0	–	7.5	18.8
Streptopelia decaocto Collared turtle dove[c,d]	16	189	6.5	52.7	29.46	15.53	2.84
Streptopelia senegalensis Laughing gull[c,d]	1	58	1.8	52.26	40.22	6.18	1.34
PSITTACIFORMES							
Melopsittacus undulatus Budgerigar[c,d]	6	36.4	1.03	46.56	47.46	4.02	1.96
Budgerigar[m]	5	36.8	1.11	47.8	–	8.2	22.9
CUCULIFORMES							
Chrysococcyx klaas Klaa's cuckoo[c,d]	3	27.0	0.66	53.0	37.03	8.4	0.16
APODIFORMES							
Colibri coruscans Violet-eared hummingbird[m]	3	3	7.3	0.27	46.4	–	13.6 11.4
COLIIFORMES							
Colius striatus Spectacled mousebird[c,d]	1	50.5	0.71	50.81	37.43	9.78	1.98

Table 1. (Continued)

Order, scientific and common name	n	Body mass (g)	Total lung volume (cm³)	Exchange tissue (%)	Lumina of parabronchi and secondary bronchi including atria (%)	Blood vessels larger than capillaries (%)	Primary bronchus (%)
PICIFORMES							
Jynx ruficollis							
Red-breasted wryneck[c,d]	2	54.3	1.09	45.7	47.65	5.44	1.20
Pogoniulus bilineatus							
Golden-rumped tinkerbird[c,d]	1		14.9	0.23	54.20	35.22	8.641.94
CASUARIIFORMES							
Dromaius novaehollandiae							
Emu[n]	1	30000	1100	17.76	49.11	28.57	4.56
PASSERIFORMES							
Ambryospiza albifrons							
Grosbeak weaver[c,d,o]	5	37.4	0.97	48.83	44.1	6.17	0.90
Cercotrichas leukophrys							
White-winged robin[c,d,o]	1	16.0	0.38	42.39	47.13	9.48	1.00
Chloropeta natalensis							
Yellow flycatcher[c,d,o]	1	10.9	0.31	–	–	–	–
Cisticola cantans							
Singing cisticola[c,d,o]	4	15.0	0.31	48.61	41.57	7.74	2.08
Cossypha cafra							
Robin chat[c,d,o]	1	25.1	0.58	46.50	44.95	6.85	1.70
Estrilda astrid							
Waxbill[c,d,o]	3	7.0	0.15	59.85	30.06	8.40	1.69

4.2 Volume of the Lung (VL)

Species	n				
Estrilda melanotis					
Yellow-bellied waxbill[c,d,o]	1	5.4	0.13	—	—
Hirundo fuligula					
African rock martin[c,d,o]	1	13.7	0.33	51.15	6.69	0.82
Laganosticta senegala					
Red-billed firefinch[c,d,o]	1	8.0	0.223	56.92	10.02	1.11
Lanius aethiopicus					
Tropical boubou[c,d,o]	1	42.8	0.89	54.57	6.62	0.91
Lanius collaris					
Fiscal shrike[c,d,o]	6	32.5	0.72	52.99	6.43	0.92
Lonchura cucullata					
Bronze mankin[c,d,o]	7	9.4	0.23	55.92	8.64	1.51
Nectarina kilimensis					
Bronze sunbird[c,d,o]	1	16.6	0.50	56.34	8.70	2.34
Nectarina reichenowi					
Golden-winged sunbird[c,d,o]	3	13.7	0.36	58.56	8.67	0.86
Passer domesticus					
House sparrow[c,d,o]	12	25.5	0.76	55.68	6.45	0.96
House sparrow[m]	5	26.5	0.80	52.60	10.3	11.80
Ploceus baglafecht					
Baglafecht weaver[c,d,o]	6	32.5	0.88	49.43	6.72	1.42
Ploceus cucullatus					
Black-headed weaver[c,d,o]	3	34.9	0.87	7.61	5.67	1.47
Ploceus ocularis					
Spectacled weaver[c,d,o]	2	26.9	0.72	47.05	6.30	1.61
Ploceus xanthops					
Holub's golden weaver[c,d,o]	3	39.3	1.03	49.13	7.87	1.91
Prinia subflava					
Twany prinia[c,d,o]	1	9.1	0.21	51.86	9.08	1.49
Serinus canaria					
Canary[c,d,o]	6	23.9	0.71	—	—	—

Table 1. (Continued)

Order, scientific and common name	n	Body mass (g)	Total lung volume (cm³)	Exchange tissue (%)	Lumina of parabronchi and secondary bronchi including atria (%)	Blood vessels larger than capillaries (%)	Primary bronchus (%)
Serinus mozambicus Yellow-fronted canary[c,d,o]	1	13.5	0.27	–	–	–	–
Sturnus vulgaris Common starling[c,d,o]	10	72.60	2.02	51.76	41.16	5.19	1.89
Turdus iliacus Redwing[c,d,o]	1	51.0	1.19	45.63	5.76	7.11	1.47
Turdus olivaceus Olive thrush[c,d,o]	2	65.1	1.40	55.84	36.38	5.92	1.86

[a] Maina and Nathaniel (2001);
[b] Maina and King (1987)
[c] Maina (1989a)
[d] Maina et al. (1989a)
[e] Maina and King (1982a)
[f] Duncker (1972)
[g] Duncker (1973)
[h] Vidyadaran et al. (1987)
[i] Abdalla et al. (1982)
[j] Abdalla and Maina (1981)
[k] Maina et al. (1982a)
[l] Maina (1987b)
[m] Dubach (1981)
[n] Maina and King (1989)
[o] Maina (1984)

4.2 Volume of the Lung (VL)

Table 2. Volume densities (%) of the components of the exchange tissue of the avian lung. *RJF* Red jungle fowl

Order, scientific and common name	Air capillaries	Blood capillaries	Tissue of the BGB	Tissue not involved in gas exchange
STRUTHIOFORMES				
Struthio camelus[a]	61.19	20.41	12.69	5.71
SPHENISCIFORMES				
Spheniscus humboldti[b]	34.36	51.04		
ANSERIFORMES				
Anas platyrhynchos[c,d,e]	59.21	32.98	5.74	2.08
Anser anser[c,d,e]	62.03	32.43	4.24	1.30
Branta canadensis[f,g]	48.5	41.5	10.0	–
Cygnus olor[f,g]	46.0	43.1	10.9	–
Cairina moschata (dom.)[h]	57.24	29.63	5.88	7.25
FALCONIFORMES				
Falco tinnunculus[d,e]	53.02	25.56	17.28	4.14
GALLIFORMES				
Gallus gallus (dom.)[i]	60.90	27.92	6.30	4.88
Gallus gallus (dom.)[f,g]	55.8	28.3	15.9	–
Gallus gallus (dom.)[h]	55.59	27.89	7.44	9.09
Gallus gallus (RJF)[h]	64.82	21.29	6.01	7.88
Numida meleagris (dom.)[j]	53.51	33.80	10.90	1.89
GRUIFORMES				
Amaurornis phoenicurus[h]	63.20	23.29	5.89	7.61
Fulica atra[f,g]	33.3	56.3	10.4	–
CHARADRIIFORMES				
Alca torda[d,e,k]	49.54	36.89	10.19	3.38
Cephus carbo[d,e,k]	47.49	39.75	9.46	3.30
Larus argentatus[d,e,k]	53.68	35.92	7.04	3.36
Larus canus[d,e,k]	58.23	30.45	8.74	2.58
Larus ridibundus[d,e,k]	62.52	28.92	5.96	2.60
COLUMBIFORMES				
Columba livia[d,e]	58.70	33.63	5.90	1.77
Columba livia[d,e]	43.0	44.0	13.0	–
Streptopelia decaocta[d,e]	62.96	22.81	10.42	3.81
Streptopelia senegalensis[d,e]	45.13	34.63	16.77	3.46
PSITTACIFORMES				
Melopsitacus undulatus[d,e]	52.47	33.34	11.44	2.75
Melopsitacus undulatus[l]	67.3	26.1	6.6	–
CUCULIFORMES				
Chrysococcyx klaas[d,e]	55.09	34.50	8.26	2.15
APODIFORMES				
Colibri coruscans[l]	53.9	36.7	9.4	–

Table 2. (*Continued*)

Order, scientific and common name	Air capillaries	Blood capillaries	Tissue of the BGB	Tissue not involved in gas exchange
COLIIFORMES				
Colius striatus[d,e]	53.75	36.43	7.49	2.33
PICIFORMES				
Pogoniulus bilineatus[d,e]	38.21	39.02	19.51	3.25
CASUARIIFORMES				
Dromaius novaehollandiae[m]	79.15	14.2	3.4	3.25
PASSERIFORMES				
Ambryospiza albifrons[d,e,n]	44.84	42.15	9.88	3.13
Cisticola cantans[d,e,n]	48.20	38.32	9.92	3.56
Hirundo fuligula[d,e,n]	45.20	44.35	7.74	2.08
Lanius collaris[d,e,n]	49.49	37.13	10.40	2.98
Passer domesticus[d,e,n]	45.68	38.04	12.87	3.41
Passer domesticus[l]	54.10	36.80	9.20	–
Ploceus baglafecht[d,e,n]	50.96	36.22	8.99	3.83
Prinia subflava[d,e,n]	50.47	37.38	9.35	2.80
Sturnus vulgaris[d,e,n]	51.68	32.55	12.55	3.22
Turdus iliacus[d,e,n]	47.11	39.33	11.19	2.37
Turdus olivaceus[d,e,n]	50.45	39.04	7.81	2.70

[a] Maina and Nathaniel (2001)
[b] Maina and King (1987)
[c] Maina and King (1982a)
[d] Maina et al. (1989a)
[e] Maina (1989a)
[f] Duncker (1972)
[g] Duncker (1973)
[h] Vidyadaran et al. (1987)
[i] Abdalla et al. (1982)
[j] Abdalla and Maina (1981)
[k] Maina (1987b)
[l] Dubach (1981)
[m] Maina and King (1989)
[n] Maina (1984)

4.2 Volume of the Lung (VL)

Table 3. Surface areas (m²) of the air capillaries (S_a), blood-gas (tissue) barrier (S_t), capillary endothelium (S_c), red blood cells (S_e) and blood plasma (S_p). *RJF* Red jungle fowl

Order, scientific and common names	S_a	S_t	S_c	S_e	S_p
STRUTHIOFORMES					
Struthio camelus[a]	182.51	120.34	146.06	119.51	133.06
SPHENISCIFORMES					
Spheniscus humboldti[b]	10.4	8.2	10.5	13.3	11.9
ANSERIFORMES					
Anas platyrhynchos[c,d,e]	3.65	2.97	3.31	3.23	3.27
Anser anser[c,d,e]	11.3	8.87	10.4	9.6	10.0
Cairina moschata (dom.)[f]	6.4	4.71	6.15	5.37	5.63
FALCONIFORMES					
Falco tinnunculus[c,d]	0.61	0.53	0.58	0.28	0.43
GALLIFORMES					
Gallus gallus (dom.)[g]	2.77	2.16	2.42	3.59	3.01
Gallus gallus (dom.)[h,i]	3.33	2.28	2.71	2.91	2.81
Gallus gallus (RJF)[f]	0.95	0.62	0.72	0.86	0.79
Numida meleagris (dom.)[j]					
GRUIFORMES					
Amaurornis phoenicurus[j]	0.76	0.51	0.65	0.52	0.61
CHARADRIIFORMES					
Alca torda[c,d,k]	2.60	2.40	1.82	1.52	0.67
Cephus carbo[c,d,k]	2.39	1.93	2.79	1.78	2.29
Larus argentatus[c,d,k]	1.58	1.46	1.81	0.91	1.36
Larus canus[c,d,k]	0.71	0.63	0.83	0.64	0.74
Larus ridibundus[c,d,k]	0.70	0.61	0.75	0.57	0.66
COLUMBIFORMES					
Columba livia (dom.)[c,d]	1.20	0.86	1.06	1.03	1.05
Streptopelia decaocta[c,d]	1.14	0.84	0.93	0.65	0.79
Streptopelia senegalensis[c,d]	0.32	0.27	0.37	0.17	0.27
PSITTACIFORMES					
Melopsitacus undulatus[c,d]	0.18	0.15	0.19	0.16	0.18
Melopsitacus undulatus[l]	0.22	0.17	0.19	–	–
CUCULIFORMES					
Chrysococcyx klaas[c,d]	0.10	0.09	0.11	0.09	0.59
APODIFORMES					
Colibri coruscans[l]	0.06	0.05	0.07	–	–
COLIIFORMES					
Colius striatus[c,d]	0.13	0.10	0.15	0.13	0.15

Table 3. (*Continued*)

Order, scientific and common names	S_a	S_t	S_c	S_e	S_p
PICIFORMES					
Pogoniulus bilineatus[c,d]	0.04	0.03	0.04	0.05	0.05
CASUARIIFORMES					
Dromaius novaehollandiae[m]	22.56	6.28	19.69	23.16	22.86
PASSERIFORMES					
Ambryospiza albifrons[c,d,n]	0.18	0.14	0.21	0.19	0.20
Cisticola cantans[c,d,n]	0.05	0.04	0.07	0.60	0.07
Hirundo fuligula[c,d,n]	0.15	0.12	0.15	0.15	0.15
Lanius collaris[c,d,n]	0.15	0.12	0.16	0.15	0.15
Passer domesticus[c,d,n]	0.18	0.17	0.20	0.12	0.16
Passer domesticus[l]	0.18	0.15	0.17	–	–
Ploceus baglafecht[c,d,n]	0.15	0.12	0.15	0.12	0.14
Prinia subflava [c,d,n]	0.03	0.03	0.04	0.03	0.30
Sturnus vulgaris[c,d,n]	0.42	0.36	0.42	0.31	0.36
Turdus iliacus[c,d,n]	0.17	0.17	0.22	0.16	0.19
Turdus olivaceus[c,d,n]	0.30	0.24	0.24	0.22	0.23

[a] Maina and Nathaniel (2001)
[b] Maina and King (1987)
[c] Maina (1989a)
[d] Maina et al. (1989a)
[e] Maina and King (1982a)
[f] Vidyadaran et al. (1987)
[g] Abdalla et al. (1982)
[h] Duncker (1972)
[i] Duncker (1973)
[j] Abdalla and Maina (1981)
[k] Maina (1987b)
[l] Dubach (1981)
[m] Maina and King (1989)
[n] Maina (1984)

4.2 Volume of the Lung (VL)

Table 4. Certain pulmonary morphometric structural parameters of the avian lung. S_t Surface area of the blood gas (tissue) barrier per unit body mass; S_t V^{-1} surface area of the blood gas (tissue) barrier per unit volume of exchange tissue; V_c S_t^{-1} volume of the pulmonary capillary blood per unit surface area of the blood gas (tissue) barrier; V_L M^{-1} volume of the lung per unit body mass; V_e volume of the red blood cells; V_{ec} volume of the red blood cell cytoplasm; *PCH* pulmonary capillary hematocrit; *RJF* red jungle fowl

Order/scientific name	S_t M^{-1} (cm^2 g^{-1})	S_t V_x^{-1} (mm^2 mm^{-1})	V_c S_t^{-1} (cm^3 m^{-2})	V_L M^{-1} (cm^3 kg^{-1})	V_e (cm^3)	V_{ec} (cm^3)	PCH (%)
STRUTHIOFORMES							
Struthio camelus[a]	30.10	98.32	2.08	39.08	–	–	–
SPHENISCIFORMES							
Spheniscus humboldti[b]	18.1	116.1	4.4	30.4	14.6	12.9	61
ANSERIFORMES							
Anas platyrhynchos[c,d,e]	28.7	240	1.4	30	2.2	1.7	52
Anser anser[c,d,e]	23.1	253	1.4	25	6.5	5.5	52
Branta canadensis[f,g]	34.8	245	–	–	–	–	–
Cygnos olor[f,g]	39.8	263	–	–	–	–	–
Ca8irina moschata (dom.)[h]	30.0.	200	1.5	30	–	–	56
FALCONIFORMES							
Falco tinnunculus[c,d]	43.4	324	0.8	18.7	0.2	0.1	36
GALLIFORMES							
Gallus gallus (dom.)[i]	8.7	172	1.6	12.6	2.3	1.8	64
Gallus gallus (dom.)[f,g]	13.6	192	–	–	–	–	–
Gallus gallus (dom.)[h]	12.5	173	1.7	14.7	1.9	1.6	53
Gallus gallus (RJF)[h]	13.0	135	1.6	18.1	0.5	0.4	51
Numida meleagris[j]	7.0	195	1.7	14.0	–	–	–
Meleagris gallopavo[f,g]	11.8	188	–	–	–	–	–
Cortunix cortunix[f,g]	–	288	–	–	–	–	–

Table 4. (Continued)

Order/scientific name	$S_t\ M^{-1}$ (cm² g⁻¹)	$S_t\ V_x^{-1}$ (mm² mm⁻¹)	$V_c\ S_t^{-1}$ (cm³ m⁻²)	$V_L\ M^{-1}$ (cm³ kg⁻¹)	V_e (cm³)	V_{ec} (cm³)	PCH (%)
GRUIFORMES							
Amaurornis phoenicurus[h]	34.8	180	1.3	35.1	0.4	0.3	60
Fulica atra[f,g]	45.8	248	–	–	–	–	–
CHARADRIIFORMES							
Alca torda[c,d,k]	29.9	237	1.6	37.2	1.2	1.0	53
Cephus carbo[c,d,k]	27.0	240	1.7	32.7	1.4	1.2	41
Larus argentatus[c,d,k]	22.1	236	1.5	27.8	0.6	0.4	27
Larus canus[c,d,k]	20.8	292	1.1	23.5	0.3	0.2	41
Larus ridibundus[c,d,k]	27.6	238	1.3	29.6	0.4	0.3	48
COLUMBIFORMES							
Columba livia[c,d]	39.8	254	1.3	34.3	0.7	0.6	64
Columba livia[f,g]	40.3	302	–	–	–	–	–
Streptopelia decaocta[c,d]	43.1	253	1.0	34.4	0.4	0.3	49
Streptopelia senegalensis[c,d]	47.2	296	1.2	31.0	0.1	0.1	31
PSITTACIFORMES							
Melopsitacus undulatus[c,d]	42.6	317	1.1	28.3	0.1	0.1	62
Melopsitacus undulatus[i]	48.1	301	0.9	29.6	0.3	–	–
CUCULIFORMES							
Chrysococcyx klaas[c,d]	31.9	274	1.4	24.4	0.1	0.1	57
APODIFORMES							
Colibri coruscans[l]	87.1	389	1.0	42.9	–	–	–

4.2 Volume of the Lung (VL)

COLIIFORMES					
Colius striatus[c,d]	19.9	260	1.4	14.1	0.1	0.1	56
PICIFORMES					
Pogoniulus bilineatus[c,d]	22.5	272	1.5	15.4	0.03	0.03	71
CASUARIIFORMES					
Dromaius novaehollandiae[m]	5.4	82	1.70	36.7	–	–	–
PASSERIFORMES					
Ambryospiza albifrons[c,d,n]	40.7	314	1.5	25.9	0.12	0.1	57
Cisticola cantans[c,d,n]	29.3	271	1.36	1.4	0.3	0.03	48
Hirundo fuligula[c,d,n]	86.5	353	0.7	24.1	0.1	0.1	62
Lanius collaris[c,d,n]	37.4	300	1.2	22.2	0.1	0.1	55
Passer domesticus[c,d,n]	63	389	0.9	29.8	0.1	0.1	52
Passer domesticus[l]	59.4	326	1.1	30.2	–	–	–
Corvus corone[c,d,n]	58.6	298	–	–	–	–	–
Ploceus baglafecht[c,d,n]	36.6	317	1.3	27.1	0.1	0.1	56
Prinia subflava[c,d,n]	29.2	247	1.5	23.1	0.02	0.12	50
Sturnus vulgaris[c,d,n]	49.3	342	1.0	27.8	0.2	0.15	53
Turdus iliacus[c,d,n]	33.4	313	1.3	23.3	0.1	0.1	49
Turdus olivaceus[c,d,n]	37.1	303	1.3	21.5	0.2	0.13	58

[a] Maina and Nathaniel (2001)
[b] Maina and King (1987)
[c] Maina (1989a)
[d] Maina et al. (1989a)
[e] Maina and King (1982a)
[f] Duncker (1972)
[g] Duncker (1973)
[h] Vidyadaran et al. (1987)
[i] Abdalla et al. (1982)
[j] Abdalla and Maina (1981)
[k] Maina (1987b)
[l] Dubach (1981)
[m] Maina and King (1989)
[n] Maina (1984)

Table 5. Thickness of the blood–gas (tissue) barrier and the plasma layer in the avian lung. τ_{ht} Harmonic mean thickness of the blood–gas (tissue) barrier; $\tau_{ht(min)}$ minimum harmonic mean thickness of the blood–gas barrier; τ_{hp} harmonic mean thickness of the plasma layer; τ_t arithmetic mean thickness of the blood–gas (tissue) barrier; *RJF* red jungle fowl

Order/scientific name	τ_{ht} (µm)	$\tau_{ht(min)}$ (µm)	τ_{hp} (µm)	τ_t (µm)	$\tau_t \tau_{ht}^{-1}$
STRUTHIOFORMES					
Struthio camelus[a]	0.560	–	0.140	0.690	1.23
SPHENISCIFORMES					
Spheniscus humboldti[b]	0.530	–	0.213	–	–
ANSERIFORMES					
Anas platyrhynchos[c,d,e]	0.113	0.062	0.369	0.903	6.80
Anas platyrhynchos[f]	0.133	0.075	–	0.240	1.80
Anser anser[c,d,e]	0.118	0.050	0.322	0.887	7.85
Cairina moschata (dom.)[g]	0.199	0.051	0.337	0.303	1.52
FALCONIFORMES					
Falco tinnunculus[c,d]	0.210	0.099	0.252	1.662	7.91
GALLIFORMES					
Gallus gallus (dom.)[h]	0.318	–	0.306	1.24	3.90
Gallus gallus (dom.)[f]	0.346	0.159	–	0.494	1.43
Gallus gallus (dom.)[g]	0.322	0.144	0.300	0.459	1.44
Gallus gallus (RJF)[g]	0.252	0.119	0.463	0.452	1.79
Numida meleagris (dom.)[i]	0.320	–	–	1.12	3.50
Meleagris gallopavo (dom.)[f]	0.385	0.177	–	0.637	1.65
GRUIFORMES					
Amaurornis phoenicurus[g]	0.204	0.099	0.384	0.395	1.94
CHARADRIIFORMES					
Alca torda[c,d,j]	0.230	–	0.254	0.803	3.49
Cephus carbo[c,d,j]	0.193	–	0.280	0.850	4.40
Larus argentatus[c,d,j]	0.153	0.075	0.399	1.27	8.33
Larus canus[c,d,j]	0.116	–	0.272	0.684	5.90
Larus ridibundus[c,d,j]	0.146	0.071	0.306	0.925	6.34
COLUMBIFORMES					
Columba livia (dom.)[c,d]	0.161	–	0.197	0.804	4.99
Columba livia (dom.)[f]	0.172	0.098	–	0.315	1.83
Streptopelia decaocta[c,d]	0.218	–	0.160	0.801	3.67
Streptopelia senegalensis[c,d]	0.227	–	0.166	0.986	4.34
PSITTACIFORMES					
Melopsitacus undulatus[c,d]	0.117	0.068	0.260	0.976	8.34
Melopsitacus undulatus[f]	0.118	0.069	0.018	0.210	1.78
CUCULIFORMES					
Chrysococcyx klaas[c,d]	0.157	–	0.236	0.679	0.433
APODIFORMES					
Colibri coruscans[f]	0.099	0.062	0.017	0.183	1.85

4.2 Volume of the Lung (VL)

Table 5. (*Continued*)

Order/scientific name	τ_{ht} (μm)	$\tau_{ht(min)}$ (μm)	τ_{hp} (μm)	τ_t (μm)	$\tau_t \, \tau_{ht}^{-1}$
COLIIFORMES					
Colius striatus[c,d]	0.148	–	0.265	0.761	5.14
PICIFORMES					
Pogoniulus bilineatus[c,d]	0.165	–	0.191	1.01	6.12
CASUARIIFORMES					
Dromaius novaehollandiae[k]	0.232	–	0.103	–	–
PASSERIFORMES					
Ambryospiza albifrons[c,d,l]	0.121	–	0.201	0.584	4.83
Cisticola cantans[c,d,l]	0.122	–	0.219	0.608	4.98
Hirundo fuligula[c,d,l]	0.090	–	0.172	0.613	6.81
Lanius collaris[c,d,l]	0.170	–	0.216	0.638	3.75
Passer domesticus[c,d,l]	0.096	0.052	0.217	1.03	10.77
Passer domesticus[f]	0.118	0.069	0.016	0.218	1.85
Ploceus baglafecht[c,d,l]	0.151	–	0.215	0.762	5.05
Prinia subflava[c,d,l]	0.124	–	0.197	0.675	5.44
Sturnus vulgaris[c,d,l]	0.141	0.065	0.226	1.124	7.87
Turdus iliacus[c,d,l]	0.120	0.060	0.457	1.012	8.43
Turdus olivaceus[c,d,l]	0.127	–	0.234	0.599	4.72

[a] Maina and Nathaniel (2001)
[b] Maina and King (1987)
[c] Maina (1989a)
[d] Maina et al. (1989a)
[e] Maina and King (1982b)
[f] Dubach (1981)
[g] Vidyadaran et al. (1987)
[h] Abdalla et al. (1982)
[i] Abdalla and Maina (1981)
[j] Maina (1987)
[k] Maina and King (1989)
[l] Maina (1984)

Table 6. Morphometric diffusing capacities (ml O_2 s^{-1} $mbar^{-1}$) (1 mbar=10^2 Pa) of the blood–gas (tissue) barrier (D_tO_2), plasma layer (D_pO_2), red blood cell (D_eO_2), membrane (D_mO_2) and total pulmonary capacity (D_LO_2)

Order/scientific name	D_tO_2	D_pO_2	D_eO_2	D_mO_2	D_LO_2
STRUTHIOFORMES					
Struthio camelus[a]	8.81	37.94	5.32	7.34	2.93
SPHENISCIFORMES					
Spheniscus humboldti[b]	0.636	2.616	0.473	0.512	0.231
ANSERIFORMES					
Anas platyrhynchos[c,d,e]	0.930	0.424	0.062	0.289	0.050
Anser anser[c,d,e]	3.360	1.506	0.211	1.018	0.172
Cairina moschata (dom.)[f]	1.009	0.834	0.123	0.404	0.093
FALCONIFORMES					
Falco tinnunculus[c,d]	0.103	0.080	0.006	0.024	0.005
GALLIFORMES					
Gallus gallus (dom.)[g]	0.279	0.488	0.041	0.168	0.032
Gallus gallus (dom.)[f]	0.300	0.452	0.041	0.178	0.033
Gallus gallus (red-jungle fowl)[f]	0.101	0.080	0.011	0.044	0.008
GRUIFORMES					
Amaurornis phoenicurus[f]	0.103	0.080	0.010	0.044	0.008
CHARADRIIFORMES					
Alca torda[c,d,h]	0.256	0.309	0.038	0.139	0.029
Cephus carbo[c,d,h]	0.419	0.394	0.058	0.198	0.043
Larus argentatus[c,d,h]	0.407	0.161	0.033	0.110	0.025
Larus canus[c,d,h]	0.224	0.126	0.012	0.080	0.010
Larus ridibundus[c,d,h]	0.180	0.102	0.011	0.062	0.011
COLUMBIFORMES					
Columba livia (dom.)[c,d]	0.221	0.249	0.025	0.116	0.020
Streptopelia decaocta[c,d]	0.168	0.236	0.014	0.096	0.012
Streptopelia senegalensis[c,d]	0.050	0.076	0.006	0.030	0.005
PSITTACIFORMES					
Melopsitacus undulatus[c,d]	0.054	0.032	0.003	0.020	0.002
Melopsitacus undulatus[i]	0.058	0.363	–	0.050	–
CUCULIFORMES					
Chrysococcyx klaas[c,d]	0.026	0.020	0.002	0.011	0.002
APODIFORMES					
Colibri coruscans[i]	0.026	0.148	–	0.022	–
COLIIFORMES					
Colius striatus[c,d]	0.028	0.030	0.003	0.014	0.002
PICIFORMES					
Pogoniulus bilineatus[c,d]	0.008	0.012	0.001	0.005	0.001

4.3 Respiratory Surface Area (RSA)

Table 6. (Continued)

Order/scientific name	D_tO_2	D_pO_2	D_eO_2	D_mO_2	D_LO_2
CASUARIIFORMES					
Dromaius novaehollandiae[j]	2.88	8.25	0.48	–	0.4
PASSERIFORMES					
Ambryospiza albifrons[c,d,k]	0.048	0.047	0.003	0.024	0.003
Cisticola cantans[c,d,k]	0.015	0.014	0.001	0.007	0.001
Hirundo fuligula[c,d,k]	0.055	0.041	0.003	0.023	0.003
Lanius collaris[c,d,k]	0.033	0.033	0.003	0.016	0.002
Passer domesticus[c,d,k]	0.074	0.035	0.003	0.024	0.003
Ploceus baglafecht[c,d,k]	0.033	0.031	0.002	0.016	0.002
Prinia subflava[c,d,k]	0.009	0.009	0.001	0.004	0.001
Sturnus vulgaris[c,d,k]	0.105	0.076	0.006	0.027	0.005
Turdus iliacus[c,d,k]	0.059	0.019	0.004	0.014	0.003
Turdus olivaceus[c,d,k]	0.074	0.056	0.006	0.032	0.005

The values of D_pO_2, D_mO_2, D_eO_2 and D_LO_2 are the mean values of the minimum and maximum diffusing capacities
[a] Maina and Nathaniel (2001)
[b] Maina and King (1987)
[c] Maina (1989a)
[d] Maina et al.(1989a)
[e] Maina and King (1982a)
[f] Vidyadaran et al. (1987)
[g] Abdalla et al. (1982)
[h] Maina (1987b)
[i] Dubach (1981)
[j] Maina and King (1989)
[k] Maina (1984)

On average, the ET (the parenchyma) of the avian lung constitutes 48 % of the VL (e.g. Maina et al. 1982a; Vitali and Richardson 1998). The lowest value (18 %) has been observed in the lung of the emu, *Dromaius novaehollandiae* (Maina and King 1989), while the highest one occurs in the ostrich, *Struthio camelus* (Maina and Nathaniel 2001; Table 1). In the mammalian lung, the parenchyma forms over 80 % of the VL (e.g. Gehr et al. 1981; Maina and King 1984). The ACs and BCs on average constitute 53 and 34 % of the ET (Table 2).

4.3
Respiratory Surface Area (RSA)

In the gas exchangers, extensive RSA is typically generated by internal subdivision (compartmentalization) or hierarchical (stratified) construction (e.g. Maina 1998). In invaginated respiratory organs, the space within which the surface area is generated is set by the volume of the lung which is in turn

Lung Volume vs Body Mass

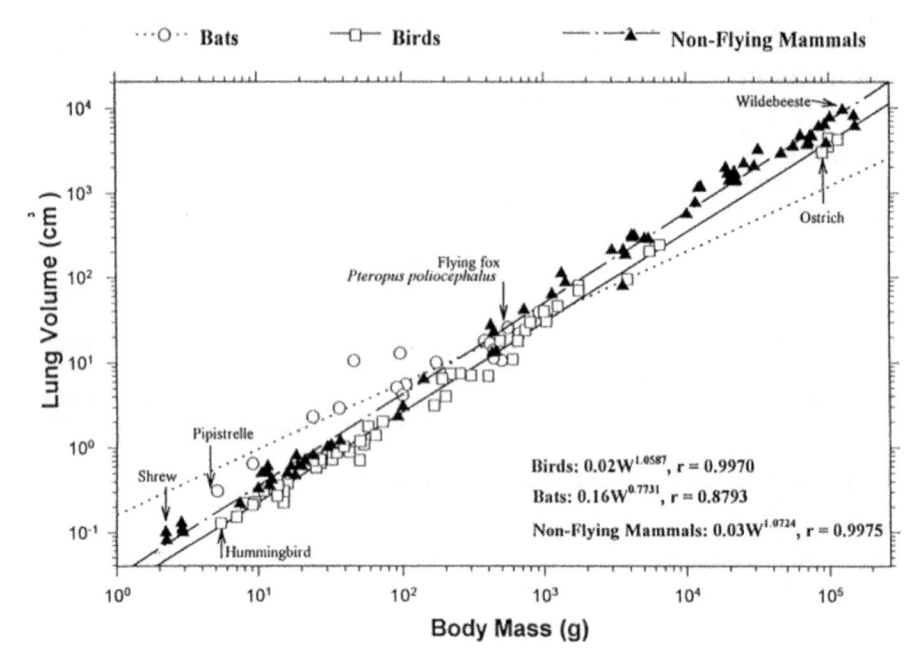

Fig. 83. Allometric comparison of the volume of the lung between birds, bats, and nonflying mammals. Bats have larger lungs when compared to those of the nonflying mammals while the lungs of birds are smaller than those of nonflying mammals and bats. Data for nonflying mammals are from Gehr et al. (1981), while those for bats are given in Maina and King (1984) and Maina et al. (1991). (Maina 2000b)

Birds: $0.02W^{1.0587}$, $r = 0.9970$
Bats: $0.16W^{0.7731}$, $r = 0.8793$
Non-Flying Mammals: $0.03W^{1.0724}$, $r = 0.9975$

determined by the size of the thoracic cavity in mammals and, in the case of birds, where a diaphragm is lacking, the coelomic cavity. For teleost gills (invaginated gas exchangers), the volume of the opercular cavity affixes the space in which RSA is generated by hierarchical construction of the gills into branchial arches, hundreds of gill filaments, and thousands of secondary lamellae (e.g. Hughes and Morgan 1973).

Increasing RSA by subdividing the ET occurs at a cost. Minute terminal respiratory units confer high surface tension at the air–tissue interface: more energy is required to inflate narrow respiratory units with air. Such divisions have a great propensity of collapsing into themselves. While in mammals the compliance of the respiratory system (except for the thoracic walls) resides in the terminal parts of the respiratory tree (mainly the alveolar spaces; e.g. Dubois et al. 1956), in the avian respiratory system compliance is restricted to the ASs (Scheid and Piiper 1989). The avian lung is firmly fixed to the ribs and is virtually rigid (Figs. 6B–D, 20C,D, 38, 39B and 43B). It changes in volume by a mere 1.4% between respiratory cycles (Jones et al. 1985). Consequently, the

4.3 Respiratory Surface Area (RSA)

Surface Area vs Body Mass

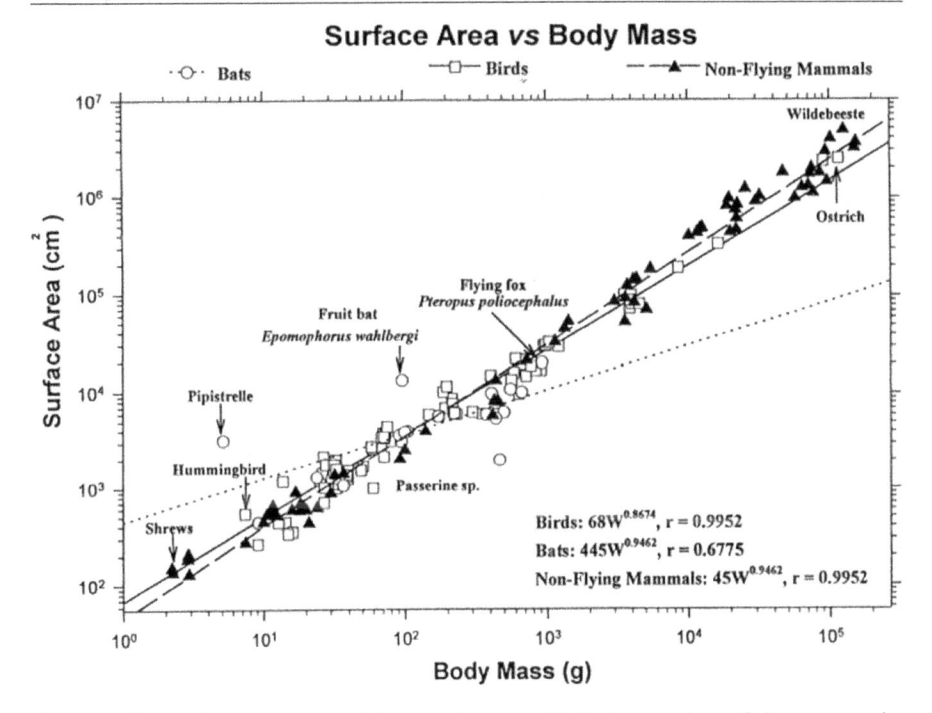

Fig. 84. Allometric comparison of the RSA between birds, bats, and nonflying mammals. Bats have relatively extensive RSA compared with nonflying mammals and birds. Small birds have more extensive RSA compared with small nonflying mammals: in heavier animals, the relationship is reversed. Data for nonflying mammals are from Gehr et al. (1981) while those for bats are given in Maina and King (1984) and Maina et al. (1991). (Maina 2000b)

degree of the internal subdivision of the ET of the avian lung has not been restricted by movements of the lung, i.e. the forces that would be needed to overcome surface tension forces in order to ventilate the ET.

Although birds have relatively small lungs (Fig. 83) and the parenchymal volume fraction (density) of the lung is about one-half (46%) that of a mammalian lung (e.g. Gehr et al. 1981; Maina et al. 1982a; Table 1), the RSA of the avian lung is interestingly 15% greater than that of a mammal, especially for the small species of birds (Fig. 84, Table 3). This stems from very intense internal subdivision of the ET, a property (as explained above) that is allowed by the rigidity of the lung itself. High surface density of the BGB (S_{VBGB}), i.e. the RSA per unit volume of the ET, eventuates in the avian lung because surface tension and size of the terminal gas-exchange units, the ACs, is not a limiting factor to the intensity of internal subdivision in a fixed lung (Table 4, Fig. 85). The ACs of the avian lung range in diameter from about 3 μm in the lungs of the smallest species (e.g. Duncker 1974) to 20 μm in that of the ostrich (Maina and Nathaniel 2001). To drive the point home, such exceptionally small functional terminal respiratory units could only form in a fixed, noncompliant lung, where surface tension was not a limiting factor during ventilatory activity. In the mammalian lung, reportedly, the narrowest alveoli (50 μm) occur in the lung of the shrew, *Suncus etruscuns*, and a bat (Tenney and Remmers 1963). In the mammalian lung, ranging from 45 mm² mm⁻³ in the genet cat, *Genetta tigrina*, to 307 mm² mm⁻³ in the shrew, *Sorex minutus* (Gehr et al. 1981), the values of the S_{VBGB} are about one-tenth those of birds (Maina 1989b; Maina et al. 1989a; Table 4): the S_{VBGB} ranges from 82 mm² mm⁻³ in the emu, *Dromaius noveahollandiae* (Maina and King 1989), to 389 mm² mm⁻³ in the hummingbird, *Colibri coruscans* (Dubach 1981). The corresponding values of the S_{VBGB} in the heterogeneous reptilian lungs range from 1.34 mm² mm⁻³ in the lizard, *Tupinambis nigropunctus*, to 1.80 mm² mm⁻³ in the turtle, *Pseudemys scripta* (Perry 1978, 1992).

In birds, the highest mass-specific RSA (MSRSA) of about 90 cm² g⁻¹ has been reported in the small and highly energetic violet-eared hummingbird, *Colibri coruscans* (Dubach 1981), and the African rock martin, *Hirundo*

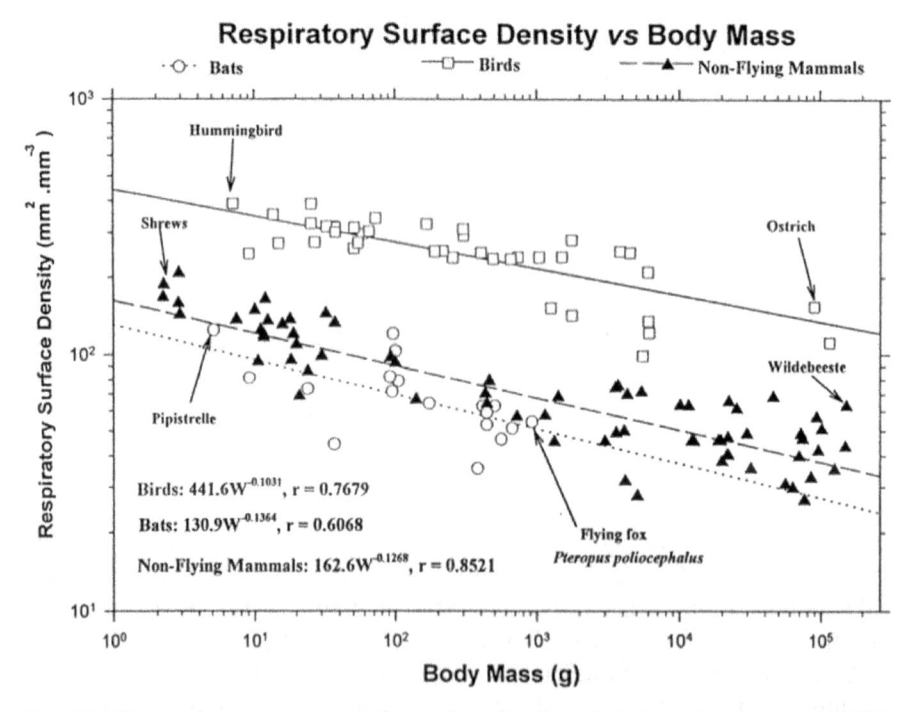

Fig. 85. Allometric comparison of the surface density of the blood–gas barrier (BGB) between birds, bats, and nonflying mammals. Birds have higher values followed by nonflying mammals and bats. The surface density of the BGB decreases with increasing body mass. Data for nonflying mammals are from Gehr et al. (1981), while those for bats are given in Maina and King (1984) and Maina et al. (1991). (Maina 2000b)

fuligula (Maina 1984, 1989b; Table 4). In nonflying mammals, the highest MSRSA of 43 cm² g⁻¹ occurs in the shrews, *Crocidura flavescens*, and *Sorex* spp. (Gehr et al. 1980). The shrews are the smallest and most highly metabolically active mammals (e.g. Fons and Sicart 1976). The astoundingly high MSRSA value of 800 cm² g⁻¹ reported for an unnamed species of hummingbird by Stanislaus (1937) should be treated with caution: the method(s) used in the determination of the value are not explained by the investigator and it is quite unlikely that it is not as reliable as the values determined stereologically. Among birds, the lowest MSRSA (5.4 cm² g⁻¹) has been reported in the lung of the emu, *Dromaius novaehollandiae* (Maina and King 1989), a large Australian bird that evolved in a habitat with low predation, especially from placental mammals. Subsequent to the fact that smaller animals have higher MRs, in birds, nonflying mammals, and bats, small animals have greater MSRSA compared with the larger ones (Fig. 84).

4.4
Thickness of the Blood–Gas Barrier

Structurally, the BGB comprises of a squamous (thin) epithelial cell and an endothelial one that are literally 'glued' back-to-back across an extracellular matrix space, commonly called BL (Chap. 3.3; Figs. 63 and 64). The epithelium comprises 12.28%, the BL 21%, and the endothelium 66.73% of the BGB (Maina and King 1982a). Among the species of birds on which data are available, the violet-eared hummingbird, *Colibri coruscans* (7.3 g; Dubach 1981), and the African rock martin, *Hirundo fuligula* (13.7 g; Maina 1984, 1989b), have the thinnest BGB that is 0.090 μm thick (harmonic mean thickness, τht). The thickest BGBs (τht) occur in the ostrich (0.56 μm; Maina and Nathaniel 2001) and the Humboldt penguin, *Spheniscus humboldti* (0.53 μm; Maina and King 1987). Welsch and Aschauer (1986) attributed the greater thickness of the BGB and presence of connective tissue elements in its interstitium/BL to need for strength in order to withstanding hydrostatic pressures during dives. Compared with bats and nonflying mammals, the BGB in the avian lung is 56–67% thinner (Maina et al. 1989a; Fig. 86; Table 5).

In the course of the evolution of the vertebrate lung, the thickness of the BGB appears to have been optimized and may be the least adaptable structural parameter (reflected in the low slopes of the regression lines; Fig. 86). Moreover, among the nonflying mammals, the thickness of the BGB in a 2.2-g shrew, *Suncus etruscus*, is 0.230 μm (Gehr et al. 1980) while the value in the lung of the bow-head whale, *Balaena mysticetus*, is only 0.35 μm (Henk and Haldiman 1990) – the factorial difference of the thickness of the BGB is only 1.5 while that of the body mass is colossal; in bats, the thickness of the BGB in a 5-g specimen of *Pipistrellus pipistrellus* is 0.206 μm (Maina and King 1984), while that in a 900-g flying fox, *Pteropus poliocephalus* (the heaviest bat reportedly weighs about 1.5 kg), is 0.303 μm (Maina et al. 1991) – the body

Harmonic Mean Thickness vs Body Mass

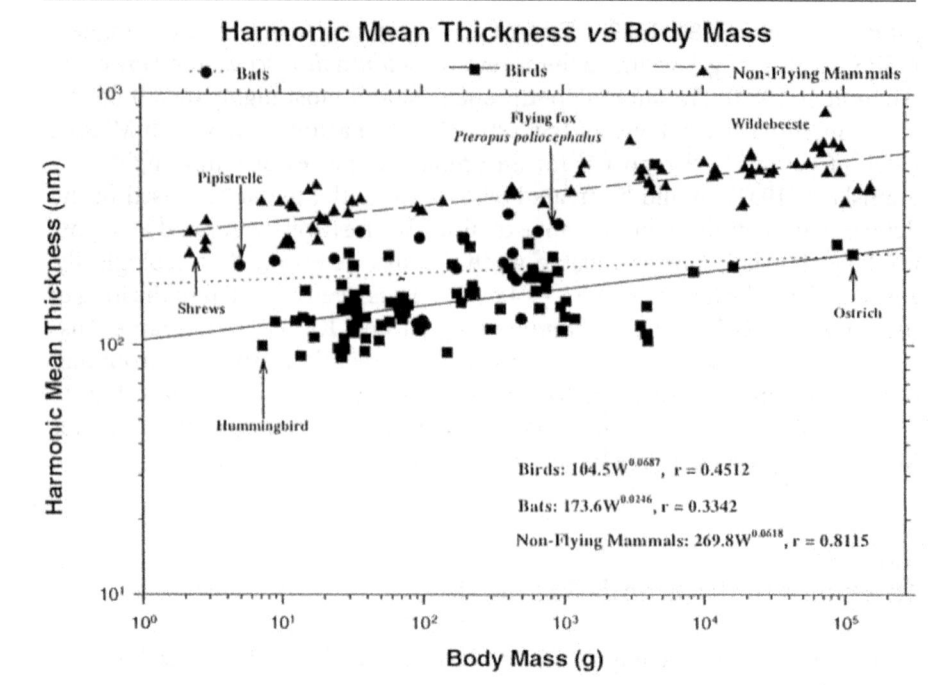

Fig. 86. Allometric comparison of the harmonic mean thickness of the BGB between birds, bats, and nonflying mammals. Birds have the thinnest barriers followed by bats and nonflying mammals. Data for nonflying mammals are from Gehr et al. (1981), while those for bats are given in Maina and King (1984) and Maina et al. (1991). (Maina 2000b)

mass factorial difference is 1.8×10^2 while that of the thickness of the BGB is only 1.5; in birds, the thickness of the BGB ranges from 0.09 μm in a hummingbird (Dubach 1981) to 0.56 μm in the ostrich (50 kg; Maina and Nathaniel 2001) – the body mass factorial difference is 7×10^3 while that of the thickness of the BGB is 5.7.

In the avian lung, sporadic attenuation is a common structural feature of the design of the BGB (Figs. 63 and 64). It is envisaged that irregular design, where extremely thin parts occur between thicker ones, is a compromise design that optimizes gas exchange, without sacrificing the mechanical integrity of the barrier (e.g. Weibel and Knight 1964). Expressed as the ratio of τ_{ht} to the arithmetic mean thickness (τ_t; Table 5; Maina and King 1982a), the highest value (10.8) occurs in the house sparrow, *Passer domesticus* (12 g), and the lowest (1.2) in the ostrich, *Struthio camelus* (45 kg). The minimum τ_{ht} (Table 5) shows the extent to which the BGB attenuates. In the lung of the graylag goose, *Anser anser*, e.g. while the τ_{ht} is 0.112 μm, the thinnest parts of the BGB are 0.050 μm (Table 5).

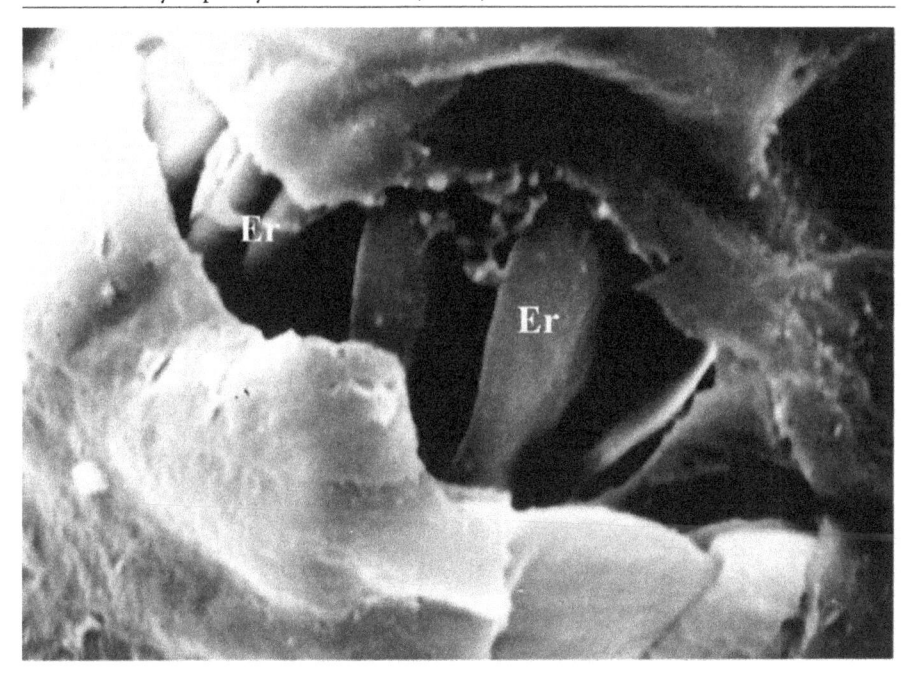

Fig. 87. Erythrocytes squeeze themselves and pass in a single file through the pulmonary blood capillaries. *Scale bar* 10 μm. (Maina 1989a)

4.5
Pulmonary Capillary Blood Volume (PCBV)

Enhanced exposure of the pulmonary capillary blood (PCB) to air is vital to efficient gas exchange. Intense vascularity and bulging of the BCs into the surrounding air spaces increases both the PCBV as well as the RSA. The total volume of blood in the avian lung constitutes as much as 36% of the VL, with 58–80% of it being located in the BCs themselves (e.g. Duncker and Güntert 1985ab; Maina et al. 1989a; Tables 1 and 2). In the African rock martin, *Hirundo fuligula*, a particularly energetic passerine bird, 29% of the VL comprises of blood, with 79% of it being located in the BCs (Maina 1984). In the lung of the rock martin, 0.075 cm^3 of blood is spread over a RSA of 0.12 m^2, giving an estimated thickness of the film of blood, a sheet (Maina 2000c), of 6.3×10^{-4} μm: erythrocytes squeeze through the blood capillaries in a single file (Figs. 58A and 82). In the avian lung, the PCBV is 2.5 to 3 times greater than that in the ET (parenchyma) of the lung of a mammal of similar body mass, where only 20% of blood is found in the alveolar capillaries (Weibel 1963). Compared with that of a nonflying mammal, the avian lung has a PCBV that is 22% greater (Maina et al. 1989a; Fig. 89; Table 4).

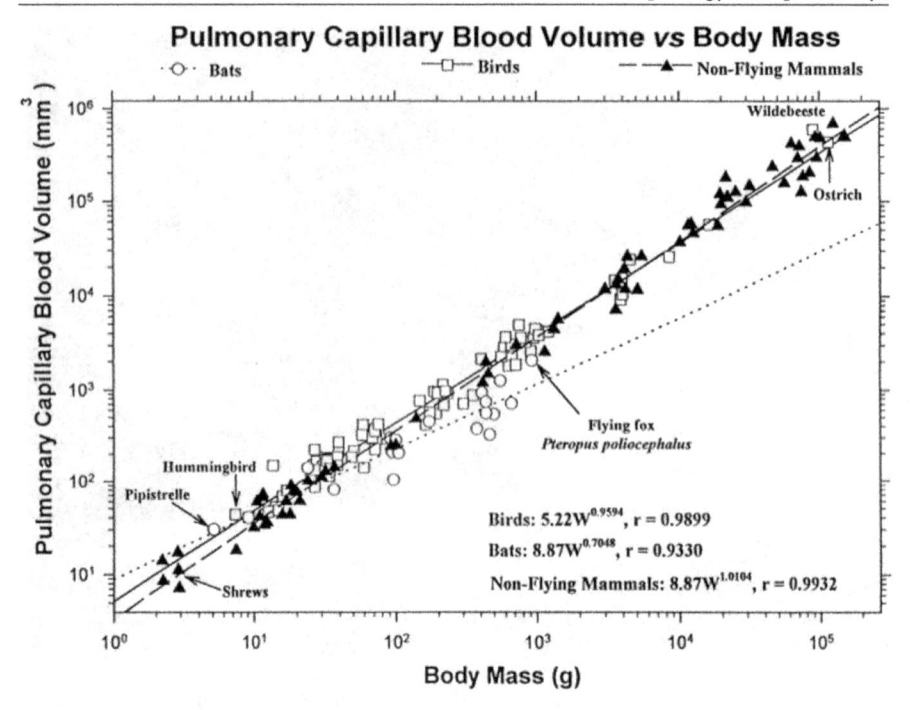

Fig. 88. Allometric comparison of the pulmonary capillary blood volume between birds, bats, and nonflying mammals. Bats have higher volumes followed by birds and nonflying mammals. Data for nonflying mammals are from Gehr et al. (1981), while those for bats are given in Maina and King (1984) and Maina et al. (1991). (Maina 2000b)

Capillary loading (CL), the ratio of the PCBV to the RSA ($V_c \cdot S_t^{-1}$, Table 4), is an indicator of the degree of exposure of the PCB to air. In the avian lung, the values range from 0.7 cm³ m⁻² in the African rock martin (Maina 1984) to 4.4 cm³ m⁻² in the Humboldt penguin (Maina and King 1987): low values are indicative of superior exposure of PCB to air. In ectotherms, values of CL as high as 13 cm³ m⁻² have been reported in the lung of the turtle, *Pseudemys scripta elegans* (Perry 1978), and 12–19 cm³ m⁻² in the lungfish, *Lepidosiren paradoxa* (Hughes and Weibel 1976). A 'double capillary system', an arrangement where blood capillaries are exposed to air only on one side of a supporting intrapulmonary septum, occurs in the lungs of amphibians, reptiles, and lungfish while a 'single capillary system', where blood capillaries are exposed to air on two sides, occurs in the mammalian lung, and a 'diffuse capillary system', where blood capillaries are exposed to air practically all around, occurs in the avian lungs (e.g. Maina 1998, 2002a; Figs. 58A and 62B). To a large extent, these refinements correspond with the metabolic capacities of the various taxa.

4.6 Modeling a Gas Exchanger: Integrative Morphometry

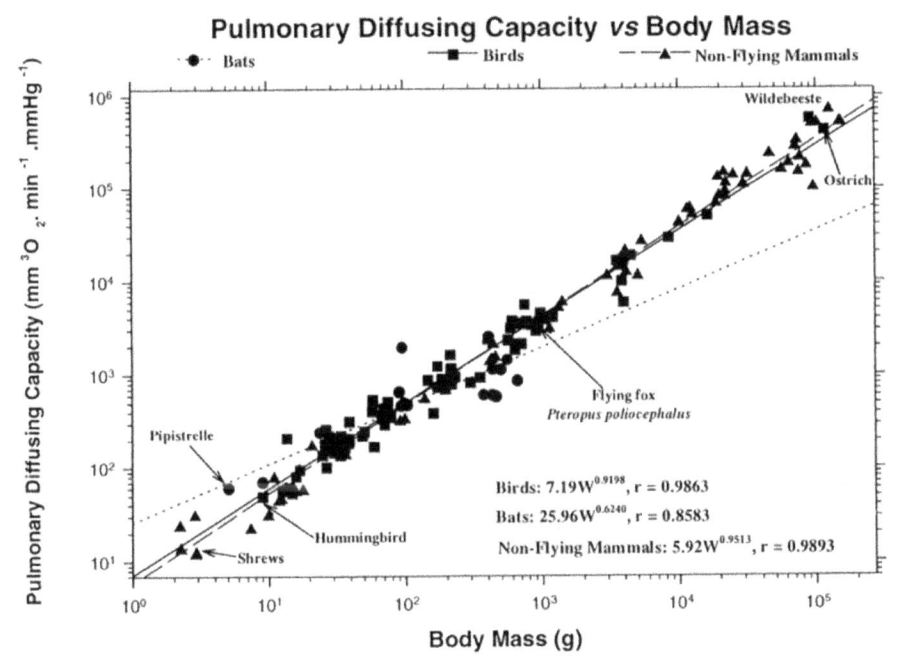

Fig. 89. Allometric comparison of the total morphometric pulmonary diffusing capacity between birds, bats, and nonflying mammals. Bats in general have higher values followed by birds and nonflying mammals. Data for nonflying mammals are from Gehr et al. (1981) while those for bats are given in Maina and King (1984) and Maina et al. (1991). (Maina 2000b)

4.6
Modeling a Gas Exchanger: Integrative Morphometry

4.6.1
General Principles

For a thorough understanding of the essence of a biological design, it is instructive not only to know how the structural elements individually influence the system, but also how they collectively engage and drive it. Utilizing the mathematical and conceptual devices of engineers underpins the basis of the form and function in biology of what may at a glance appear like complex, insoluble structural designs or phenomena. Gas exchangers manifest elaborate multilevel construction in which the integral structural elements are functionally coupled. Since gas exchangers possess a high degree of functional/adaptive plasticity (e.g. Maina 1998), inferences based on assessments, measurements, and tests performed on individual parameters such as VL, RSA, PCBV, and thickness of the BGB are useful only to a certain point since changes and refinements can occur at any level of the air-hemoglobin pathway (AHP). Perceptive analysis and understanding of composite systems call for utilization of an appropriate model. Fundamentally, a mathematical simplification of a highly complex system, a model separates a structure into its individual functional parts, isolates those parts that are central to understanding pertinent questions, and then reassembles it. Gutman and Bonik (1981) defined a model as *"an abstraction of a real situation which describes only the essential aspects of the situation"* while Scheid (1987) defined it as *"an image of part of the physical or mental world apt to explain or predict observations"*. Emphasizing the utility of models in biology, Scheid (1987) observed that *"models are not only helpful but often indispensable in quantitative biology"*. Instructively, Powell and Scheid (1989) pointed out that *"in an attempt at deriving a functional model for gas exchange from morphologic evidence, the physiologist has to identify the simplest functional subunit in the gas exchange organ"*. They went on to add that in creating a morphological model *"many anatomical details have to be omitted to allow quantitative treatment of gas exchange"*.

Assumptions are inevitable in modeling. The inclusions and exclusions determine the noise inherent in a model. It is imperative that a model captures as much of the essential attributes of a structure as possible: it should meaningfully describe the system that it endeavors to represent. Most importantly, it must be simple and easy to conceptualize, and must be theoretically and practically testable. The robustness of a model is measured in terms of its predictive value. By changing or isolating one or a number of factors while holding another or others constant, negative and positive controls as well as redundant features/factors can be identified. Overall, biological models underpin the mechanisms that regulate the performance of whole biological systems.

4.6.2
Modeling the Avian Lung

In general, pulmonary modeling is based on the fundamental fact that gas exchangers comprise of a construction in which external (water/air) and internal (hemolymph/blood) respiratory fluid media are contained in different compartments and are exposed to each other across a tissue barrier (Fig. 81A). Existing PO_2 and that of CO_2 (PCO_2) drives the gases across a cascade of spaces (Fig. 81A). Utilization/consumption of O_2 at the mitochondrial level for generation of energy (ATP) drives the flow of the gas across the AHP from where it is transported by the circulatory system. According to Fick's law, the diffusing capacity (DO_2; i.e. the conductance of a respiratory organ for O_2) is directly proportional to the surface area (Sa), the permeation coefficient across the BGB (K_tO_2), and the ΔPO_2 prevailing across the barrier. DO_2 correlates inversely with the thickness of a barrier (τht). Thus:

4.6 Modeling a Gas Exchanger: Integrative Morphometry

$$DO_2 = KtO_2 \cdot \Delta PO_2 \cdot Sa \cdot \tau ht^{-1} \tag{1}$$

Fick's law is the physiologists equivalent of Ohm's law of electricity where,

$$I = U R^{-1} \tag{2}$$

where I is the electric current, R the resistance, and U the potential difference (i.e. the voltage) between two points of a conductor.

The AHP is the distance that an O_2 molecule travels from the respiratory surface to the erythrocyte cytoplasm (EC), where it is biochemically bound to hemoglobin. The AHP comprises of the BGB, the plasma layer (PL), and the EC (Fig. 81): molecular O_2 diffuses in succession through the three tissue compartments. To satisfy the unique design of the avian lung, certain modifications to the model, which was advanced by Weibel (1970/71) and has since been extensively applied (e.g. Weibel 1973), were necessary (Maina 1989b; Maina et al. 1989a). Individual diffusing capacities of the components of the AHP, namely the tissue barrier (D_tO_2), the plasma layer (D_pO_2), and the erythrocyte (D_eO_2), and the aggregative ones like the membrane (D_mO_2) and the total morphometric diffusing capacity ($D_LO_{2\,m}$) were calculated from relevant morphometric measurements and the physicochemical coefficients. Regarding D_tO_2 and D_pO_2, respective surface areas, harmonic mean thicknesses, and Krogh's permeation coefficients (K) for O_2 were used (e.g. Weibel 1970/1971; Maina et al. 1989a). The permeation coefficient (K) is a product of the solubility (α) and diffusion (D) coefficients of O_2. Since α and D are influenced by temperature in opposite directions, i.e. at a higher temperature solubility is reduced while diffusion is enhanced, K is not significantly affected by temperature.

The diffusing capacity of the BGB (D_tO_2) is mathematically estimated as:

$$D_tO_2 = S_t K_tO_2 \, \tau ht^{-1}, \tag{3}$$

where S_t is the surface area of the BGB, K_tO_2 the Krogh's permeation coefficient of the BGB for O_2 (4.1×10^{-8} cm^2 s^{-1} Pa^{-1}; 1 mbar=10^2 Pa) and τ_{ht}, the harmonic mean thickness of the BGB.

The diffusing capacity of the PL (D_pO_2) is estimated as:

$$D_pO_2 = S_p \cdot K_pO_2 \cdot \tau hp^{-1}, \tag{4}$$

where S_p is the surface area of the PL [estimated as the mean of the surface area of the capillary endothelium (S_c) and that of the erythrocytes (S_e); Table 3], K_pO_2 the Krogh's permeation coefficient for O_2 across the PL (minimum value: 4.0×10^{-8} cm^2 s^{-1} Pa^{-1}; maximum value 5.4×10^{-8} cm^2 s^{-1} Pa^{-1}), and τhp the harmonic mean thickness of the PL.

The conductance of the erythrocytes (D_eO_2) is calculated from the O_2 uptake coefficient of the whole blood (θO_2; minimum value: 1.13 mLO$_2$ s^{-1} Pa^{-1}; maximum value 3.13 mLO$_2$ s^{-1} Pa^{-1}) and the PCBV. Thus:

$$D_eO_2 = \theta O_2 \cdot PCBV \tag{5}$$

Since the BGB, the PL, and the EC are arranged in series (Fig. 80B) and thus an O_2 molecule passes sequentially through all the sections of the AHP, the overall resistance (or its reciprocal the conductance) is the sum of the individual resistances. D_mO_2 is the total diffusing capacity (conductance) of the BGB and the PL, i.e. the reciprocal of D_tO_2 and D_pO_2.

$$D_mO_2^{-1} = D_tO_2^{-1} + D_pO_2^{-1} \tag{6}$$

The total D_LO_{2m} is estimated from the diffusing capacities of the BGB (D_tO_2), the PL (D_pO_2), and that of the erythrocyte (D_eO_2). Thus:

$$D_LO_{2m}^{-1} = D_tO_2^{-1} + D_pO_2^{-1} + D_eO_2^{-1} \tag{7}$$

D_LO_{2m} is an inclusive parameter that expresses the overall capacity of the lung or for that matter any gas exchanger to transfer (conduct) O_2. The physiological diffusing capacity (D_LO_{2p}) is defined as the ratio of the flow rate of oxygen (VO_{2f}) to the driving force, i.e. the ΔPO_2 (Bohr 1909). Thus:

$$D_LO_{2p}^{-1} = VO_{2f} \cdot \Delta PO_2^{-1} \tag{8}$$

where ΔPO_2 is the difference in P between alveolar and mean capillary O_2 tensions.

In mammals, D_LO_{2p} is consistently lower than D_LO_{2m} (e.g. Siegwart et al. 1971; Crapo and Crapo 1983; Weibel et al. 1983). There are indications that a similar relationship applies in birds (e.g. Meyer et al. 1977; Scheid et al. 1977; Burger et al. 1979). It is envisaged that D_LO_{2p} underestimates the diffusing capacity of a gas exchanger mainly on account of regional inhomogeneities of ventilation and perfusion. Moreover, in birds, the particularly high VO_{2c} of the nucleated erythrocytes (e.g. Lutz et al. 1973) may lower D_LO_{2p}. In the mammalian lung, during the fixation of the lung by intratracheal instillation for morphometric analysis, the alveolar surface area is expanded to that at functional residual capacity (e.g. Siegwart et al. 1971; Weibel 1973). Consequently, combined with the assumption that at any moment the entire BGB is utilized for gas exchange, D_LO_{2m} predicts the maximal achievable conductance of the gas exchanger under ideal conditions, e.g. where inequalities of ventilation and perfusion are eliminated. In a healthy mammalian lung, at $VO_{2\,max}$, D_LO_{2p} approaches D_LO_{2m} (Weibel 1990). Compared with nonflying mammals, small birds have greater D_LO_{2m} (Fig. 88; Table 6): bats have higher D_LO_{2m} than birds and nonflying mammals. Predictably, small animals, be they birds, nonflying mammals, or bats, have higher D_LO_{2m}.

In modeling the avian lung, application of mammalian-derived physical constants/coefficients is unavoidable, since appropriate values for birds are largely lacking. Nguyen-Phu et al. (1986) reported a value of 2.75 ml O_2 s^{-1} Pa^{-1}

for the θO_2 of the whole blood of the domestic fowl and 2.79 ml O_2 s⁻¹ Pa⁻¹ for that of the domestic muscovy duck. Before that publication, θO_2 values were based on the mammalian blood. The avian venous haematocrit (AVH) values were adjusted by mathematically 'subtracting' the volume of the nucleus (estimated morphometrically by point counting) from that of the whole cell (e.g. Abdalla et al. 1982; Maina et al. 1989a), an adjustment that gave a value that corresponded to the volume of the cytoplasm in the anucleate mammalian RBC. The 'modified' AVH was used to adjust the mammalian θO_2, a value that is based on an average venous hematocrit of 45%. The value of $D_L O_{2m}$ determined using the avian θO_2 (Nguyen-Phu et al. 1986) was lower than that determined by adjusting the mammalian θO_2 (Maina et al. 1989a).

There is a need to streamline the methodology of modeling gas exchangers. Unsettling differences occur in the approaches applied by different investigators: Gehr et al. (1981), Abdalla et al. (1982), and Maina (1989b) have determined the individual and inclusive diffusing capacities; Perry (1978) and Stinner (1982) have only reported $D_t O_2$; and the ratio of the RSA to the thickness of the BGB has been termed 'anatomical diffusion factor' (ADF; e.g. Perry 1983). Another unsatisfactory aspect that affects comparison between diffusing capacities of various gas exchangers is that of lack and discrepancy of the physical constants/coefficients, i.e. $K_p O_2$, and θO_2: presently, ranges rather than specific values of $D_p O_2$, $D_e O_2$, $D_m O_2$, and $D_L O_2$ are available.

4.6.3
Pros and Cons of Pulmonary Modeling

While the effort involved in generating morphometric data and the uncertainties that exist on the physical constants may dissuade investigators from carrying out comprehensive modeling of gas exchangers, it important to take note of the fact that inferences made from individual morphometric parameters, e.g. surface area, volume, and thickness and partial diffusing capacities, i.e. $D_t O_2$, $D_p O_2$ and $D_e O_2$, may be incorrect especially when the functional capacities of different gas exchangers are compared. For example, the Humboldt penguin, *Spheniscus humboldti*, has an exceptionally thick BGB (Maina and King 1987; Table 5), a feature that gives low mass-specific $D_t O_2$. However, emanating from a particularly large PCBV, the mass-specific $D_e O_2$ is exceptionally high (Table 6). Overall, for a flightless bird, high $D_e O_2$ boosts the $D_L O_{2m}$ to equal or exceed that of certain other supposedly more energetic birds. In the particular case of the Humboldt penguin, if one examined only the thicknesses of the BGB and the PL or estimated partial conductances, i.e. $D_t O_2$, $D_p O_2$, and $D_m O_2$, the conclusion would inevitably be that the bird's lung is structurally inferior, compared to those of most other birds of similar body mass.

5
Comparative Respiratory Morphology

> *Amongst animals, diversity of form and of environmental circumstances has given rise to a multitude of different adaptations subserving the relatively unified patterns of cellular metabolism. Nowhere else is this state of affairs better exemplified than in the realm of respiration.* Jones (1972)

5.1
General Observations

Essentially, respiratory organs grant an interface between O_2 in the external environment and the aerobic machinery of the body. The design and construction of a gas exchanger, i.e. the assembly and sizing of its formative components, determine the efficiency of procurement of O_2 and hence the metabolic capacity of an animal. Factors such as body size, sex, season, habitat occupied, respiratory medium utilized, phylogenetic status, and lifestyle pursued collectively prescribe the structure and function of the respiratory organs. Conspicuous morphological heterogeneity occurs in the modern vertebrate lungs. However, at their actual point of operation, i.e. at the gas exchange level, remarkable structural and functional similarities occur. For example, the surfactant lines the respiratory surface (e.g. Pattle 1976), a three-ply (tripartite laminated) design of the BGB is ubiquitous (e.g. Maina and King 1982a; Maina and West 2005), and the flux of respiratory gases (O_2 and CO_2) occurs entirely by passive diffusion under P maintained by ventilation and perfusion of the respiratory organ (e.g. Forster 1996). The uniformity between the morphologies of the gas exchangers, i.e. structural convergence, in remarkably phylogenetically diverse animals intimates that: (1) similar selective pressures obligated and directed the evolution of respiratory organs, and (2) only specific structural designs can efficiently exchange respiratory gases. Additionally, the structural-functional correlations in the designs of gas exchangers can be explained by the facts that: (1) at ordinary temperatures and pressures under which animals evolved, only two respiratory fluid media, water and air, have existed – gas exchangers were dedicated to utilize one or the other and rarely both of the media, and (2) since the movement and transfer of respiratory gases in both water and air are fundamentally governed by immutable laws of physics and chemistry, the structural requirements for gas exchange must basically be the same, irrespective of existing phylogenetic differences. Structurally, gas exchangers must have: (1) a vast RSA, (2) large PCBV, and (3) thin BGB separating the respiratory media: these properties optimize the diffusing capacity for the respiratory gases. Extensive RSA is generated by programmed branching of the pulmonary airway and vascular systems (branching morphogenesis); large PCBV by intense anastomoses and vascularization of the terminal respiratory units, the BCs; and thin BGB by qualitative and quantitative changes of the tissue and cellular components lining the airway and vascular systems (e.g. Weibel 1984).

At all levels of biological organization, from cellular through organic to organismal levels, morphology is the outward manifestation of form, size, and arrangement of the constitutive structural elements while function is the expression of multiple minuscule events generated at different levels of structure. Structure and function impact strongly on each other and hence are inextricably interrelated (e.g. Weibel 1984, 2000; Maina 2002b). The enterprise of developing optimal, i.e. cost-effective, structures through evolution and adaptation has not been easily achieved: about 99.99 % of the animal species that ever evolved in the about 4 billion years that life has existed on earth are now extinct (e.g. Pough et al. 1989). Regardless of the factor(s) that directly or indirectly precipitated their demise, unprepared as they were, evolutionary, such animals may be considered to have been failed experiments. In both human and biological engineering, only certain structural designs can perform optimally under a given set of conditions. To perpetuate fitness for survival, once novel states are founded, they are passionately defended and ultimately genomically conserved.

The objective of this chapter is to sketchily illustrate the structural and functional similarities and differences between the avian respiratory system (the lung-air sac system) and those gas exchangers that have evolved in some air-breathers. Comparisons with the mammalian lung have been made at different points in the earlier chapters and are hence not repeated here. It will be emphasized that the similarities that exist between the designs of gas exchangers are not solely determined by phylogenetic status: features such as body mass, lifestyle, and habitat are meaningfully consequential.

5.2
Comparison of the Structure of the Avian Respiratory System with Those of Some Other Animals

5.2.1
Dipnoan Lung

From their bimodal mode of respiration, the focal systematic position that they occupy, and a fascinating natural history, lungfish (Dipnoi) are important animal models in studies of tetrapod evolution and respiratory adaptive biology. Evolved some 300 million years ago (Thompson 1971; Marshall 1986a), the discovery of the South American lungfish, *Lepidosiren paradoxa*, and the African lungfish, *Protopterus*, almost concurrently nearly two centuries ago, i.e. 1830s (see Marshall 1986a), and later that of the Australian lungfish, *Neoceratodus forsteri*, about one and a half centuries ago (Krefft 1870), prompted intense interest and controversy on tetrapod evolution, a scientific debate that has since only been matched by the much later discovery in the 1920s of the archaic coelacanth, *Latimeria chalumnae*, dubbed a 'living fossil' (Smith 1956). Out of an estimated 55 extinct genera and 112 species, now there are only 3 genera and 6 extant species of lungfish (Marshall 1986a,b).

Shift from water to land is one of the momentous events that have taken place in the evolution of the animal life (e.g. Schmalhausen 1968; Little 1990). The realization of air-breathing was a decisive event in a sequence of purposeful preadaptations that occurred for preparation for life on land. Transactions that entailed change from gill to lung breathing and establishment of intricate neural coordination between respiration and circulation were an imperative. Ancient fish like the lungfish and the bichir and three-quarters of the modern fish that live in the tropical and subtropical waters to various extents breath air (e.g. Munshi and Hughes 1992; Graham 1997). This indicates that the selective pressures that compelled air-breathing were most severe in the tropical and subtropical regions of earth or under such conditions. Increasing environmental temperatures reduced the solubility of O_2 in water, increased the rate of decay of plant organic matter, and caused the shallow and extensive continental shelves to shrink and ultimately dry up. The consequence of this cycle of events was congestion and competition for fast diminishing resources. With increasing levels of O_2 above (in the atmosphere) and a hypoxic crisis below (in water), especially in standing plant infested waters of the hot tropical regions (e.g. Carter and Beadle 1930), the switch from water- to air-breathing became of the essence rather than a choice. When accompanied by hypercapnia, hypoxia constitutes a very strong driving force that induces air-breathing (e.g. Jensen and Weber 1985).

The structure and function of the lung of *Lepidosiren paradoxa* was studied by, e.g., Hughes and Weibel (1976), those of *Protopterus* by, e.g., De Groodt et al. (1960), Klika and Lelek (1967), Kimura et al. (1987), Maina and Maloiy

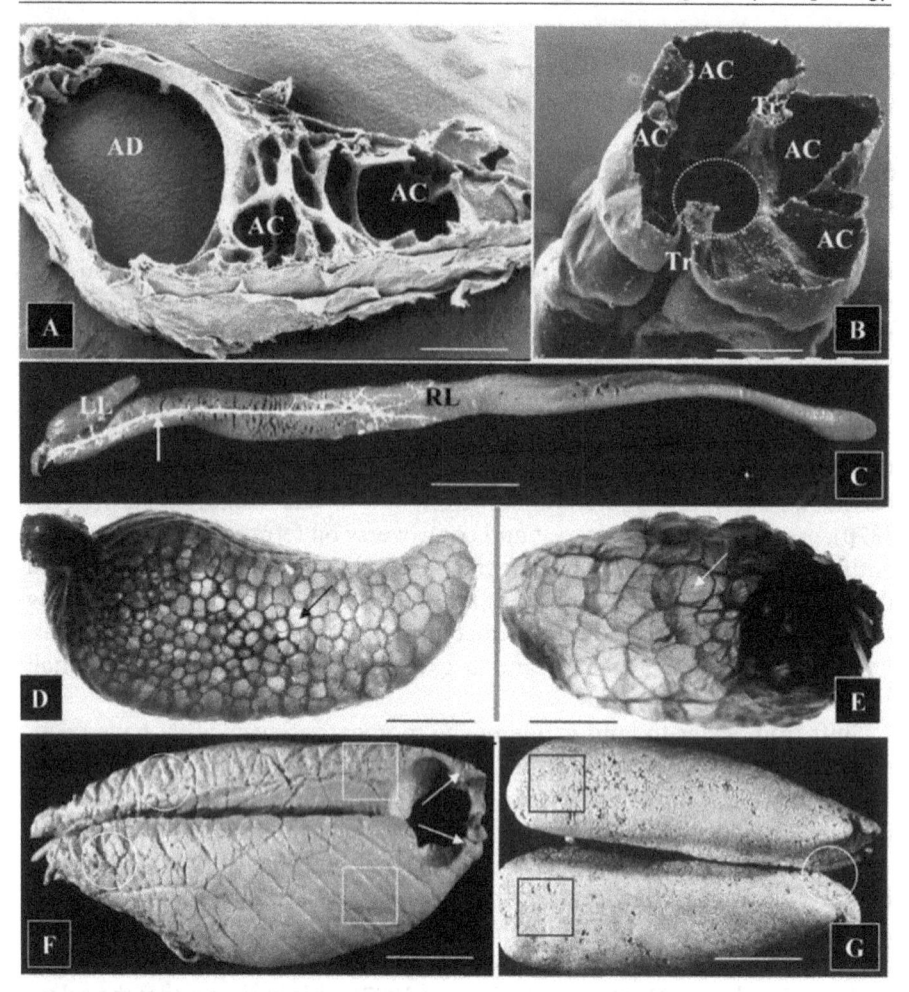

Fig. 90. A,B Lung of the lungfish, *Protopterus aethiopicus* and the caecilian, *Boulengerula taitanus*, showing an air duct (*AD*) and circle (**B**) surrounded by air cells (*AC*). *Tr* Trabeculae. *Scale bar* **A**, 2 mm; **B**, 0.5 mm. **C** Cast of a lung of *B. taitanus* showing a well-developed right lung (*RL*) and a vestigial left one (*LL*). *Arrow* Pulmonary artery. *Scale bar* 1 mm. **D,E** Lungs of frogs, *Xenopus laevis* and *Chiromantis petersi*, showing internal subdivision (arrows). *Scale bars* 1 mm. **F,G** Lungs of a chameleon, *Chameleo chameleon*, and monitor lizard, *Varanus exanthematicus*, showing greater intensity of subdivision cranially (*circle*) and poor subdivision caudally (*square*). *Arrow* Distal dilatations of the lung. *Scale bars* 2 mm. **A** from Maina (1987); **B** from Maina and Maloiy (1988); **C** from Maina (1998); **E** from Maina (2002a); **G** from Maina et al. (1989b)

5.2 Comparison of the Structure of the Avian Respiratory

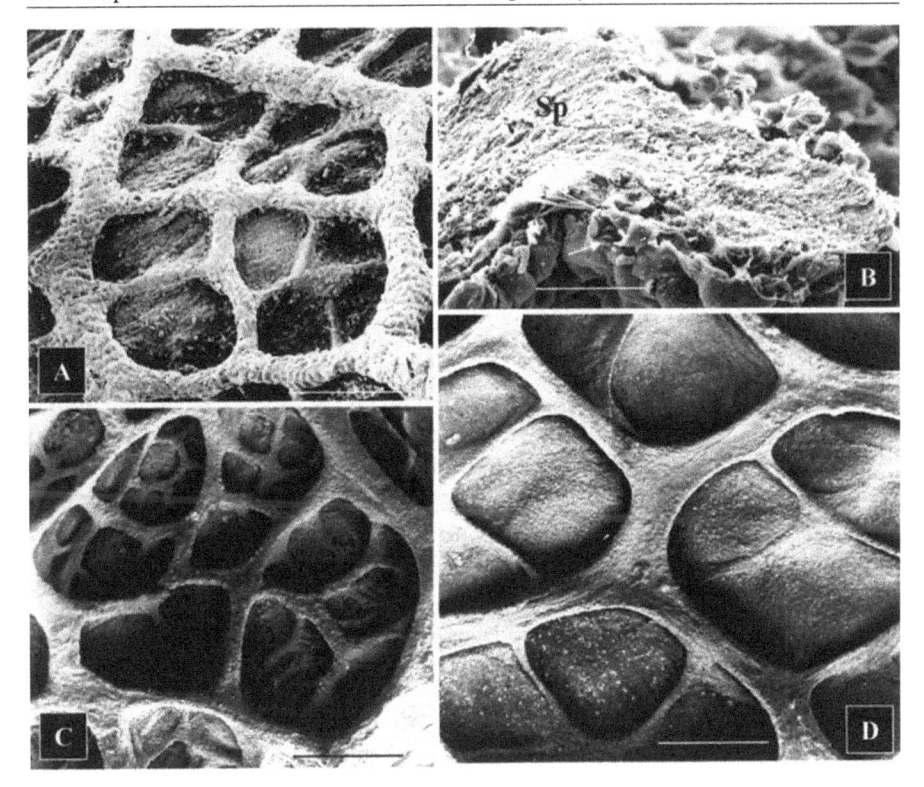

Fig. 91. Views of the respiratory surfaces of the lungs of the lungfish, *Protopterus aethiopicus* (**A, B**), pancake tortoise, *Malacochersus tornieri* (**C**), and the monitor lizard, *Varanus exanthematicus* (**D**). The hierarchical pattern of internal subdivision is similar in the lungs of the three groups of animals. As in the lungs of the tortoise and the lizard, the blood capillaries are exposed to air only on one side of the septa that divide the air spaces, a double capillary arrangement. *Sp* Septum. *Scale bars* **A, C, D** 0.5 mm; **B** 50 µm. **A, B** from Maina (1987); **C, D** from Maina et al. (1989b)

(1985) and Maina (1987), and those of *Neoceratodus forsteri* by, e.g. Grigg (1965), Gannon et al. (1983), and Power et al. (1999). Generally, the lungs are tubular in shape and internally subdivided (Figs. 90A and 91A). The air cells (faveoli) are delineated by hierarchically arranged septa and open into an eccentrically located air-duct (Fig. 90A). In *P. aethiopicus*, the intensity of internal subdivision of the lung decreases proximal-distally (Maina and Maloiy 1985), with the terminal end being less well vascularized. To a certain degree, the somewhat cylindrical lungs of the Dipnoi resemble an individual PR of the avian lung. Moreover, the morphological heterogeneity of the lung, where the proximal part is intensely subdivided while the distal one is relatively smooth, corresponds with that of the avian respiratory system where the intensely compartmentalized lung (parenchyma) and the ASs that are smooth and transparent have been effectively separated. In the dipnoan lung, the blood capillaries are exposed to air only on one side (e.g. Maina and Maloiy 1985), a configuration that constitutes a 'double capillary arrangement' (Fig. 91B). The lungs are lined by a surfactant (e.g. Hughes and Weibel 1976; Maina and Maloiy 1985) that contains both surfactant A and surfactant B like proteins (Power et al. 1999) and the pneumocytes are not differentiated into type I and II cells (e.g. Maina and Maloiy 1985; Maina 1987).

5.2.2
Amphibian Lung

Life at the air–water interface obligates compromise physiological and morphological adaptations. Multiple respiratory structures correspond with the diverse habitats occupied and the metamorphosing amphibian pattern of development. During embryonic and larval stages, gills are the singular gas exchangers. The lungs take over the respiratory role during juvenile and adult life stages. The amphibian lungs are generally simple, saccular (Fig. 90B–E; e.g. Burggren 1989; Maina 1989c). They are satisfactory in supplying O_2 to ectothermic animals with characteristically low MR (e.g. Feder 1976; Guimond and Hutchison 1976). Furthermore, O_2 can be acquired across an assortment of respiratory sites that include skin and buccal cavity. The degree of vascularization of the amphibian lungs and skin correlates with the level of terrestrialness and behavior (e.g. Stinner and Shoemaker 1987; McClanahan et al. 1994). In the predominantly aquatic species, the skin is the primary respiratory pathway while in the more terrestrial ones it is has been downgraded or rendered totally superfluous. In the latter case, the lung serves as the primary respiratory organ.

Among the modern amphibians, three orders occur. These are the highly elusive (fossorial or aquatic), vermiform, tropical caecilians (Gymnophiona=Apoda=caecilians), the frogs (Salentia=Anura), and salamanders (Caudata=Urodela). The morphologies of the lungs vary between taxa and species. The caecilians have simple, particularly long, tubular, internally divided lungs (e.g. Maina and Maloiy 1988; Fig. 90B,C). In the African caecilian, *Boulengerula taitanus*, the left lung is vestigial (Maina and Maloiy 1988; Fig. 90C) while, in the aquatic *Typhlonectes compressicauda*, as many as three functional lungs develop (Toews and MacIntyre 1977). The lungs of *B. taitanus* are supported by diametrically placed trabeculae from which septa attach to delineate respiratory air spaces (Maina and Maloiy 1988; Fig. 90B). The lungs of *Necturus* and *Cryptobranchus* are thin-walled, poorly vascularized, and nonsepted (e.g. Guimond and Hutchison 1973). Generally, the lungs of anurans and apodans are more complex than those of the urodeles (e.g. Smith and Rapson 1977; Meban 1980; Fig. 90D,E). For example, the highly terrestrial species, e.g. the toad, *Bufo marinus* (Smith and Rapson 1977; Meban 1980), and the tree frogs, *Hyra arborea* (Goniakowska-Witalinska 1986) and *Chiromantis petersi* (Maina 1989c), have relatively well internally subdivided lungs (Fig. 90D, E): a central air duct opens into hierarchically arranged septa that delineate the air spaces. The air cells range in diameter from 1.45 mm in *Rana pipiens* to 2.3 mm in *Bufo marinus* and *Rana catesbeiana* (Tenney and Tenney 1970).

The RSA in the lungs of the more terrestrial amphibian species is greater than that in the lungs of the more aquatic ones (Tenney and Tenney 1970). The newt, *Triturus alpestris*, has rather smooth-surfaced lungs (e.g. Czopek 1965) with 569 capillary meshes per cm² (Matsumura and Setoguti 1984), while the relatively more metabolically active tree frog, *Hyla arborea* (Goniakowska-Witalinska 1986), has more complex lungs with 652 capillary meshes per cm² (Czopek 1965). Plethodontid (lungless) salamanders, a taxon that constitutes the largest family among the Caudata, derive their O_2 needs from the cold, well-oxygenated water that they live in entirely across a highly vascularized skin: the length of the BCs in the skin forms about 90 % of all blood vessels on the respiratory surfaces, with the remainder being in the buccal cavity (Czopek 1965). In two species of Salentia that subsist in well-oxygenated high mountain lakes, *Telmatobius* and *Batrachophrynus*, the lungs are very small, the skin is highly vascularized, and the epidermis is very thin (e.g. Martin and Hutchison 1979). The lungs of *Pipa pipa* and *Xenopus laevis* are strengthened by septal cartilages that preserve the patency of the air-passages (e.g. Marcus 1937; Goniakowska-Witalinska 1995). Differentiated pneumocytes as well as dust cells (free=surface phagocytes) occur on the respiratory surface of some amphibian lungs (e.g. Welsch 1983; Maina and Maloiy 1988). On average, the thickness of the BGB in the lungs of the urodeles is 2.59 µm, in apodans 2.35 µm, and in anurans 1.89 µm (Meban 1980). Some parts of the BGB of the lungs of the caecilians, *Chthonerpoton indistinctum* and *Ichthyophis paucesulcus*, are only 1 µm thick (Welsch 1981) while in the tree frog, *Hyla arborea*, the BGB is as thin as 0.6 µm (Meban 1980).

5.2.3
Reptilian Lung

Through multiple adaptations that included development of an impermeable surface covering that averted risk of desiccation on land, reptiles were the first vertebrates to become adequately adapted for terrestrial habitation. With the skin rendered redundant as a respiratory site, the lung became the sole gas exchanger. The lungs of reptiles display remarkable morphological heterogeneity (e.g. Perry 1983, 1989a, 1992; Maina 1989a; Maina et al. 1989b, 1999). In the more advanced species of snakes, e.g. Colubridae, Viperidae, and Elapidae, the left lung is vestigial and in some cases is totally missing while, in the primitive species, e.g. the boas and the pythons, the left and right lungs are equally well developed (e.g. Luchtel and Kardong 1981; Maina 1989a). In the Amphisbenia, the right lung is very small (Gibe 1970) while, in the order Squamata, single-chambered lungs preponderate, particularly in families such as Teiidae (Perry 1989b), Scindae (Klemm et al. 1979), Lacertidae (Meban 1978), and

Gekkonidae (Perry et al. 1989). Exceptionally simple lungs occur in the family Angioidea (Gibe 1970). In view of the fact that the lungs of the more primitive species of reptiles are more homogenous, morphological heterogeneity of the lung appears to confer certain functional advantage. The gas ET is more intensely subdivided in the proximal parts of the lung while the distal ones are sparsely subdivided and in some cases are smooth and saccular (Fig. 90F,G). While the greatest extent of separation of the respiratory (the gas exchange) and the gas storage/ventilatory sites occurs in the avian respiratory system – the lung-air sac system, such dissociation is generally evident in the reptilian lung.

Based to a large extent on the degrees of internal subdivision, various categories of the reptilian lungs have been formulated (e.g. Duncker 1979b; Hlastala et al. 1985): the multicameral lungs are intensely subdivided and occur in, e.g., turtles, monitor lizards, crocodiles, and snakes (e.g. Perry and Duncker 1980; Perry 1988; Maina 1989a; Maina et al. 1989b, 1999); the paucicameral lungs are less elaborate and occur, e.g. in the chameleons and the iguanids; while the unicameral ones are simple, saccular, and smooth-walled and occur in, e.g. the lungs of the teju lizard, *Tupinambis nigropunctatus* (e.g. Klemm et al. 1979; Perry 1983). While to a certain extent useful, this classification is very simplistic since transitional forms exist. The land-based chelonians and lacertids have paucicameral lungs: the lungs lack an intrapulmonary bronchus and have two or three hierarchical air-cells that open into a central air-conduit (Fig. 91C). In a certain manner (Fig. 92A–C), such lungs resemble the individual PR of the avian lung (Figs. 52A; 55; 92D). Marine reptiles have multichambered, bronchiolated lungs (e.g. Solomon and Purton 1984; Pastor et al. 1989). The elongated lungs of the snakes (Ophidia) and the amphisbaenids are divided into two zones: the anterior (respiratory) region is highly vascularized while the posterior one is saccular and avascular (e.g. Kardong 1972; Klemm et al. 1979; Stinner 1982; Maina 1989a; Pastor 1995). In the crocodilian lung, much of the parenchyma (the gas ET) is found in the anterior two-thirds of the lung where blood comprises 38–50% of the total volume of the region (Perry 1988). The less vascularized posterior part of the lung is envisaged to store air (e.g. Heatwole 1981), serve a hydrostatic role (Graham et al. 1975), and conceivably mechanically ventilate the ET in the anterior region of the lung. Although conclusions must be made cautiously, from the above-mentioned morphological attributes of the reptilian lung, it is valid to surmise that, to an extent, structurally, the avian lung is more-or-less a complex reptilian lung: stacks of parabronchi (=many multicameral reptilian lungs; Fig. 92A,B) form the bulk of the avian lung and complete disengagement of the respiratory and nonrespiratory parts has been accomplished. It remains unclear whether this outcome arose entirely by evolutionary progression or was a functional imperative.

The BALu (mammalian) lung and the PRLu (avian) one are presumed to have evolved from transformation of the reptilian multicameral lung (e.g. George and Shah 1956; Duncker 1978; Becker et al. 1989; Perry 1989b, 1992).

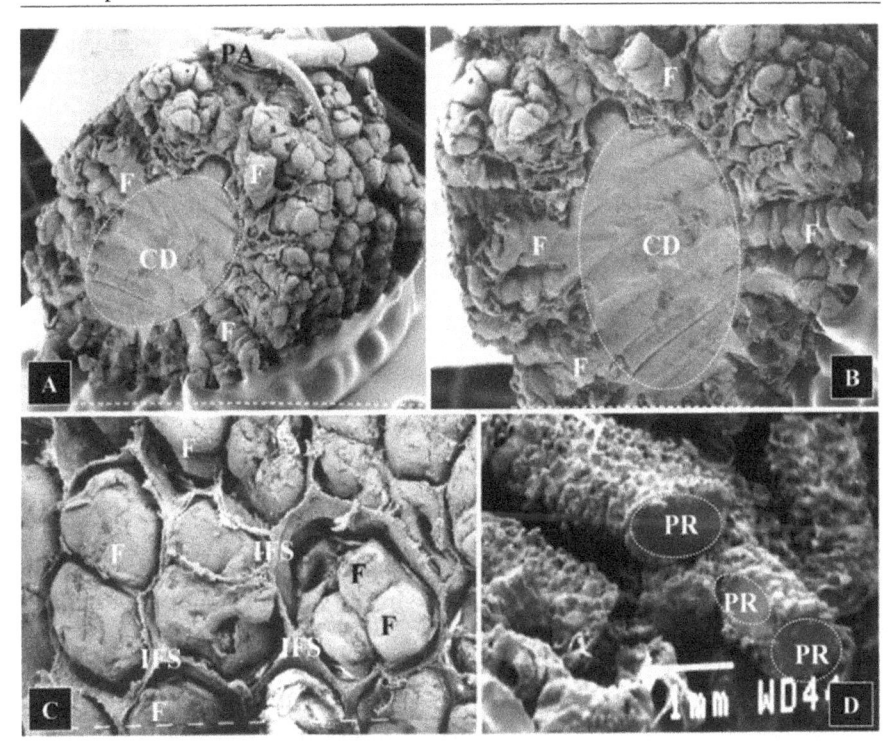

Fig. 92. A, B Cross-sectional views of cast of the lung of the sand boa, *Eryx colubrinus*, showing a central air duct (*CD*; *circled*). *F* Faveolus; *PA* pulmonary artery. *Scale bars* 1 mm; **C** Faveoli (*F*) of the lung of the sand boa separated by interfaveolar septa (*IFS*). *Scale bar* 0.5 mm. **D** Starks of parabronchi (*PR*; *circled*) of the lung of the domestic fowl, *Gallus gallus* variant *domesticus*, showing the similarity of the lung of the snake (**A, B**) with that of a single parabronchus of the bird lung. *Scale bar* 3 mm. **A–C** from Maina et al. (1999)

Having probably arisen from thecodonts or coelurosaurs in the Triassic (e.g. de Beer 1954; Ostrom 1975), birds have comparatively recent evolutionary connection with reptiles. It is therefore reasonable to expect the anatomy of birds to include reptilian features. Among the living reptiles, crocodiles are closest to birds (e.g. Pough et al. 1989): among reptiles, the crocodilian lungs are the most complex (Perry 1988). As the mammalian line departed from the reptilian-bird line at a very early stage in the evolution of the higher vertebrates (e.g. Romer 1966; Bakker 1971), the phylogenetic relationship of mammals to modern reptiles is relatively more remote. Consequently, anatomically, birds and mammals show only general similarities. It has been speculated that the innate structural-functional inadequacies of the reptilian lungs may have largely hindered reptiles from attaining endothermic-homeothermy (Perry 1989a,b), consigning their metabolic capacities to well below those of birds and mammals. At a temperature of 37 °C, a lizard weighing 1 kg consumes O_2

at a rate of 122 ml O_2 h^{-1}, a value that is only 18 % of the VO_{2c} of an equivalent-sized mammal (Bennett and Dawson 1976). Varanids (monitor lizards) present the greatest degree of pulmonary structural complexity in the suborder Sauria: *Varanus exanthematicus* and the pancake tortoise, *Malacochersus tornieri*, have multichambered lungs with bifurcated intrapulmonary bronchi and profuse internal subdivision (Perry and Duncker 1978; Maina et al. 1989b). The simple lungs in Sphenodontia structurally correspond to those of amphibians and lungfish. Such lungs have a central air-duct that opens into peripherally located, shallow respiratory air spaces that are poorly vascularized. The upper air passages of the reptilian lungs are lined by ciliated and mucus-secreting epithelial cells (Luchtel and Kardong 1981; Maina 1989a). In the more advanced species, the pulmonary epithelial cells are conspicuously differentiated into type I (squamous=smooth), type II (cuboidal=granular; e.g. Luchtel and Kardong 1981; Maina 1989a; Perry et al. 1989; Daniels et al. 1990), and type III (brush) pneumocytes (Gomi 1982). A rare mitochondria-rich cell has been described in the lung of the turtle, *Pseudemys scripta* (Bartels and Welsch 1984). Dust cells (surface macrophages) occur in the reptilian lungs, e.g. in the turtle, *Testudo graeca* (Gomi 1982). Reptilian lungs are lined by a surfactant (Daniels et al. 1990), have a 'double capillary arrangement', i.e. blood capillaries are exposed to air on one side (e.g. Maina 1989a), and a preponderance of smooth muscle tissue (e.g. Perry 1988). In the tegu and the monitor lizards, respectively, pulmonary smooth muscle tissue constitutes 7.4 % and 1.3 % of the nontrabecular tissue (e.g. Perry 1981; Pastor et al. 1989). Smooth muscle tissue is involved in promoting intrapulmonary convective movement of air (Perry and Duncker 1980; Carrier 1987).

5.2.4
Insectan Tracheal System

Among the air-breathing animals, the insects have evolved an exceptional respiratory system. The tracheal system is astonishing both for its structural simplicity and functional efficiency. In fine-tuning almost past belief, the circulatory and respiratory systems have been totally disengaged, with the former being relegated from meaningful respiratory role: O_2 is delivered from the atmosphere directly to the body tissues and cells (e.g. Buck 1962). The ΔPO_2 between the tracheoles (terminal trachea), structures analogous to vertebrate blood capillaries, and the tissue cells is about 5.3 kPa (39.8 mmHg; Weis-Fogh 1964a, 1967) compared with that of no more than 0.3 kPa (2.3 mmHg) at the mitochondrial levels of the mammalian tissues (e.g. Wittenberg and Wittenberg 1989). In adult *Aphelocheirus*, between the spiracles and the tracheoles, the PO_2 drops by only 0.3 kPa (2.3 mmHg; Thorpe and Crisp 1941). The tracheal system can supply ten times more O_2 per gram tissue than the blood capillary system (Steen 1971). Ectodermal invaginations, spiracles, are the portals of entry of air into the body. In mechanical terms, the

5.2 Comparison of the Structure of the Avian Respiratory

spiracular valve corresponds to a carburetor while the trachea match a compressor and an exhaust pipe.

Although best developed in insects, tracheal respiration has evolved in various animal taxa. These include the Onychophora (Peripatus), Solifugae, Phalangidae, some Acarina, Myriapoda, and Chilopoda. The bodies of tracheates are suffused by fine air-filled conduits. In small and relatively inactive insects and arachnids, the tracheal system may be simple but, in larger and more energetic species (e.g. wasps and bees), it is highly developed (e.g. Wigglesworth 1965). In such cases, the system comprises of a maze of longitudinal and transverse branches that connect to ASs (Fig. 93A,B). Deter-

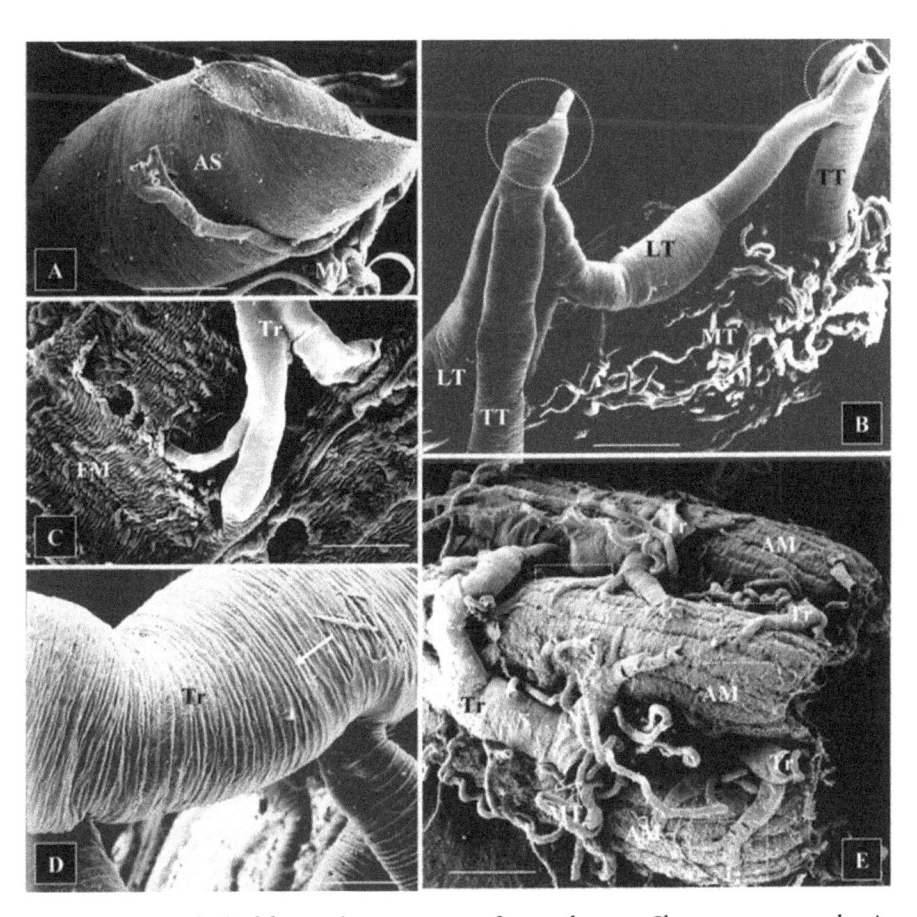

Fig. 93. A Air sac (*AS*) of the respiratory system of a grasshopper, *Chrotogonus senegalensis*. *MT* Malphigian tubules. *Scale bar* 2 mm. **B** Trachea of the grasshopper showing longitudinal trachea (*LT*) and transverse trachea (*TT*). The trachea arise from the spiracles (*dashed circles*). *MT* Malphigian tubules. *Scale bar* 1 mm. **C** Trachea directly supplying air to the flight muscle (*FM*). *Scale bar* 1 mm. **D** Close-up of a trachea (*Tr*) showing the supporting taenidia (*arrow*). *Scale bar* 0.5 mm. **E** Tracheal (*Tr*) air supply to the abdominal muscles (*AM*). *Rectangles* Trachea entering muscle. *Scale bar* 1 mm. (Maina 1989d)

mined by factors such as age and developmental stage, the entire respiratory system in insects (trachea, tracheoles, and ASs) may constitute as much as 50% of the whole body volume (Steen 1971). In the silkworm, *Bombyx mori*, the tracheoles are 1.5 m long (Buck 1962): in a 5.7-g worm with a volume of 49 l g^{-1}, about 5% of the body volume comprises of the trachea (Bridges et al. 1980). In the adult cockchafer, *Melolotha*, the trachea comprise a volume of 585 l g^{-1} (Demoll 1927). The volume of trachea in a 5-g *Cecropia* pupae is about 250 mm^3 (Kanwisher 1966). In the flight muscle of the locust, between 10^{-1} and 10^{-3} (volume of the trachea per volume of muscle) is formed by the tracheal system (Weis-Fogh 1967). In much the same process as that between capillarization of tissues and metabolic activity in vertebrates, in insects, the development of the tracheal system is largely determined by factors inherent in the target tissues (Locke 1958a–c). These include the metabolic requirements of different organs/tissues (Edwards et al. 1958; Locke 1958b) and the degree of hypoxia prevailing in particular organs and parts of the body (Edwards et al. 1958; Wigglesworth 1965; Steen 1971). In the legs of the spiders of the family Uloboridae that are actively used for web monitoring (Opell 1987), the trachea are particularly well developed. In larval mealworms, *Tenebrio molitor*, hypoxia influences tracheal growth and development. At an ambient PO$_2$ of below ~10 kPa, wider trachea form (Loudon 1989).

The tracheas are simple, noncollapsible hollow tubes that are strengthened by endocuticular spiral or annular chitinous thickenings, the taenidia (Fig. 92C,D). The smallest divisions of the trachea, the tracheoles, may be as small as 0.2 µm in diameter as they approach the tissue cells (Fig. 94). In highly metabolically active tissues, the terminal tracheoles reportedly indent cells more-or-less in a manner of jabbing a finger onto the surface of a balloon (Steen 1971). In the flight muscle, the tracheoles are never more than 0.2–0.5 µm from a mitochondrion and in some tissues they may lie as close as 0.005 µm (Wigglesworth and Lee 1982; Maina 1989d). Commonly, mitochondria cluster around terminal tracheoles (Fig. 93C), forming what was termed 'mitochondrial continuum' by Edwards et al. (1958). In the flight muscle, the tracheoles may surround single muscle fibrils (Krogh 1941). The tracheoles terminate blindly (Richards and Korda 1950) but anastomosis was reported by Buck (1948). Estimations made on the tracheal system of the giant *Cossus* larva (mass 3.4 g, length 60 mm) gave a total cross-sectional surface area of all trachea supplying the tissues of 6.7 mm^2, with an average length of 6 mm (Krogh 1920a,b): O$_2$ diffuses at a rate of 0.3 mm^3 s^{-1} at a ΔPO$_2$ of 1.5 kPa (11.3 mmHg), a value considered to be more than adequate even during muscular exercise. The terminal tracheoles contain fluid (Wigglesworth 1953, 1965). The extent of filling depends on the state of activity: endotracheal fluid is removed osmotically by increased concentration of end-products of metabolism in the interstitial spaces. During exercise and exposure to hypoxia, the air–fluid interface advances closer to the tissue cells as the interstitial fluid is drawn upstream of the peripheral tracheoles and into the cytoplasm of the

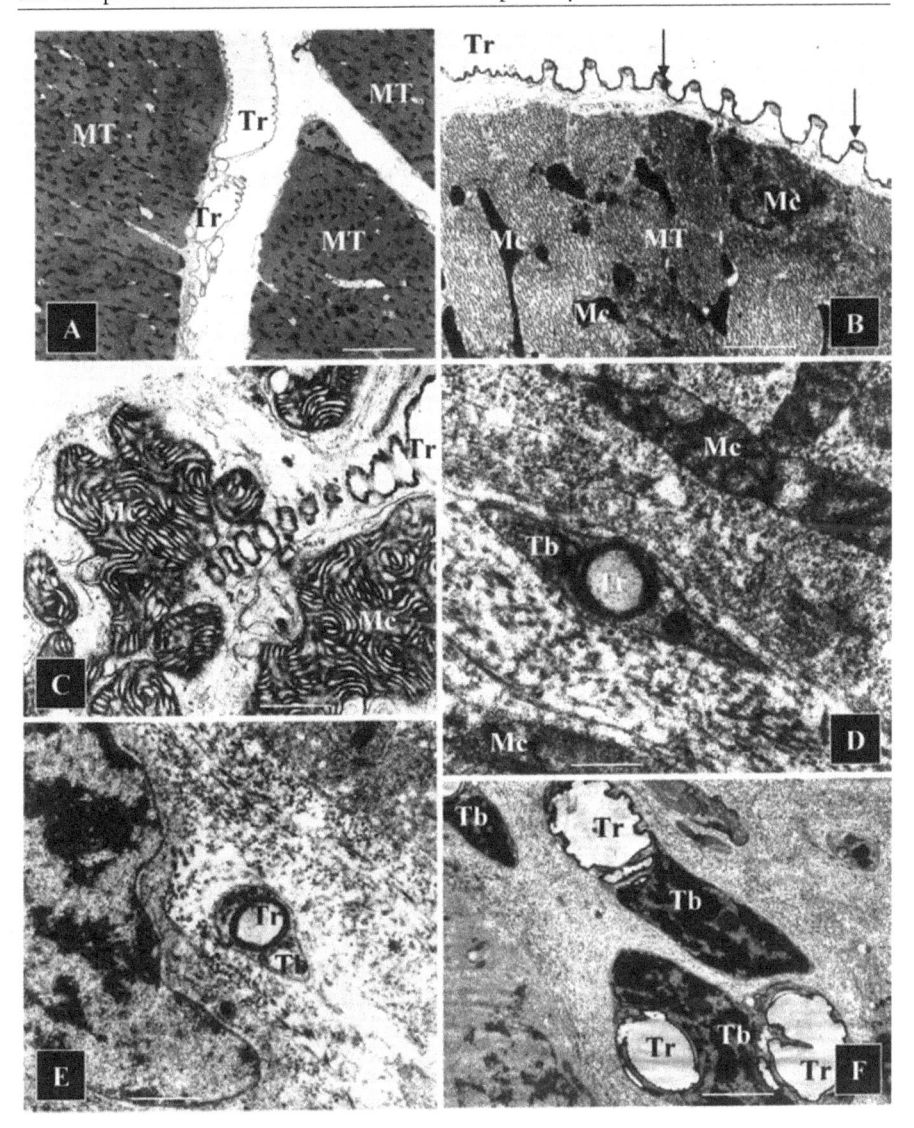

5.2 Comparison of the Structure of the Avian Respiratory

Fig. 94. A Trachea (*Tr*) supplying air to flight muscle (*MT*) of a locust, *Locusta migratoria*. *Scale bar* 20 μm. **B** Close-up of a trachea overlying the flight muscle (*MT*). *Mc* Mitochondria. *Scale bar* 0.5 μm. **C** Mitochondria (*Mc*) clustered around a tracheole (*Tr*). *Scale bar* 0.25 μm. **D–F** Trachea (*Tr*) are formed by tracheoblasts (*Tb*). *Mc* Mitochondria. *Scale bar* 0.5 μm. **A, F** from Maina (1998)

surrounding cells by active ion transport. The movement shortens the diffusional pathway for the respiratory gases, enhancing gas-exchange efficiency. During resting conditions, the process is reversed when the acidic metabolites are eliminated.

The design of the respiratory system of insects, the tracheal-air sac system, provides a novel mechanism of delivering O_2 to the body tissues and cells. It is incredibly cost-effective. In trade-off, however, insects paid a heavy price. The intrinsic limitations of diffusion and the large mechanical ventilatory forces needed to move air at high rates through copious, narrow conduits have consigned insects to small body sizes of which adaptive progress to states like endothermy were untenable. The average tracheolar length for optimal diffusion is 5–10 mm and the minimum diameter is 0.2 µm (Krogh 1920a,b; Weis-Fogh 1964a). Tracheates that utilize diffusion as the only mode of moving respiratory gases include the Onychophora (Peripatus), the tracheate Arachnoidea, Myriapoda, and Chilopoda, almost all terrestrial insect larvae, and all pupae and most of the small imagines. This is made possible by the relatively rapid diffusion of O_2 in air: the process would be inadequate if the trachea were fluid-filled. The largest insect that ever lived is reportedly the tropical dragonfly-like *Meganeura*, of the Carboniferous. It reached a body length of 30 cm, a wingspan of 60 cm, and a body thickness of 3 cm (Krogh 1941). The largest modern insects are the tropical beetles which may be as long as 15 cm. The stick insect, *Dixippus morosus*, suitably displays the trade-offs and compromises, mainly regarding size and shape, that have been compelled by the diffusional distance and ventilatory capacities. While a house-fly that weighs about 15–20 mg does not need to ventilate the tracheal system, a bee that is more energetic and weighs about 100 mg has to do so constantly. In insects such as locusts, dragon-flies, and cockroaches, at rest, well-synchronized abdominal and to a lesser extent thoracic ventilations occur (Brocher 1931). Whereas at rest no ventilatory movements take place in the cockroaches *Peripaneta* and *Blatella*, during flight, when VO_{2c} increases 10–100 times, wing movements compress the thorax, ventilating the trachea and the ASs (Brocher 1920; Portier 1933). During steady flight, in the desert locust, about 320l kg^{-1} h^{-1} of air with an average tidal volume of 167 cm^3 and frequencies of 30– 60 times min^{-1} is pumped into the tracheal system by abdominal and thoracic pumping, with the intratracheal pressure increasing from 0.9 to 3.3 kPa (6.8–24.8 mmHg) at the peak of abdominal contraction (Miller 1960; Weis-Fogh 1967). The giant beetle, *Petrognatha gigas*, has a ventilatory rate of about 2000l kg^{-1} h^{-1} (Miller 1966).

Through synchronized activity of the spiracles, particularly among the Orthoptera, the trachea are ventilated unidirectionally (e.g. Fraenkel 1932; Weis-Fogh 1964a,b, 1967). In the honey bee, the flow is unidirectional during flight (Bailey 1954) and, in *Sphodromantis*, 95 % of the inhaled air passes unidirectionally while only 5 % of it passes tidally (Miller 1974). In the cockroaches *Periplaneta* and *Blatella*, tidal ventilation only occurs during stress while in other roaches, *Byrsotria*, *Blaberus*, and *Nyctobra*, anteroposterior ventilation occurs during rest (Buck 1962). In the avian lung (Chap. 3.11), unidirectional and continuous ventilation occurs in the PPPR while the air flow in the NPPR is bidirectional (e.g. Scheid 1979; Fedde 1980). In both birds and insects, unidirectional air flow minimizes or eliminates dead space, enhancing respiratory efficiency. In some large insects, even abdominal pumping is inadequate for supplying O_2 to the long muscles of the legs. In the grasshopper, the concentration of O_2 in the tibial tracheae is relatively high in the resting state (16%) but drops to 5% during physical exertion (Krogh 1913). In the harvestmen (Opiliones), special spiracles have developed on the legs (Hansen 1893) to overcome the diffusion and convective limitations. Akin to birds, ASs are an important part of the insectan respiratory system: in insects, they increase the tidal volume by as much as 70% of the total air capacity (Bursell 1970) and reduce the longitudinal diffusion gradient for O_2 through the gas-exchange pathway. The ASs are well developed in Diptera and Hymenoptera but are absent in the subclass Apterygota. In cicada, *Fidicina monnifera*, together with the tracheal system, the ASs constitute 45% of the body volume (Bartholomew and Barnhart 1984).

5.3
Conclusions

The similarity between the development (the two gas exchangers form by invagination), structural (ASs increase tidal volume, ASs are actively involved in ventilating the gas exchangers, and the terminal tracheoles to a certain extent morphologically resemble the ACs), and functional (unidirectional ventilation at the gas exchange level) designs of the avian and the insectan respiratory systems, animals separated by some 100 million years of evolution (e.g. Miller and Orgel 1974), is astounding. Clearly, this is a classical case of convergence where analogous structural and functional solutions were engineered to accomplish and support a common enterprise – flight. To supply the large amounts of O_2 for volancy, exceptionally efficient respiratory systems were an imperative (Maina 1997, 2002b). Owing to their disparate phylogenies, birds and insects have utilized different resources to fabricate efficient respiratory devises.

In closing, it is important to underscore the fact that the designs that have evolved in birds and insects were not a prerequisite for flight. Some 100 million years after birds (e.g. Yalden and Morris 1975) and some 300 million years after insects (e.g. Wigglesworth 1965; Miller and Orgel 1974), bats (Order: Chiroptera) evolved flight (e.g. Thewissen and Babcock 1992) by structurally highly refining the mammalian lung (e.g. Maina et al. 1982b, 1991; Maina and King 1984). In addition these refinements, parameters/features like one-to-one synchronization between wing-beat frequency and respiratory cycles (e.g. Suthers et al. 1972; Thomas 1981; Carpenter 1986), enormous hearts that provided large cardiac output (e.g. Snyder 1976; J¸rgens et al. 1981), and high hemoglobin concentration, hematocrit, and O_2-carrying capacity (e.g. Riedesel 1977; Black and Wiederhielm 1976; Wolk and Bodgdanowicz 1987), contributed to delivery of the large amounts of O_2 needed for active flight.

References

Abdalla MA (1989) The blood supply to the lung. In: King AS, McLelland J (eds) Form and function in birds, vol 4. Academic Press, London, pp 281–306
Abdalla MA, King AS (1975) The functional anatomy of the pulmonary circulation of the domestic fowl. Respir Physiol 23:267–290
Abdalla MA, King AS (1976a) Pulmonary arteriovenous anastomoses in the avian lung: do they exist? Respir Physiol 27:187–191
Abdalla MA, King AS (1976b) The functional anatomy of the bronchial circulation of the domestic fowl. J Anat 121:537–550
Abdalla MA, King AS (1977) The avian bronchial arteries: species variations. J Anat 123:697–704
Abdalla MA, Maina JN (1981) Quantitative analysis of the exchange tissue of the avian lung (Galliformes). J Anat 134:677–680
Abdalla MA, Maina JN, King AS, King DZ, Henry J (1982) Morphometrics of the avian lung. 1. The domestic fowl, *Gallus domesticus*. Respir Physiol 47:267–278
Abraham JA, Mergia A, Wang JL, Tumolo A, Friedman KA, Hyerild D, Gospodarowicz D, Fiddes JC (1986) Nucleotide sequence of a bovine cDNA clone encoding the angiogenetic protein, basic fibroblast growth factor. Science 233:545–548
Acarregui MJ, Penisten ST, Goss KL, Ramirez K, Snyder JM (1999) Vascular endothelial growth factor gene expression in human fetal lung in vitro. Am J Respir Cell Mol Biol 20:14–23
Adamson IYR (1997) Development of the lung. In: Crystal RD, West JB, Weibel ER, Barnes PJ (eds) The lung: scientific foundations. Lippincott-Raven, Philadelphia, pp 993–1001
Affolter M, Bellusci S, Itoh N, Shilo B, Thiery JP, Werb Z (2003) Tube or not tube: remodeling epithelial tissues by branching morphogenesis. Dev Cell 4:11–18
Agraves WS, Larue AC, Fleming PA, Drake CJ (2002) VEGF signaling is required for the assembly but not the maintenance of embryonic blood vessels. Dev Dyn 225:298–304
Akeson AL, Wetzel B, Thompson FY, Brooks SK, Paradis H, Gendron RL, Greenberg JM (2000) Embryonic vasculogenesis by endothelial precursor cells derived from lung mesenchyme. Anat Rec 217:11–23
Akester AR (1970) Osmiophilic inclusion bodies as the sources of laminated membrane in the epithelial lining of avian tertiary bronchi. J Anat 107:189–190
Alsberg E, Moore K, Huang S, Polte T, Inger DE (2004) The mechanical and cytoskeletal basis of lung morphogenesis. In: Massaro DJ, Massaro GC, Chambon P (eds) Lung development and regeneration. Marcel Dekker, New York, pp 247–274
Arman E, Haffner-Krausz R, Gorivodsky M, Lonai P (1999) FGF2 is required for limb outgrowth and lung branching morphogenesis. Proc Natl Acad Sci USA 96:11895–11899
Asahara T, Murohara T, Sullivan A, Silver M, van der Zee R, Li T, Witzenbichler B et al. (1997) Isolation of putative progenitor endothelial cells for angiogenesis. Science 275:964–967
Aschoff J, Pohl H (1970) Rhythmic variations in energy metabolism. Fed Proc 29:1541–1552
Atwal OS, Singh B, Staempfli H, Minhas KJ (1992) Presence of pulmonary intravascular macrophages in the equine lung: some structural-functional properties. Anat Rec 234:530–540

Auerbach A, Auerbach W (1997) Profound effects on vascular development caused by perturbations during organogenesis. Am J Pathol 151:1183–1186
Bailey L (1954) The respiratory currents in the tracheal system of the adult bee. J Exp Biol 31:589–595
Bakker RT (1971) Dinosaur physiology and origin of mammals. Evolution 25:636–658
Banzett RB, Nations CS, Barnas JL, Lehr JL, Jones JH (1987) Inspiratory aerodynamic valving in goose lungs depends on gas density and velocity. Respir Physiol 70:287–300
Banzett RB, Nations CS, Wang N, Fredberg JJ, Butler PJ (1991) Pressure profiles show features essential to aerodynamic valving in geese. Respir Physiol 84:295–309
Banzett RB, Nations CS, Wang N, Butler JP, Lehr JL (1992) Mechanical interdependence of wing beat and breathing in starlings. Respir Physiol 89:27–36
Bard JB (2002) Growth and death in the developing mammalian kidney, signals, receptors and conversations. Bioessays 24:72–82
Barker FK, Cibois A, Schikler P, Fenstein J, Cracraft J (2004) Phylogeny and diversification of the largest avian radiation. Proc Natl Acad Sci USA 101:11040–11045
Barnas G, Mather FB, Fedde MR (1978) Response of avian intrapulmonary smooth muscle to changes in carbon dioxide concentration. Poult Sci 57:1400–1407
Bartels H, Welsch U (1984) Freeze-fracture study of the turtle lung. 2. Rod-shaped particles in the plasma membrane of a mitochondria-rich pneumocyte in *Pseudemys (Chrysemys) scripta*. Cell Tissue Res 236:453–467
Bartholomew GA, Barnhart CM (1984) Tracheal gases, respiratory gas exchange, body temperature and flight in some tropical cicadas. J Exp Biol 111:131–144
Bartholomew GA, Lighton JRB (1986) Oxygen consumption during hover-feeding in free-ranging Anna hummingbirds. J Exp Biol 123:191–199
Becker HO, Bohme W, Perry SF (1989) The lung morphology of lizards (Reptilia: Varaniidae) and its taxonomic-phylogenetic meaning. Bonn Zool Beitr 40:27–56
Bellavite P, Dri P, Bisiachi B, Patricia P (1977) Catalase deficiency in myeloperoxidase deficient polymorphonuclear leukocytes from chicken. Fed Exp Biol Soc Lett 81:73–76
Bellairs A, d'A, Attridge J (1975) Reptiles. Hutchison University Library, London
Bellusci S, Henderson R, Winnier G, Oikawa T, Hogan BL (1996a) Evidence from normal expression and targeted misexpression that bone morphogenetic protein-4 (BMP-4) plays a role in mouse embryonic lung morphogenesis. Development 122:1693–1702
Bellusci S, Furuta Y, Rush MG, Handerson R, Winnier G, Hogan BL (1996b) Involvement of sonic hedgehog (shh) in mouse embryonic lung growth and morphogenesis. Development 124:53–63
Bellusci S, Grindley J, Emoto H, Itoh N, Hogan BL (1997) Fibroblast growth factor-10 (FGF-10) and branching morphogenesis in the embryonic mouse lung. Development 124:4867–4878
Bennett AF, Dawson WR (1976) Metabolism. In: Gans C, Dawson WR (eds) Biology of reptilia, vol 5. Academic Press, New York, pp 127–223
Berger AJ (1960) Some anatomical characters of the Cuculidae and the Musophagidae. Wilson Bull 72:60–104
Berger AJ (1961) Bird study, 1st edn. Dover Publications, New York
Bernfield MR (1977) The periphery in morphogenesis. In: Littlefield JW, de Grouchy J, Ebling FJG (eds) Birth defects, vol 432. Excerpta Medica, Amsterdam, pp 111–125
Betram TA, Overby LH, Brody AR, Eling TE (1989) Comparison of pulmonary intravascular and alveolar macrophages exposed to particulate and soluble stimuli. Lab Invest 61:457–466
Bezuidenhout AJ, Groenewald HB, Soley JT (2000) An anatomical study of the respiratory air sacs in ostriches. Onderst J Vet Res 66:317–325
Biggs PM, King AS (1957) A new experimental approach to the problem of the air pathway within the avian lung. J Physiol Lond 138:282–289

Bikfalvi A, Klein S, Guiseppe P, Rifkin D (1997) Biological roles of fibroblast growth factor-2. Endocr Rev 18:26–45

Black CP, Tenney SM (1980) Oxygen transport during progressive hypoxia in high altitude and sea level water-fowl. Respir Physiol 39:217–239

Black CP, Tenney SM, Kroonenburg MV (1978) Oxygen transport during progressive hypoxia in bar-headed geese (*Anser anser*) acclimated to sea level and 5600 m. In: Piiper J (ed) Respiratory function in birds adult and embryonic. Springer, Berlin Heidelberg New York, pp 79–83

Black LL, Wiederhielm CA (1976) Plasma oncotic pressures and hematocrit in the intact, anaesthetized bat. Microvasc Res 12:55–58

Bohr C (1909) Ueber die spezifische Tätigkeit der Lungen bei der respiratorischen Gasaufnahme. Scand Arch Physiol 22:221–280

Borges M, Linnoila RI, van de Velde HJ, Chen H, Nelkin BD, Mabry M, Baylin SB, Ball DW (1997) An *achaete-scute* homologue essential for neuroendocrine differentiation in the lung. Nature 386:852–855

Bowden DH (1987) Macrophages, dust, and pulmonary diseases. Exp Lung Res 12:89–107

Brackenburry JH (1984) Physiological responses of birds to flight and running. Biol Rev 59:559–575

Brackenburry JH (1989) Functions of the syrinx and the control and sound production. In: King AS, McLelland J (eds) Form and function in birds, vol 4. Academic Press, London, pp 193–220

Bramwell CD (1971) Aerodynamics of *Pteranodon*. J Linn Soc Biol 3:313–328

Brazelton TR, Blau HM (2004) Plasticity of circulating adult stem cells. In: Massaro DJ, Massaro GC, Chambon P (eds) Lung development and regeneration. Marcel Dekker, New York, pp 217–245

Bremer JL (1912) The development of the aorta and aortic arches in rabbits. Am J Anat 13:111–128

Bridges CR, Kester P, Scheid P (1980) Tracheal volume in the pupa of the saturniid moth *Hyalophora cecropia* determined with inert gases. Respir Physiol 40:281–291

Brocher F (1920) Étude expérimentale sur le fonctionnement du vaisseau dorsal et sur lar circulation du sang chez le Insectes. III. Le Sphinx convolvuli. Arch Zool Exp Gén 60:1–45

Brocher F (1931) Le mécanisme de la respiration et celui de la circulation du sang chez les insectes. Arch Zool Exp Gén 74:25–32

Brown RE, Kovacs CE, Butler JP, Wang N, Lehr J, Banzett RB (1995) The avian lung: is there an aerodynamic expiratory valve? J Exp Biol 198:2349–2357

Bruce MC, Honaker CE, Ross RJ (1999) Lung fibroblasts undergo apoptosis following alveolization. Am J Respir Cell Mol Biol 20:228–236

Buck JB (1948) The anatomy and physiology of the light organs in fire flies. Ann NY Acad Sci 9:397–482

Buck J (1962) Some physical aspects of insect respiration. Ann Rev Entomol 7:27–56

Buffon GLL (1777) Essai d'arithmétique morale. Supplément à l'Histoire Naturelle (Paris) 4

Burger RE, Meyer M, Werner G, Scheid P (1979) Gas exchange in the parabronchail lung of birds: experiments in unidirectionally ventilated ducks. Respir Physiol 36:19–37

Burggren WW (1989) Lung structure and function. In: Wood SC (ed) Comparative pulmonary physiology: current concepts. Marcel Dekker, New York, pp 153–192

Burri PH (1997) Structural aspects of pre- and postnatal development and growth of the lung. In: McDonald J (ed) Growth and development of the lung. Marcel Dekker, New York, pp 1–35

Burri PH, Tarek MR (1990) A novel mechanism of capillary growth in the rat pulmonary circulation. Anat Rec 228:35–45

Burri PH, Weibel ER (1977) Ultrastructure and morphometry of the developing lung. In: Hodson WA (ed) Development of the Lung. Marcel Dekker, New York, pp 215–268.

Bursell E (1970) An introduction to insect physiology. Academic Press, London
Butler JP, Banzett RB, Fredberg JJ (1988) Inspiratory valving in avian bronchi: aerodynamic considerations. Respir Physiol 73:241–256
Cadigan KM, Nusse R (1997) Wnt signaling: a common term in animal development. Genes Dev 11:3286–3305
Caduff JH, Fischer LC, Burri PH (1986) Scanning electron microscopic study of the developing microvasculature in the postnatal rat lung. Anat Rec 216:154–164
Cardoso WN (2000) Lung morphogenesis revisited: old facts, current ideas. Dev Dyn 219:121–130
Cardoso WV (2001) Molecular regulation of lung development. Annu Rev Physiol 63:471–494
Cardoso WV, Itoh A, Nogawa H, Mason I, Brody JS (1997) FGF-1 and FGF-7 induce distinct patterns of growth and differentiation in embryonic lung epithelium. Dev Dyn 208:398–405
Carey SW (1976) The expanding earth. Elsevier, New York
Carlson CW (1960) Aortic rupture. Turkey Producer (January issue)
Carlson CW, Beggs EC (1973) Ultrastructure of the abdominal air sac of the fowl. Res Vet Sci 14:148–150
Carmeliet P (2000) Mechanisms of angiogenesis and arteriogenesis. Nat Med 6:389–395
Carmeliet P, Ferrara V, Breier G, Pollefeyt S, Kieckens L, Gertsenstein M, Fahrig M et al. (1996) Abnormal blood vessel development and lethality in embryos lacking a single vascular endothelial growth factor allele. Nature 380:435–439
Carpenter FL, Paton DC, Hixon MA (1983) Weight gain and adjustment of feeding territory size in migrant hummingbirds. Proc Natl Acad Sci USA 80:7259–7263
Carpenter RE (1985) Flight physiology of flying foxes, *Pteropus poliocephalus*. J Exp Biol 114:619–747
Carpenter RE (1986) Flight physiology of intermediate sized fruit-bats (Family: Pteropodidae). J Exp Biol 120:79–103
Carrier DR (1987) Lung ventilation during walking and running in four species of lizards. J Exp Biol 47:33–42
Carter GS, Beadle LC (1930) Notes on the habits and development of *Lepidosiren paradoxa*. J Linn Soc Zool 37:327–368
Casey TM, May ML, Morgan KR (1985) Flight energetics of euglossine bees in relation to morphology and wing stroke frequency. J Exp Biol 116:271–289
Chandler DB, Brannen AL (1990) Interstitial macrophage subpopulations: responsiveness to chemotatctic stimuli. Tissue Cell 22:427–434
Chandler DB, Bayless G, Fuller WC (1988) Prostaglandin synthesis and release by subpopulations of rat interstitial macrophages. Am Rev Respir Dis 138:901–907
Chen WT, Chen JM, Mueller SC (1986) Coupled expression and colocalization of 140 K cell adhesion molecules, fibronectin, and laminin during morphogenesis and cytodifferentiation of chick lung cells. J Cell Biol 103:1073–1090
Cheng CW, Smith SK, Charnock-Jones DS (2003) Wnt-1 signaling inhibits human umbilical vein endothelial cell proliferation and alters cell morphology. Exp Cell Res 291:415–425
Choi K, Kennedy M, Kazarov A, Papadimitriou JC, Keller G (1998) A common precursor for hematopoietic and endothelial cells. Development 125:725–732
Clark GA (1979) Body weights of birds. Condor 18:193–302
Cleaver O, Krieg PA (1999) Molecular mechanisms of vascular development. In: Harvey RP, Rosenthal N (eds) Heart development. Academic Press, San Diego, pp 221–252
Coffin JD, Poole TJ (1988) Embryonic vascular development: immunohistochemical identification of the origin and subsequent morphogenesis of the major vessel primordia in the quail embryos. Development 102:735–748

Cohen ED, Mariol MC, Wallace RM, Weyers J, Kamberov YG, Pradel J, Wilder EL (2002) DWnt4 regulates cell movement and focal adhesion kinase during *Drosophila* ovarian morphogenesis. Dev Cell 2:437–448

Coitier V (1573, cit by Campana 1875) Anatomia avium. In: Externum et internarum praecipalium humani corporis partium tabulae arque anatomicae exercitationes. Nuremberg

Colvin JS, Feldman B, Nadeau JH, Goldfab M, Ornitz DM (1999) Genomic organization and embryonic expression of the mouse fibroblast growth factor 9 gene. Dev Dyn 216:72–88

Colvin JS, White AC, Pratt SJ, Ornitz DM (2001) Lung hypolasia and neonatal death in FGF-9-null mice identify this gene as an essential regulator of lung mesenchyme. Development 128:2095–2106

Cook RD, King AS (1970) Observations on the ultrastructure of the smooth muscle and its innervation in the avian lung. J Anat 106:273–283

Cook RD, Vaillant CR, King AS (1987) The structure and innervation of the saccopleural membrane of the domestic fowl, *Gallus gallus*: an ultrastructural and immunohistochemical study. J Anat 150:1–9

Constable G (1990) How things work: flight. Time Life Books, Alexandria, Virginia

Coraux C, Meneguzzi G, Rousselle P, Puchelle E, Gaillard D (2002) Distribution of laminin5, integrin receptors, and branching morphogenesis during human fetal lung development. Dev Dyn 225:176–185

Costa DP, Prince PA (1987) Foraging energetics of grey-headed albatrosses, *Diomedea chrysostoma*, at Bird Island, South Georgia. Ibis 129:149–158

Crapo JD, Crapo RO (1983) Comparison of total lung diffusion capacity and the membrane component of diffusion capacity as determined by physiologic and by morphometric techniques. Respir Physiol 51:183–194

Cutts CJ, Speakman JR (1994) Energy savings in formation flight of pink-footed geese. J Exp Biol 189:251–261

Czopek J (1965) Quantitative studies of the morphology of respiratory surfaces in amphibians. Acta Anat 62:296–323

Dale TC (1998) Signal transduction by the Wnt family of ligands. Biochem J 329:209–223

D'Angio CT, Maniscalco WM (2002). The role of vascular growth factors in hyperoxia-induced injury to the developing lung. Front Biosci 7:1609–1623

Danchakoff V (1918) Cell potentialities and differential factors considered in relation to erythropoiesis. Am J Anat 24:1–30

Daniels CB, Barr HA, Power JHT, Nicholas TE (1990) Body temperature alters the lipid composition of pulmonary surfactant in the lizard *Ctenophorus nuchalis*. Exp Lung Res 16:435–449

Davis GE, Bayless KJ (2003) An integrin and Rho GTPase-dependent pinocytotic vacuole mechanism controls capillary lumen formation in collagen and fibrin matrices. Microcirculation 10:27–44

de Beer G. (1954). *Archeopteryx lithographica*. British Museum of Natural History, London

del Corral JPD (1995) Anatomy and histology of the lung and air sacs of birds. In: LM Pastor (ed) Histology, ultrastructure and immunohistochemistry of the respiratory organs in non-mammalian vertebrates. Publicaciones de la Universitatd de University of Murcia, Murcia (Spain), pp 179–233

De Groodt M, Lagasse A, Sebruyns M (1960) Electronenmikroskopische morphologie der lungenalveolen des *Protopterus* und *Amblystoma*. Proc Int Congress Electr Microsc, Springer, Berlin Heidelberg New York, 1960, p 418A

Dehring DM, Wismar BL (1989) Intravascular macrophages in pulmonary capillaries of humans. Am Rev Respir Dis 139:1027–1029

Dejours P (1989) Current concepts in comparative physiology of respiration. In: Wood SC (ed) Comparative pulmonary physiology: current concepts. Marcel Dekker, New York, pp 1–10

Delessé MA (1846) Procédé mécanique pour déterminer la composition des roches. Ann Mines, Ser 4. 13:379-388
De Mello DE, Sawyer D, Galvin N, Reid LM (1997) Early fetal development of lung vasculature. Am J Respir Cell Mol Biol 16:568-581
Demoll R (1927) Untersuchungen über die Atmung der Insekten. Z Biol 87:8-22
Desai TJ, Cardoso WV (2002) Growth factors in lung development and disease: friends or foe. Respir Res 3:2
Diamond JM, Karasov WH, Phan D, Carpenter FL (1986) Digestive physiology is a determinant of foraging bout frequency in hummingbirds. Nature 320:62-63
Dotterweich H (1930) Versuch über den Weg der Atemluft in der Vogellunge. Zeitsch Vergl Physiol 11:271-284
Dotterweich H (1934) Ein weiterer Beitrag zur Atmungsphysiologie der Vögel. Z Vergl Physiol 18:803-809
Dotterweich H (1936) Die Atmung der Vögel. Zeitscht vergleich Physiol 23:744-770
Downs KM (2003) Florence Sabin and the mechanism of blood vessel lumenization during vasculogenesis. Microcirculation 10:5-25
Drake CJ, Brandt SJ, Trusk TC, Little CD (1997) TAL/SCL is expressed in endothelial progenitor cells/angioblasts and defines a dorsal-to-ventral gradient of vasculogenesis. Dev Biol 192:17-30
Dubach M (1981) Quantitative analysis of the respiratory system of the house sparrow, budgerigar, and violet-eared hummingbird. Respir Physiol 46:43-60
Dubois A, Brody AW, Lewis DH, Burgess F (1956) Oscillation mechanics of lungs and chest in man. J Appl Physiol 8:587-594
Dumont DJ, Gradwohl G, Fong GH, Puri MC, Gertsenstein M, Auerbach A, et al. (1994) Dominant-negative and targeted null mutations in the endothelial receptor tyrosine kinase, tek, reveal a critical role in vasculogenesis of the embryo. Genes Dev 8:1897-1909
Duncker HR (1971) The lung-air sac system of birds. A contribution to the functional anatomy of the respiratory apparatus. Ergeb Anat Entwicklung 45:1-171.
Duncker H-R (1972) Structure of the avian lung. Respir Physiol 14:4-63
Duncker H-R (1973) Der quantitative Aufbau des Lungenluftsacksystems der Vögel. Verhandlungen Anat Ges 67:197-204
Duncker H-R (1974) Structure of the avian respiratory tract. Respir Physiol 22:1-34
Duncker H-R (1978) Development of the avian respiratory and circulatory systems. In: Piiper J (ed) Respiratory function in birds, adult and embryonic. Springer, Berlin Heidelberg New York, pp 260-273
Duncker H-R (1979a) Die funktionelle Anatomie des Lungen-Luftsack-Systems der Vögel - mit besonderer Berücksichtigung der Greivögel. Der Prakt Tier 60:209-218
Duncker H-R (1979b) General morphological principles of amniotic lungs. In: Piiper J (ed) Respiratory function in birds, adult and embryonic. Springer, Berlin Heidelberg New York, 1979:1-15
Duncker H-R, Guntert M (1985a) The quantitative design of the avian respiratory system: from hummingbird to the mute swan. In: Nachtigall W (ed) BIONA Report No 3. Gustav-Fischer, Stuttgart, pp 361-378
Duncker H-R, Guntert M (1985b) Morphometric analysis of the avian respiratory system. In: Duncker HR, G Fleischer G (eds) Vertebrate morphology. Gustav-Fischer, Stuttgart, pp 383-387
Dunnill MS (1962) Quantitative methods in the study of pulmonary morphology. Thorax 17:320-328
Durbeej M, Ekblom P (1997) Dystroglycan and laminins: glycoconjugates involved in branching epithelial morphogenesis. Exp Lung Res 23:109-118
Edwards GA, Ruska H, Harven de E (1958) The fine structure of insect tracheoblasts, tracheae and tracheoles. Arch Biol 69:351-369

References

Elkins N (1983) Weather and bird behaviour. T and AD Poyser, Stoke on Trent
Ellington CE (1999) Limitations of animal flight performance. J Exp Biol 160:71–91
Evans HE (1909) On the development of the aortae, cardinal veins, and other blood vessels of vertebrate embryos from capillaries. Anat Rec 3:498–518
Farner DS (1970) Some glimpses of comparative avian physiology. Fed Proc 29:1649–1663
Fedde MR (1980) The structure and gas flow pattern in the avian lung. Poult Sci 59:2642–2653
Fedde MR, Orr JA, Shams H, Scheid P (1989) Cardiopulmonary function in exercising barheaded geese during normoxia and hypoxia. Respir Physiol 77:239–262
Feder ME (1976) Lungless, body size, and metabolic rate in salamanders. Physiol Zool 49:398–418
Fedi P, Bafico A, Nieto-Soria A, Burgess WH, Miki T, Bottaro DP, Kraus MH, Aaronson SA (1999) Isolation and biochemical characterization of the human Dkk-1 homologue, a novel inhibitor of mammalian Wnt signaling. J Biol Chem 274:19465–19472
Fenton MB, Bringham RM, Mills AM, Rautenbach IL (1985) The roosting and foraging areas of *Epomophorus wahlbergi* (Pteropodida) and *Scotophilus viridis* (Vespertilionidae) in Kruger National Park, South Africa. J Mammal 66:461–468
Ferrara N (1999) Molecular and biological properties of vascular endothelial growth factor. J Mol Med 77:527–543
Ferrara N (2000) The role of vascular endothelial growth factor in angiogenesis. In: Ware JA, Simons M (eds) Angiogenesis in health and disease: basic mechanisms and clinical applications. Marcel Dekker, New York, pp 47–73
Ferrara N, Davis-Smyth T (1997) The biology of vascular endothelial growth factor. Endocr Rev 18:4–25
Ferrara N, Gerber HP (1999) The vascular endothelial growth factor family. In: Ware JA, Simons M (eds) Angiogenesis and cardiovascular disease. Oxford University Press, New York, pp 101–127
Ferrara N, Houck K, Jakeman L, Leung DW (1992) Molecular and biological properties of the vascular endothelial growth factor family of proteins. Endocr Rev 13:18–32
Ferrara N, Carver-Moore K, Chen H, Dowd M, Lu L, O'Shea KS, Powell-Braxton L, et al. (1996) Heterozygous embryonic lethality induced by targeted inactivation of the VEGF gene. Nature 380:439–442
Ficken MD, Edwards JF, Lay JC (1986) Induction, collection, and partial characterization of induced respiratory macrophages of the turkey. Avian Dis 30:766–771
Fisher HI (1946) Adaptations and comparative anatomy of the locomotor apparatus of the New World Vultures. Am Midl Nat 35:545–727
Fisher HI (1955) Avian anatomy 1925–1950, and some suggested problems. In: Wolfson A (eds) Recent studies in avian biology. University of Illinois Press, Urbana, pp 57–104
Fisher J, Peterson RT (1988) World of birds. Crescent Books, New York
Fletcher OJ (1980) Pathology of the avian respiratory system. Poult Sci 59:2666–2679
Fong GH, Rossant J, Gertsenstein M, Breitman BL (1995) Role of the flt-1 receptor tyrosine kinase in regulating the assembly of vascular endothelium. Nature 376:66–70
Fons R, Sicart R (1976) Contribution à la connaissance du métabolisme énergétique chez deux Crocidurinae: *Suncus etruscus* (Savi 1822) et *Crocidura russula* (Herman 1780), Insectivora, Soricidae, Mammalia. Mammalia 40:229–311
Forster RE (1996) Transfer of gas by diffusion and chemical reaction in pulmonary capillaries. In: West JB (ed) Respiration physiology: peoples and ideas. Oxford University Press, New York, pp 49–74
Fraenkel G (1932) Der Atmungsmechanismus des Skorpions. Z Vergl Physiol 11:656–661
Franch-Marro X, Casanova J (2002) *Spalt*-induced specification of distinct dorsal and ventral domains is required for *Drosophila* tracheal patterning. Dev Biol. 250:374–382

French MJ (1988) Invention and evolution: design in nature and engineering. Cambridge University Press, Cambridge

Fu YM, Spirito P, Yu ZX, Biro S, Sasse J, Lei J, Ferrans VJ, Epstein SE, Casscells W (1991) Acidic fibroblast growth factor in the developing rat embryo. J Cell Biol 114:1261-1273

Fujiwara T, Adams FH, Nozaki M, Dermer GB (1970) Pulmonary surfactant phospholipids from turkey lung: comparison with rabbit lung. Am J Physiol 218:218-225

Gannon BJ, Randall DJ, Browning J, Lester RJG, Rogers LJ (1983) The microvascular organization of the gas exchange organs of the Australian lungfish, *Neoceratodus forsteri* (Krefft). Aust J Zool 31:651-673

Gebb SA, Shannon JM (2000) Tissue interactions mediate early events in pulmonary vasculogenesis. Dev Dyn 217:159-169

Gehr P, Sehovic S, Burri PH, Classen H, Weibel ER (1980) The lung of shrews: morphometric estimation of diffusion capacity. Respir Physiol 44:61-86

Gehr P, Mwangi DK, Amman A, Maloiy GMO, Taylor CR, Weibel ER (1981) Design of the mammalian respiratory system. V. Scaling morphometric diffusing capacity to body mass: wild and domestic animals. Respir Physiol 44:41-86

George JC, Shah RV (1956) Comparative morphology of the lung in snakes with remarks on the evolution of the lung in reptiles. J Anim Morphol Physiol 3:1-7

Gerritsen ME, Soriano R, Yang S, Zlot C, Ingle G, Toy K, Williams PM (2003) Branching out: a molecular fingerprint of endothelial differentiation into tube-like structures generated by affymetrix oligonucleotide arrays. Microcirculation 10:63-81

Gibe J (1970) L'appareil respiratoire. In: Grasse PP (ed) Traité de zoologie, tome XIV, fascicule III. Masson and Cie, Paris, pp 499-520

Gier HT (1952) The air sacs of the loon. Auk 69:40-49

Gloe T, Pohl U (2002) Laminin binding conveys mechanosensing in endothelial cells. News Physiol Sci 17:166-169

Goldin GV (1980) Towards a mechanism for morphogenesis in epithelio-mesenchymal organs. Q Rev Biol 55:215-265

Goldin GV, Opperman LA (1980) Induction of supernumerary tracheal buds and the stimulation of DNA synthesis in the embryonic chick lung and trachea by epidermal growth factor. J Embryol Exp Morphol 60:235-243

Gomi T (1982) Electron microscopic studies of the alveolar brush cell of the striped snake (*Elaphe quadrivirgata*). J Med Soc Toho Jpn 29:481-102

Goniakowska-Witalinska L (1986) Lung of the tree frog, *Hyla arborea*: a scanning and transmission electron microscope study. Anat Embryol 174:379-389

Goniakowska-Witalinska L (1995) The histology and ultrastructure of the amphibian lung. In: Pastor LM (ed) Histology, ultrastructure and immunohistochemistry of the respiratory organs in non-mammalian vertebrates. Publicaciones Universidad de Murcia, Murcia (Spain), pp 77-112

Gonzalez AM, Buscaglia M, Ong M, Baird A (1990) Distribution of fibroblast growth factor in the 18-day rat fetus: localization in the basement laminas of diverse tissues. J Cell Biol 110:753-765

Gonzalez AM, Hill DJ, Logan A, Maher PA, Baird A (1996) Distribution of fibroblast growth factor (FGF-2) and FGF receptor-1 messenger RNA expression and protein presence in the mid-trimester human fetus. Pediatr Res 39:375-385

Gonzalez-Crussi F (1971) Vasculogenesis in the chick embryo. An ultrastructural study. Am J Anat 130:441-460

Gory-Faure S, Prandini MH, Pointu H, Roullot V, Pignot-Paintrand I, Vernet M, Huber P (1999) Role of vascular endothelial-cadherin in vascular morphogenesis. Development 126:2093-2102

Gospodarowicz D (1991) Biological activities of fibroblast growth factors. In: Baird A, Klagsbrun M (eds) The fibroblast growth factor family. New York Academy of Science, New York, pp 1-8

Graham JB (1994) An evolutionary perspective of bimodal respiration: a biological sythesis of fish air breathing. Am Zool 34:229-237
Graham JB (1997) Air breathing fishes: evolution, diversity and adaptation. Academic Press, San Diego
Graham JB, Gee JH, Robinson FS (1975) Hydrostatic and gas exchange functions of the lung of the sea snake, *Pelamis platurus*. Comp Biochem Physiol 50:477-482
Grant MM, Brain JD, Vinegar A (1981) Pulmonary defense mechanisms in *Boa constrictor*. J Appl Physiol 50:979-983
Grigg GC (1965) Studies on the Queensland lungfish, *Neoceratodus forsteri* (Krefft). Aust J Zool 13:243-257
Grindley JC, Bellusci S, Perkins D, Hogan BLM (1997) Evidence for the involvement of the Gli gene family in embryonic mouse lung development. Dev Biol 188:337-348
Groebbels F (1932). Die Vögel. Bau, Funktion, Lebenserscheinung, Einpassung, vol 1. Borntraeger, Berlin
Grubb BR (1982) Cardiac output and stroke volume in exercising ducks and pigeons. J Appl Physiol 53:203-211
Gruson ES (1976) Checklist of birds of the world. William Collins, London
Guimond RW, Hutchison VH (1973) Trimodal gas exchange in the large aquatic salamander, *Siren lacertina*. Comp Biochem Physiol 46A:249-268
Guimond RW, Hutchison VH (1976) Gas exchange of the giant salamanders of North America. In: Hughes GM (ed) Respiration of amphibious vertebrates. Academic Press, New York, pp 313-338
Gutman WF, Bonik K (1981) Kritische Evolutionstheorie. Gerstenberg, Hildesheim
Gyles NR (1989) Poultry, people and progress. Poult Sci 68:1-8
Hackett BP, Binge CD, Gitlin JD (1996) Mechanisms of gene expression and cell fate determination in the developing pulmonary epithelium. Annu Rev Physiol 58:51-71
Hacohen N, Kramer S, Sutherland D, Hiromi Y, Krasnow M (1998) *Sprouty* encodes a novel antagonist of FGF signaling that patterns apical branching of the *Drosophila* airways. Cell 92:253-263
Hahn H (1909) Experimental studies on the development of the blood and the first blood vessels in the chick. Arch Entwicklungsmechanik Organ 4:140-143
Hainsworth FW (1981) Locomotion. In: F.W. Hainsworth (ed) Animal physiology: adaptation in function. Addison-Wesley, Reading (MA), pp 259-292
Haldane JS, Priestley JG (1935) Respiration. Oxford University Press, London
Hall SM, Hislop AA, Harworth SG (2002) Origin, differentiation, and maturation of human pulmonary veins. Am J Respir Cell Mol Biol 26:333-340
Hamburger V, Hamilton HL (1951) A series of normal stages in the development of the chick embryo. J Morphol 88:49-92
Han RNN, Liu J, Tanswell AK, Post M (1992) Expression of basic fibroblast growth factor and receptor: immunolocalization studies in developing rat fetal lung. Pediatr Res. 31:435-440
Hanahan D (1997) Signaling vascular morphogenesis and maintenance. Science 277:48-50
Hansen HJ (1893) Organs and characters in different orders of Arachnida. Entomol Medd 4:135-144
Hazelhoff EH (1951) Structure and function of the lung of birds. Poult Sci 30:3-10
Healy AM, Morgenthau L, Zhu X, Farber HW, Cardoso WV (2000) VEGF is deposited in the subepithelial matrix at the leading edge of branching airways and stimulates neovascularization in murine embryonic lung. Dev Dyn 219:341-352
Heatwole H (1981) Role of the saccular lung in the diving of the sea krait, *Laticuda colubrina* (Serpentes: Laticaudidae). Aust J Herpetol 1:11-16
Henk WG, Haldiman JT (1990) Microanatomy of the lung of the bowhead whale *Balaena mysticetus*. Anat Rec 226:187-197

Hilfer SR (1996) Morphogenesis of the lung: control of embryonic and fetal branching. Annu Rev Physiol 58:93–113

His W (1900) Lecithoblast und Angioblast der Wirbeltiere. Abh Math-Phys Gen Wiss 26:171–328

Hislop AA (2002) Airway and blood vessel interaction during lung development. J Anat 201:325–333

Hirakow R, Hiruma T (1981) Scanning electron microscopic study on the development of primitive blood vessels in chick embryos at the early somite stage. Anat Embryol 163:299–306

Hlastala MP, Standaert TA, Pierson DJ, Luchtel DL (1985) The matching of ventilation and perfusion in the lung of the tegu, *Tupinambis nigropunctus*. Respir Physiol 60:277–294

Hochachka PW (1973) Comparative intermediary metabolism. In: Prosser CL (ed) Comparative animal physiology, 3rd edn. Saunders, Philadelphia, pp 212–278

Hof IM, Klauser M, Gehr P (1990) Phagocytic properties and organelle motility of pulmonary macrophages from smokers and nonsmokers estimated in vitro by magnetometric means. Eur Respir J 3:157–162

Hogan BLM (1999) Morphogenesis. Cell 96:225–233

Hogan BLM, Grindley J, Bellusci S, Dunn NR, Emoto H, Itoh N (1997) Branching morphogenesis of the lung: new models for classical problems. Cold Spring Harbour Symp Q Biol 62:249–256

Holle JP, Meyer M, Scheid P (1977) Oxygen affinity of duck blood determined by in vivo and in vitro techniques. Respir Physiol 29:355–361

Holle JP, Heisler N, Scheid P (1978) Blood flow distribution in the bird lung and its controls by respiratory gases. Am J Physiol 234:R146–R154

Holt PG (1979) Alveolar macrophage. I. A simple technique for the preparation of high numbers of viable alveolar macrophages from small laboratory animals. J Immunol Methods 27:189–198.

Hsieh JC, Kodjabachian L, Rebbert ML, Rattner A, Smallwood PM, Samos CH, Nusse R, Dawid IB, Nathans J (1999) A new secreted protein that binds to Wnt proteins and inhibits their activities. Nature 398:431–436

Huchzermeyer FW, de Ruyk AMC (1986) Pulmonary hypertension syndrome associated with ascites in broilers. Vet Rec 119:94

Hudlicka O, Brown MD (2000) Modulators of angiogenesis: hemodynamic factors. In: Ware JA, Simons M (eds) Angiogenesis and cardiovascular disease. Marcel Dekker, New York, pp 215–244

Hudlicka O, Tyler KR (1986) Angiogenesis. The growth of the vascular system. Academic Press, London

Hughes GM, Morgan M (1973) The structure of fish gills in relation to their respiratory function. Biol Rev 48:419–475

Hughes GM, Weibel ER (1976) Morphometry of fish lungs. In: Hughes GM (ed) Respiration of amphibious vertebrates. Academic Press, London, pp 213–232

Hui CC, Slusarski D, Platt KA, Holmgren R, Joyner AL (1994) Expression of three mouse homologs of the *Drosophila* segment polarity gene cubitus interruptus, Gli, Gli-2, and Gli-3, in ectoderm- and mesoderm-derived tissues suggests multiple roles during postimplantation development. Dev Biol 162:402–413

Huxley TH (1882) On the respiratory organs of *Apteryx*. Proc Zool Soc Lond 1882:560–569

Hyatt BA, Shangguan X, Shannon JM (2002) BMP-4 modulates fibroblast growth factor-mediated induction of proximal and distal lung differentiation in mouse embryonic tracheal epithelium in mesenchyme free culture. Dev Dyn 225:153–165

Ishikawa T, Tamai Y, Zorn AM, Yoshida H, Seldin MF, Nishikawa S, Taketo MM (2001) Mouse Wnt receptor gene fzd5 is essential for yolk sac and placental angiogenesis. Development 128:25–33

Ito T, Udaka N, Yazawa T, Okudela K, Hayashi H, Sudo T, Guillemot F, Kageyama R, Kitamura H (2000) Basic helix-loop-helix transcription factors regulate the neuroendocrine differentiation of fetal mouse epithelium. Development 127:3913-3921

James AE, Hutchins G, Bush M, Natarajan TK, Burns B (1976) How birds breathe: correlation of radiographic and anatomical and pathological studies. J Vet Radiol Soc 17:77-86

Jameson W (1958) The wandering albatross. Hart-Davis, London

Jankow RP, Luo P, Campbell A, Belcastro R, Cabacungan J, Johnstone L et al. (2003) Fibroblast growth factor receptor-1 and neonatal compensatory lung growth after exposure to 95% oxygen. Am J Respir Crit Care Med 167:1554-1561

Jarecki J, Johnson E, Krasnow MA (1999) Oxygen regulation of airway branching in *Drosophila* is mediated by branchless FGF. Cell 99:211-220

Jensen FB, Weber RE (1985) Kinetics of the acclimational responses of tench to combined hypoxia and hypercapnia. I. Respiratory responses. J Comp Physiol B 156:197-203

Johnston DW, McFarlane RW (1967) Migration and bioenergetics of flight in the Pacific golden plover. Condor 69:156-168

Jones AW, Radnor CJP (1972a) The development of the chick tertiary bronchus. I. General development and the mode of production of the osmiophilic inclusion body. J Anat 113:303-324

Jones AW, Radnor CJP (1972b) The development of the chick tertiary bronchus. II. The origin of the surface lining system. J Anat 113, 325-340

Jones G, Rayner JMV (1989) Optimal flight speed of pipistrelle bats (*Pipistrellus pipistrellus*). In: Hanak V, Horaceck I, Gaisler J (eds) European bat research 1987: Proc 4th Eur Bat Research Symposium. Charles University Press, Praha, pp 87-103

Jones JD (1972) Comparative physiology of respiration. Edward Arnold, London

Jones JH (1982) Pulmonary blood flow distribution in panting ostriches. J Appl Physiol 53:1411-1417

Jones JH (1998) Symmorphosis and the mammalian respiratory system: what is optimal design and does it exist? In: Weibel ER, Taylor CR, Bolis L (eds) Principles of animal design: the optimization and symmorphosis debate. Cambridge University Press, Cambridge, pp 241-248.

Jones JH, Effmann EL, Schmidt-Nielsen K (1985) Lung volume changes during respiration in ducks. Respir Physiol 59:15-25

Juillet A (1912) Recherches anatomiques, embryologiques, histologiques et comparatives sur le poumon des oiseaux. Arch Zool Exp Gen IX:207-371

Julian RJ (1987) The effects of increased sodium in the drinking water on right ventricular failure and ascites in broiler chickens. Avian Pathol 16:61-71

Julian RJ, Wilson JB (1986) Right ventricular failure as a cause of ascites in broiler and rooster chickens. Proc 4th Intern SympVe Lab Diag 1986: 608-611

Julian RJ, Moran ET, Revintonn W, Hunter DB (1984) Acute hypertensive angiopathy as a cause of sudden death in turkeys. Proc Am Vet Med Assn 117:47-52

Jürgens JD, Bartels H, Bartels R (1981) Blood oxygen transport and organ weight of small bats and small nonflying mammals. Respir Physiol 45:243-260

Kallianpur AR, Jordan JE, Brandt SJ (1994) The SLC/TAL-1 gene is expressed in progenitors of both the hematopoietic and vascular systems during embryogenesis. Blood 83:1200-1208

Kanwisher JW (1966) Tracheal gas dynamics in pupae of the *Cecropia* silkworm. Biol Bull 130:96-105

Kaplan F (2000) Molecular determinants of fetal lung organogenesis. Mol Genet Metab 71:321-341

Karasov WH, Phan D, Diamond JM, Carpenter FL (1986) Food passage and intestinal nutrient absorption in hummingbirds. Auk 103:453-464

Kardong KV (1972) Morphology of the respiratory system and its musculature in different snake genera. I). *Crotalus* and *Elaphe*. Gegenbaurs Morphol Jahrb 117:285-302

Kaufman SL, Burri PH, Weibel ER (1974) The postnatal growth of the rat lung. II. Autoradiography. Anat Rec 180:63-76

Kessler DS, Melton DA (1994) Vertebrate embryonic induction: mesodermal and neural patterning. Science 266:595-604

Kimura A, Gomi T, Kikuchi Y, Hashimoto T (1987) Anatomical studies of the lung of air breathing fish. I. Gross anatomical and light microscopic observations of the lungs of the African lungfish *Protopterus aethiopicus*. J Med Soc (Toho) Jpn 34:1-18

King AS (1966) Structural and functional aspects of the avian lung and its air sacs. Intern Rev Gen Exp Zool 2:171-267

King AS (1975) Aves respiratory system. In: Getty R (ed) Sisson and Grossman's The anatomy of the domestic animals, 5th edn, vol 2. Saunders, Philadelphia, pp 1011-1075

King AS (1979) Systema respiratorium. In: Baumel JJ, King AS, Lucas AM, Breazile JE, Evans HE (eds) Nomina anatomica avium. Academic Press, London, pp 227-265

King AS (1989) Functional anatomy of the syrinx. In: King AS, McLelland J (eds) Form and function in birds, vol 4. Academic Press, London, pp 105-192

King AS, Atherton JD (1970) The identity of the air sacs of the turkey (*Melleagris gallopavo*). Acta Anat 77:78-91

King AS, King DZ (1979) Avian morphology: general principles. In: King AS, McLelland J (eds) Form and function in birds, vol I. Academic Press, London, pp 1-38

King AS, McLelland J (eds) (1989) Form and function in birds, vol 4. Academic Press, London

King AS, Molony V (1971) The anatomy of respiration. In: Bell DF, Freeman BM (eds) Physiology and biochemistry of the domestic fowl, vol 1. Academic Press, London, pp 347-384

King JR (1974) Seasonal allocation of time and energy resources in birds. In: Paynter RA (ed) Avian energetics. Nuttal Ornithological Club, Cambridge (MA), pp 4-85

Klagsbrun M (1989) The fibroblast growth factor family: structural and biological properties. Prog Growth Factor Res 1:207-235

Klemm RD, Gatz RN, Westfall JA, Fedde MR (1979) Microanatomy of the lung parenchyma of a tegu lizard *Tupinambis nigropunctatus*. J Morphol 161:257-280

Klika E, Lelek A (1967) A contribution to the study of the lungs of *Protopterus annectens* and *Polypterus segegalensis*. Folia Morphol 15:168-175

Klika E, Scheuermann DW, de Groodt-Lasseel MHA, Bazantova I, Switka A (1996)Pulmonary macrophages in birds (barn owl, *Tyto tyto alba*), domestic fowl (*Gallus gallus domestica*), quail (*Coturnix coturnix*), and pigeons (*Columba livia*). Anat Rec 256:87-97

Kodama H, Sato G, Mikami T (1976) Age dependent resistance of chickens to Salmonella in vitro: phagocytic and bactericidal activities of splenic phagocytes. Am J Vet Res 37:1091-1094.

Konarzewski M, Gavin A, McDevitt TA, Wallis IR (2000) Metabolic and organ mass responses to selection for high growth rates in the domestic chicken (*Gallus domesticus*). Physiol Biochem Zool 73:237-248

Krebs JR, Harvey PH (1986) Busy doing nothing - efficiently. Nature 320:18-19

Krefft G (1870) Description of a giant amphibian allied to the genus *Lepidosiren* from the Wide Bay District, Queensland. Proc Zool Soc Lond 221-224

Krogh A (1913) On the composition of the air in the tracheal system of some insects. Skand Arch Physiol 29:29-36

Krogh A (1920a) Studien über Tracheen Respiration. II. Ueber Gasdiffusion in der Tracheen. Pflugers Arch 179:95-112

Krogh A (1920b) Studien über Tracheen Respiration. III. Die Kombination von mechanischer Ventilation mit Gasdiffusion nach Versuchen an Dytiscuslarven. Pfl Arch Ges Physiol Menschen Tiere 179:113-120

Krogh A (1941) The comparative physiology of respiratory mechanisms. University of Pennsylvania Press, Philadelphia
Kuethe DO (1988) Fluid mechanical valving of airflow in bird lungs. J Exp Biol 136:1–12
Lacaud C, Robertson S, Pallis J, Kennedy M, Keller G (2001) Regulation of hemangioblast development. Ann NY Acad Sci 938:96–107
Larrivée B, Karsan A (2000) Signaling pathways induced by vascular endothelial growth factor. Int J Mol Med 5:447–545
LaRue AC, Mirov VA, Argraves WS, Czirók A, Flemming PA, Drake CJ (2003) Patterning of embryonic blood vessels. Dev Dyn 228:21–29
Lasiewski RC (1962) The energetics of migrating hummingbirds. Condor 64:324
Lasiewski RC, Dawson WR (1967) A re-examination of the relation between standard metabolic rate and body weight in birds. Condor 69:13–23
Lasser A (1983) A mononuclear phagocytic system. A review. Hum Pathol 14:108–126
Laybourne RC (1974) Collision between a vulture and an aircraft at an altitude of 37,000 ft. Wilson Bull 86:461–462
Le ACN, Musil L (2001) FGF signaling in chick lens development. Development 233:394–411
Lebeche M, Marpel S, Cardoso WV (1999) Fibroblast growth factor interactions in the developing lung. Mech Dev 86:125–136
LeCouter J, Lin R, Ferrara N (2002) Endocrine gland-derived VEGF and the emerging hypothesis of organ-specific regulation of angiogenesis. Nat Med 8:913–917
Levy AP, Levy NS, Goldberg MA (1996) Post-transcriptional regulation of vascular endothelial growth factor by hypoxia. J Biol Chem 271:2746–2753
Li J, Hampton T, Morgan JP, Simons M (1997) Stretch-induced VEGF expression in the heart. J Clin Invest 100:18–24
Lillie FR (1918) The development of the chick, 2nd edn. Holt, New York
Lippmann M, Schlesinger RB (1984) Interspecies comparison of particle deposition and mucociliary clearance in tracheobronchial airways. J Toxicol Environ Health 13:441–469
Litingtung Y, Lei L, Westphal H, Chiang C (1998) Sonic hedgehog is essential to foregut development. Nat Genet 20:58–61
Little C (1990) The terrestrial invasion. Cambridge University Press, Cambridge
Liu Z, Xu J, Colvin JS, Ornitz DM (2002) Coordination of chondrogenesis and osteogenesis by fibroblast growth factor-18. Genes Dev 16:859–869
Locke MJ (1958a) The structure of insect tracheae. Q J Microsc Sci 98:487–492
Locke MJ (1958b) Coordination of growth in the tracheal system of insects. Q J Microsc Sci 99:373–391
Locke MJ (1958c) The formation of trachea and tracheoles in *Rhodinius prolixus*. Q J Microsc Sci 99:29–46
Lockley RM (1970) The most aerial bird in the world. Animals 13:4–7
Locy WA, Larsell O (1916a) The embryology of the bird's lung based on observations of the bronchial tree. Part I. Am J Anat 19:447–504
Locy WA, Larsell O (1916b) The embryology of the bird's lung based on observations of the domestic fowl. Part II. Am J Anat 20:1–44
López J (1995) Anatomy and histology of the lung and air sacs of birds. In: LM Pastor (ed) Histology, ultrastructure, and immunohistochemistry of the respiratory organs in non-mammalian vertebrates. Publicaciones de la Universitatd de University of Murcia, Murcia (Spain), pp 179–233
Lorz C, López J (1997) Incidence of air pollution in the pulmonary surfactant system of the pigeon (*Columba livia*). Anat Rec 249:206–212
Loudon C (1989) Tracheal hypertrophy in mealworms: design and plasticity in oxygen supply systems. J Exp Biol 147:217–235
Luchtel DL, Kardong KV (1981) Ultrastructure of the lung of the rattlesnake, *Crotalus viridis oreganus*. J Morphol 169:29–47

Lutz PL, Longmuir IS, Tuttle JV, Schmidt-Nielsen K (1973) Dissociation curve of bird blood and effect of red cell oxygen consumption. Respir Physiol 17:269-275

Macklem P, Bouverot P, Scheid P (1979) Measurement of the distensibility of the parabronchi in duck lungs. Respir Physiol 33:23-35

Madri JA, Pratt BM, Tucker AM (1988) Phenotypic modulation of endothelial cells by transforming growth factor-β depends on the composition and organization of the extracellular matrix. J Cell Biol 106:1375-1384

Magnussen H, Willmer H, Scheid P (1976) Gas exchange in the air sacs: contribution to respiratory gas exchange in ducks. Respir Physiol 26:129-146

Maina JN (1982a) A scanning electron microscopic study of the air and blood capillaries of the lung of the domestic fowl (*Gallus domesticus*). Experientia 35:614-616

Maina JN (1982b) Stereological analysis of the paleopulmo and neopulmo respiratory regions of the avian lung (*Streptopelia decaocto*). IRCS Med Sci 10:328

Maina JN (1984) Morphometrics of the avian lung. 3. The structural design of the passerine lung. Respir Physiol 55:291-309

Maina JN (1985) A scanning and transmission electron microscopic study of the bat lung. J Zool Lond 205B:19-27

Maina JN (1986) The structural design of the bat lung. Myotis 23:71-77

Maina JN (1987) The morphology of the lung of the African lungfish, *Protopterus aethiopicus*: a scanning electron microscopic study. Cell Tissue Res 250:191-196

Maina JN (1988) Scanning electron microscopic study of the spatial organization of the air- and blood-conducting components of the avian lung (*Gallus gallus domesticus*). Anat Rec 222:145-153

Maina JN (1989a) The morphology of the lung of the black mamba *Dendroaspis polylepis* (Reptilia: Ophidia: Elapidae): a scanning and transmission electron microscopic study. J Anat 167:31-46

Maina JN (1989b) The morphometry of the avian lung. In: King AS, McLelland J (eds) Form and function in birds, vol 4. Academic Press, London, pp 307-368

Maina JN (1989c) The morphology of the lung of the East African tree frog *Chiromantis petersi* with observations on the skin and the buccal cavity as secondary gas exchange organs: A TEM and SEM study. J Anat 165:29-43

Maina JN (1989d) A scanning and transmission electron microscopic study of the tracheal air-sac system in a grasshopper (*Chrotogonus senegalensis*, Kraus) - (Orthoptera: Acrididae: Pygomorphinae). Anat Rec 223:393-405

Maina JN (1994) Comparative respiratory morphology and morphometry: the functional design of the respiratory systems. In: Gilles R (ed) Advances in comparative and environmental physiology. Springer, Berlin Heidelberg New York, pp 111-232

Maina JN (1996) Principles of the structure and function of birds. In: Rosskopff E, Woepel P (eds) Petraks diseases of cage and aviary birds, vol 2. Lea and Febiger, New York, pp 167-256

Maina JN (1997) The lungs of the volant vertebrates - birds and bats: how are they relatively structurally optimized for this elite mode of locomotion. In: Weibel ER, Taylor CR, Bolis L (eds) Diversity in biological design: symmorphosis - fact or fancy? Cambridge University Press, New York, pp 177-185

Maina JN (1998) The gas exchangers: structure, function, and evolution of the respiratory processes. Springer, Berlin Heidelberg New York

Maina JN (2000a) Comparative respiratory morphology: themes and principles in the design and construction of the gas exchangers. Anat Rec 261:25-44

Maina JN (2000b) What it takes to fly: the novel respiratory structural and functional adaptations in birds and bats. J Exp Biol 203:3045-3064

Maina JN (2000c) Is the sheet-flow design a 'frozen core' (a Bauplan) of the gas exchangers? Comparative functional morphology of the respiratory microvascular systems: illustra-

tion of the geometry and rationalization of the fractal properties. Comp Biochem Physiol 126A:491–515
Maina JN (2002a) Functional morphology of the vertebrate respiratory systems. Science Publishers Inc, Enfield (NH)
Maina JN (2002b) Fundamental structural aspects in the bioengineering of the gas exchangers: comparative perspectives. Springer, Berlin Heidelberg New York
Maina JN (2002 c) Some recent advances of the study and understanding of the functional design of the avian lung: morphological and morphometric perspectives. Biol Rev 77:97–152
Maina JN (2003a). A systematic study of the development of the airway (bronchial) system of the avian lung from days 3 to 26 of embryogenesis: a transmission electron microscopic study on the domestic fowl, *Gallus gallus* variant *domesticus*. Tissue Cell 35:375–391
Maina JN (2003b) Developmental dynamics of the bronchial (airway) and air sac systems of the avian respiratory system from days 3 to 26 of life: a scanning electron microscopic study of the domestic fowl, *Gallus gallus* variant *domesticus*. Anat Embryol 207:119–134
Maina JN (2004a) A systematic study of hematopoiesis, vasculogenesis, and angiogenesis in the developing avian lung, *Gallus gallus* variant *domesticus*. Tissue Cell 36:307–322
Maina JN (2004b) Morphogenesis of the laminated tripartite cytoarchitectural design of the blood-gas barrier of the avian lung: a systematic electron microscopic study of the domestic fowl, *Gallus gallus* variant *domesticus*. Tissue Cell 36:129–139
Maina JN (2004 c) Structure and function of non mammalian vertebrate lung. In: Massaro DJ, Massaro GC, Chambon P (eds) Lung development and regeneration. Marcel Dekker, New York, pp 319–354
Maina JN, Africa M (2000) Inspiratory aerodynamic valving in the avian lung: functional morphological study of the extrapulmonary primary bronchus. J Exp Biol 203:2865–2876
Maina JN, Cowley HM (1998) Ultrastructural characterization of the pulmonary cellular defenses in the lung of a bird, the rock dove, *Columba livia*. Proc R Soc Lond B 265:1567–1572
Maina JN, King AS (1982a) The thickness of the avian blood-gas barrier: qualitative and quantitative observations. J Anat 134:553–562
Maina JN, King AS (1982b) Morphometrics of the avian lung. 2. The wild mallard (*Anas platyrhynchos*) and greylag goose (*Anser anser*). Respir Physiol 50:299–313
Maina JN, King AS (1984) The structural functional correlation in the design of the bat lung. A morphometric study. J Exp Biol 111:43–63
Maina JN, King AS (1987). A morphometric study of the lung of a humboldt penguin (*Spheniscus humboldti*). Zentralbl Vet Med C Anat Histo Embryol 16:293–297
Maina JN, King AS (1989) The lung of the emu, *Dromaius novaehollandiae*: a microscopic and morphometric study. J Anat 163:67–74
Maina JN, Madan AK (2003) Occurrence and distribution of fibroblast growth factor-2 (FGF-2) in the early development of the avian lung. FASEB J 17:779, 468.10A
Maina JN, Maloiy GMO (1985) The morphometry of the lung of the lungfish (*Protopterus aethiopicus*): its structural-functional correlations. Proc R Soc Lond B 244:399–420
Maina JN, Maloiy GMO (1988) A scanning and transmission electron microscopic study of the lung of a caecilian *Boulengerula taitanus*. J Zool Lond 215:739–751
Maina JN, Nathaniel C (2001) A qualitative and quantitative study of the lung of an ostrich, *Struthio camelus*. J Exp Biol 204:2313–2330
Maina JN, van Gils P (2001) Morphometric characterization of the airway and vascular systems of the lung of the domestic pig, *Sus scrofa*: comparison of the airway, arterial, and venous systems. Comp Biochem Physiol 130A:781–798

Maina JN, West JB (2005) Thin and strong! The bioengineering dilemma in the structural and functional design of the blood-gas barrier: comparative and evolutionary perspectives. Physiol Rev (in press)

Maina JN, Abdalla MA, King AS (1982a) Light microscopic morphometry of the lungs of 19 avian species. Acta Anat 112:264-270

Maina JN, King AS, King DZ (1982b) A morphometric analysis of the lung of a species of bat. Respir Physiol 50:1-11

Maina JN, Howard CV, Scales L (1983) Length densities and maximum diameter distribution of the air capillaries of the paleopulmo and neopulmo region of the avian lung. Acta Stereol 2:101-107

Maina JN, King AS, Settle G (1989a) An allometric study of the pulmonary morphometric parameters in birds, with mammalian comparison. Philos Trans R Soc Lond 326B:1-57

Maina JN, Maloiy GMO, Warui CN, Njogu EK, Kokwaro ED (1989b) A scanning electron microscopic study of the morphology of the reptilian lung: The savanna monitor lizard (*Varanus exanthematicus*) and the pancake tortoise (*Malacochersus tornieri*). Anat Rec 224:514-522

Maina JN, Thomas SP, Dallas DM (1991) A morphometric study of bats of different size: correlations between structure and function of the chiropteran lung. Philos Trans R Soc Lond B 333:31-50

Maina JN, Maloiy GMO, Makanya AN (1992) The morphology and morphometry of the lungs of two East African fossorial rodents: the mole rats *Tachyoryctes splendens* and *Heterocephalus glaber* (Rodentia: Rhizomyidae: Bathyergidae). Zoomorphology 112:167-179

Maina JN, Veltcamp CJ, Henry J (1999) A study of the spatial organization of the gas-exchange components of a snake lung – the sandboa *Eryx colubrinus* (Reptilia: Ophidia; Corubridae) by latex casting. J Zool Lond 247:81-90

Maina JN, Madan AK, Alison B (2003) Expression of fibroblast growth factor-2 (FGF-2) in early stages (days 3-11) of the development of the avian lung, *Gallus gallus* variant *domesticus*. J Anat 203:505-512

Makanya AN, Sparrow MP, Warui CN, Mwangi DK, Burri PH (2001) Morphological analysis of the postnatally developing marsupial lung: the quokka wallaby. Anat Rec 262:253-265

Maniscalco WM, Watkins RH, Finkelstein JN, Campbell MH (1995) Vascular endothelial growth factor mRNA increases in alveolar epithelial cells during recovery from oxygen injury. Am J Respir Cell Mol Biol 13:377-386

Maniscalco WM, Watkins RH, D'Angio C, Ryan R (1997) Hyperoxic injury decreases alveolar epithelial cell expression of vascular endothelial growth factor (VEGF) in neonatal rabbit lung. Am J Respir Cell Mol Biol 16:557-567

Marden JH (1994) From damselflies to pterosaurs: how burst and sustainable flight performance scale with size. Am J Physiol 266:R1077-R1984

Marcus H (1937) Lungen. In: Bolk L, Goppert E, Kallius E, Lubosch W (eds) Handbuch der vergleichenden Anatomie der Wirbeltiere, III. Urban and Schwarzenberg, Berlin-Wien, 1937, pp 909-1018

Marshall AJ (1962) In: Parker JT, Haswell PW (eds) A textbook of zoology, vol 2. Macmillan, London

Marshall CR (1986a) Lungfish: phylogeny and parsimony. J Morphol 1:151-162

Marshall CR (1986b) A list of fossil and extant Dipnoans. J Morphol 1:15-23

Martin BP (1987) World birds. Guiness Superlatives Ltd, London

Martin GR, Timpl R, Kuhn K (1988) Basement lamina proteins: molecular structure and function. Adv Protein Chem 39:1-50

Martin KM, Hutchison VH (1979) Ventilatory activity in *Amphiuma tridactylum* and *Siren lacertina* (Amphibia, Caudata). J Herpatol 13:427-434

Martinez del Rio C (1990) Dietary, phylogenetic and ecological correlates of intestinal sucrase and maltase activity in birds. Physiol Zool 63:987–1011

Mason DK, Collins AE, Watkins KL (1983) Exercise induced pulmonary hemorrhage in horses. In: Snow DH, Persson SGB, Rose RJ (eds) Equine exercise physiology.Cambridge University Press, Cambridge, pp 57–63

Mathieu-Costello O, Szewczak JM, Logermann RB, Agey PJ (1992) Geometry of blood-tissue exchange in bat flight muscle compared with bat hindlimb and rat soleus muscle. Am J Physiol 262:R955–R965

Matsumura H, Setoguti T (1984) Electron microscopic studies of the lung of the salamander, *Hynobius nebulosus*. I. A scanning and transmission microscopic observation. Okajimas Folia Anat Jpn 61:15–25

Maxwell MH, Robertson GW, Spence S (1986a) Studies on an ascetic syndrome in young broilers. I. Hematology and pathology. Avian Pathol 15:511–524

Maxwell MH, Robertson GW, Spence S (1986b) Studies on an ascitic syndrome in young broilers. II. Ultrastructure. Avian Pathol 15:525–538

May RM (1992) How many species inhabit Earth? Sci Am Oct:18–24

McClanahan LL, Rodolfo R, Shoemaker VH (1994) Frogs and toads in deserts. Sci Am 273:82–88

Meban C (1978) Functional anatomy of the lungs of the green turtle, *Lacerta viridis*. J Anat 125:421–436

Meban C (1980) Thicknesses of the air-blood barriers in vertebrate lungs. J Anat 131:299–307

Mensah GA, Brain JD (1982) Deposition and clearance of inhaled aerosol in the respiratory tract of chickens. J Appl Physiol 53:1423–1428

Metzger RJ, Krasnow MA (1999) Genetic control of branching morphogenesis. Science 284:1635–1639

Meyer MR, Burger RE, Scheid P, Piiper J (1977) Analyses des échanges gazeux dans les poumons d'oiseaux. J Physiol Paris 73:9A

Miller AM, McWhorter JE (1914) Experiments on the development of blood vessels in the area pellucida and embryonic body of the chick. Anat Rec 8:203–227

Miller DN, Bondurant S (1961) Surface characteristics of vertebrate lung extracts. J Appl Physiol 16:1075–1077

Miller JR, Hocking AM, Brown JD, Moon RT (1999) Mechanism and function of signal transudation by the Wnt/β-catenin and Wnt/Ca^{2+} pathways. Oncogene 18:7860–7872

Miller PL (1960) Respiration in the desert locust. III. Ventilation and spiracles during flight. J Exp Biol 37:264–278

Miller PL (1966) The supply of oxygen to the active flight muscles of some large beetles. J Exp Biol 45:285–304

Miller PL (1974) Respiration – aerial gas transport. In: Rockstein M (ed) The physiology of insects, 2nd edn. Academic Press, New York, pp 346–402

Miller SL, Orgel LE (1974) The origins of life on earth. Prentice-Hall, Englewood Cliffs

Min H, Danilenko DM, Scully SA, Bolon B, Ring BD, Tarpley JE, DeRose M, Simmonett WS (1998) FGF-10 is required for both limb and lung development and exhibits striking functional similarities to *Drosophila* branchless. Genes Dev 12:3156–3161

Minoo P, King RJ (1994) Epithelial-mesenchymal interactions in lung development. Annu Rev Physiol 56:13–45

Miquerol L, Gertsenstein M, Harpal K, Rossant J, Nagy A (1999) Multiple developmental roles of VEGF suggested by a lacZ-tagged allele. Dev Biol 212:307–322

Molony V, Graf W, Scheid P (1976) Effects of CO_2 on pulmonary flow resistance in the duck. Respir Physiol 26:333–349

Moon RT, Brown JD, Torres M (1997a) Wnts modulate cell fate and behaviour during vertebrate development. Trends Genet 14:452–162

Moon RT, Brown JD, Yang-Snyder JA, Miller JR (1997b) Structurally related receptors and antagonists compete for secreted Wnt ligands. Cell 88:725–728

Morony JJ, Bock WJ, Farrand J (1975) Reference list of the birds of the world. Department of Ornithology, American Museum of Natural History, New York

Motoyama J, Liu J, Mo R, Ding Q, Post M, Hui CC (1998) Essential function of Gli 2 and Gli 3 in the formation of lung, trachea and oesophagus. Nat Genet 20:54–57

Müller B (1908) The air sacs of the pigeon. Smithson Misc Colls 50:365–414

Munshi JSD, Hughes GM (1992) Air breathing fishes of India: their structure, function and life history. AA Balkema Uitgevers BV, Rotterdam

Muraoka RS, Bushdid PB, Brantley DM, Yull FE, Kerr LD (2000) Mesenchymal expression of nuclear factor-kappaβ inhibits epithelial growth and branching in the embryonic chick lung. Dev Biol 225:322–338

Murray PDF (1932) The development 'in vitro' of the blood of the early chick embryo. Proc R Soc Lond 11:497–521

Nagaraja KV, Emery DA, Jordan A, Sivanandan JA, Newman JA, Pomeroy BS (1984) Effect of ammonia on the quantitative clearance of *Escherichia coli* from the lungs, air sacs, and livers of turkeys aerosol vaccinated against *Echerichia coli*. Am J Vet Res 45:392–395

Neufeld G, Cohen T, Gengrinovitch S, Poltorak Z (1999) Vascular endothelial growth factor (VEGF) and its receptors. FASEB J 13:9–22

Newman PJ, Albeida SM (1992) Cellular and molecular aspects of PECAM-1. Nouv Rev Fr Hematol 34:S9-S13

Ng YS, Rohan R, Sunday ME, Demello DE, D'Amore PA (2001) Differential expression of VEGF isoforms in mouse during development and in the adult. Dev Dyn 220:112–121

Nganpiep L, Maina JN (2002) Composite cellular defense stratagem in the avian respiratory system: functional morphology of the free (surface) macrophages and specialized pulmonary epithelia. J Anat 200:499–516

Nguyen-Phu D, Yamaguchi K, Scheid P, Piiper J (1986) Kinetics of oxygen uptake and release by erythrocytes of the chicken. J Exp Biol 125:15–27

Nicosia RF, Villaschi S (1999) Autoregulation of angiogenesis by cells of the vascular wall. Int Rev Cytol 185:1–43

Noden DM (1989) Embryonic origins and assembly of blood vessels. Am Rev Respir Dis 140:1097–1103

Norberg UM (1990) Vertebrate flight: mechanics, physiology, morphology, ecology and evolution. Springer, Berlin Heidelberg New York

Nudds RL, Bryant DM (2000) The energy cost of short flights in birds. J Exp Biol 203:1561–1582

Nusse R, Varmus HE (1992) Wnt genes. Cell 69:1073–1087

Ohbayashi N, Shibayama M, Kurotaki Y, Imanishi M, Fujimori T, Itoh N, Takada S (2002) FGF-18 is required for normal cell proliferation and differentiation during osteogenesis and chondrogenesis. Genes Dev 16:870–879

Okubo T, Hogan BLM (2004) Hyperactive Wnt signaling changes the developmental potential of embryonic lung endoderm. J Biol 3:11–31

Opell BD (1987) The influence of web monitoring tactics of the tracheal systems of spiders in the family Uroboridae (Arachnida, Areneida). Zoomorphology 107:255–259

Ornitz DM, Itoh N (2001) Fibroblast growth factors. Genome Biol 2:REVIEWS 3005

Ostrom JH (1975) The origin of birds. Annu Rev Earth Planet Sci 3:55–77

Pardanaud L, Dieterlen-Lièvre F (1993) Emergence of endothelial and hemopoietic cells in the avian embryo. Anat Embryol 187:107–114

Pardanaud L, Altman C, Kitos C, Dieterlen-Lièvre F, Buck CA (1987) Vasculogenesis in the early quail blastodisc as studied with a monoclonal antibody recognizing endothelial cells. Development 100:339–349

Pardanaud L, Yasiine F, Dieterlen-Lièvre F (1989) Relationship between vasculogenesis, angiogenesis, and haemopoiesis during avian ontogeny. Development 105:473–485

Pardanaud L, Luton D, Prigent M, Bourcheix LM, Catala M, Dieterlen-Lièvre F (1996) Two distinct endothelial lineages in ontogeny, one of them related to hemopoiesis. Development 122:1363–1371

Park JE, Keller GA, Ferrara N (1993) The vascular endothelial growth factor (VEGF) isoforms: differential deposition in the subepithelial extracellular matrix and bioactivity of extracellular matrix bound VEGF. Mol Biol Cell 4:1317–1326

Park WY, Miranda B, Lebeche D, Hashimoto G, Cardoso WV (1998) FGF-10 is a chemotactic factor for distal epithelial buds during lung development. Dev Biol 201:125–134

Parry K, Yates MS (1979) Observations on the avian pulmonary and bronchial circulation using labeled microspheres. Respir Physiol 38:131–140

Pastor LM (1995) The histology of the reptilian lungs. In: Pastor LM (ed) Histology, ultrastructure and immunohistochemistry of respiratory organs in non-mammalian vertebrates. Publicaciones Universidad de Murcia, Murcia (Spain), pp 131–153.

Pastor LM, Calvo A (1995) The extrapulmonary airways in birds. In: Pastor LM (ed) Histology, ultrastructure and immunohistochemistry of the respiratory organs on nonmammalian vertebrates. Murcia University Press, Murcia (Spain), pp 159–173

Pastor LM, Ballesta J, Castells MT, Perez-Tomas R, Marin JA, Madrid JF (1989) A microscopic study of the lung of *Testudo graeca*. J Anat 162:19–33

Patt DI, Patt GR (1969) Comparative vertebrate histology. Harper and Row, New York

Pattle RE (1976) The lung surfactant in the evolutionary tree. In: Hughes GM (ed) Respiration of amphibious vertebrates. Academic Press, London, pp 233–255

Pattle RE (1978) Lung surfactant and lung lining in birds. In: Piiper J (ed) Respiratory function in birds, adult and embryonic. Springer, Berlin Heidelberg New York, pp 23–32

Paul GS (1991) The many myths, some old, some new, of dinosaurology. Mod Geol 16:69–99

Pèault B, Thiery JP, Le Douarin NM (1983) A surface marker for the hemopoietic and endothelial cell lineage in the quail species defined by a monoclonal antibody. Proc Natl Acad Sci USA 80:2976–2980

Pennial R, Spitznagel JK (1975) Chicken neutrophils: oxidative metabolism in phagocytic cells devoid of myeloperoxidase. Proc Natl Acad Sci USA 72:5012–5015

Pennycuick CJ (1975) Mechanics of flight. In: Farner DS, King JR (eds) Avian biology, vol 5. Academic Press, New York, pp 1–75

Pepicelli CV, Lewis P, McMahon A (1998) Sonic hedgehog regulates branching morphogenesis in the mammalian lung. Curr Biol 8:1083–1086

Pepper MS (1997) Manipulating angiogenesis: from basic science to the bedside. Arteriosclerosis Thromb Vasc Biol 17:605–619

Pérez-Aparicio FJ, Carretero A, Navarro M, Ruberte J (1996) Angiogenesis in the gonadal capillary network of the chick embryo. Scanning Microsc 10:859–874

Perl AK, Whitsett JA (1999) Molecular mechanisms controlling lung morphogenesis. Clin Genet 56:14–27

Perry SF (1978) Quantitative anatomy of the lungs of the red-eared turtle, *Pseudemys scripta elegans*. Respir Physiol 35:245–262

Perry SF (1981) Morphometric analysis of pulmonary structure: methods for evaluation of unicameral lungs. Microscopie 38:278–293

Perry SF (1983) Reptilian lungs: functional anatomy and evolution. Adv Anat Embryol Cell Biol 79:1–81

Perry SF (1988) Functional morphology of the lungs of the Nile crocodile *Crocodylus niloticus*: non-respiratory parameters. J Exp Biol 143:99–117

Perry SF (1989a) Mainstreams in the evolution of vertebrate respiratory structures. In: King AS, McLelland J (eds) Form and function in birds, vol V. Academic Press, London, pp 1–67

Perry SF (1989b) Structure and function of the reptilian respiratory system. In: Wood SC (ed) Comparative pulmonary physiology: current concepts. Marcel Dekker, New York, pp 193–236.

Perry SF (1992) Gas exchange strategies in reptiles and the origin of the avian lung. In: Wood SC, Weber RE, Hargens AR, Millard RW (eds) Physiological adaptations in vertebrates: respiration, circulation, and metabolism. Marcel Dekker, New York, pp 149–167

Perry SF, Duncker HR (1978) Lung architecture, volume and static mechanics in five species of lizards. Respir Physiol 34:61–81

Perry SF, Duncker HR (1980) Interrelationship of static mechanical factors and anatomical structure in lung ventilation. J Comp Physiol 138:321–334

Perry SF, Darian-Smith C, Alston D, Limpus CJ, Maloney JE (1989) Histological structure of the lungs of the loggerhead turtle, *Caretta caretta*, before and after hatching. Copeia 1989:1000–1010

Peters K, Werner S, Liao X, Wert S, Whitsett J, Williams L (1994) Targeted expression of a dominant negative FGF receptor blocks branching morphogenesis and epithelial differentiation of the mouse. EMBO J 13:3296–3301

Petrova TV, Makinen T, Alitalo K (1999) Signaling via vascular endothelial growth factor receptors. Exp Cell Res 253:117–130

Petschow D, Würdinger I, Baumann R, Duhn J, Braunitzer G, Bauer C (1977) Causes of high blood oxygen affinity of animals living at high altitude. J Appl Physiol 42:139–143

Pettingill OS (1969) Ornithology in laboratory and field. Burgess Publishing, Minneapolis

Pfeifer M, Polakis P (2000) Wnt signaling in oncogenesis and embryogenesis: a look outside the nucleus. Science 287:1606–1609

Piiper J (1978) Origin of carbon dioxide in caudal air sacs of birds. In: Piiper J (ed) Respiratory function in birds, adult and embryonic. Springer, Berlin Heidelberg New York, pp 221–248

Piiper J, Scheid P (1973) Gas exchange in the avian lung: model and experimental evidence. In: Bolis L, Schmidt-Nielsen K, Maddrell SHP (eds) Comparative physiology. Elsevier, Amsterdam, pp 161–185

Pizarro B, Salas A, Parades J (1970) Mal de altura en aves. Inst Vet Invest Trop Altura Cuarto Bol Extraordinaire 1970:147–151

Poole TJ, Coffin JD (1989) Vasculogenesis and angiogenensis: two distinct morphogenetic mechanisms establish embryonic vascular pattern. J Exp Zool 251:224–231

Poole TJ, Coffin JD (1991) Morphogenetic mechanisms in avian vascular development. In: Feinberg RN, Sherer GK, Auerbach R (eds) The development of the vascular system, vol 14. Karger, Basel, pp 25–36

Poole TJ, Finkelstein EB, Cox CM (2001) The role of FGF and VEGF in angioblast induction and migration during vascular development. Dev Dyn 200:1–17

Portier P (1933) Locomotion aérienne et respiration des lépidoptéres, un nouveau rôle physiologique des ailes et des écailles. Trav V Congr Intern Ent Paris 2:25–31

Pough FH, Heiser JB, McFarland WN (1989) Vertebrate life, 3rd edn. Macmillan, New York

Powell FL (1983) Respiration. In: Abs M (ed) Physiology and behaviours of the pigeon. Academic Press, New York, pp 73–95

Powell FL, Scheid P (1989) Physiology of gas exchange in the avian respiratory system. In: King AS, McLelland J (eds) Form and function of the avian lung, vol 4. Academic Press, London, pp 393–437

Power JHT, Doyle IR, Davidson K, Nicholas TE (1999) Ultrastructural and protein analysis of surfactant in the Australian lungfish *Neoceratodus forsteri*: evidence for conservation of composition for 300 million years. J Exp Biol 202:2543–2550

Powers DR, Naggy KA (1988) Field metabolic rate and food consumption by free-living Anna's hummingbirds (*Calypte anna*). Physiol Zool 61:500–506.

Radu C, Radu L (1971) Le dispositif vasculaire du poumon chez les oiseaux domestiques (coq, dindon, oie, canard). Revue Med Vet 122:1219–1226

Rawal UM (1976) Nerves in the avian air sacs. Pavo 14:57–60

Rayner JMV (1981) Flight adaptations in vertebrates. Symp Zool Soc Lond 48:137–172

Rayner JMV (1985) Bounding and undulating flight in birds. J Theor Biol 117:47-77
Reagan FP (1916) A further study of the origin of blood vascular tissues in chemically treated teleost embryos, with special reference to haematopoiesis in the anterior mesenchyme and in the heart. Anat Rec 10:99-118
Reagan FP, Thorington JM (1916) The vascularization of the embryonic body of hybrid teleosts without circulation. Anat Rec 10:79-98
Richards AG, Korda FH (1950) Studies on arthropod cuticle. IV. An electron microscope survey of the intima of arthropod tracheae. Ann Entomol Soc Am 43:49-71
Ricklefs RE (1985) Modification of growth and development of muscles in poultry. Poult Sci 64:1563-1576
Riedesel ML (1977) Blood physiology. In: Wimsatt WA (ed) Biology of bats, vol II. Academic Press, London, pp 485-517
Riedesel ML, Williams BA (1976) Continuous 24 hr oxygen consumption studies of *Myotis velifer*. Comp Biochem Physiol 54A:95-99.
Risau W (1997) Mechanisms of angiogenesis. Nature 386:671-674
Risau W, Flamme I (1995) Vasculogenesis. Annu Rev Cell Dev Biol 11:73-91
Romanoff AL (1960) The avian embryo. Macmillan, New York
Romer AS (1966) Vertebrate paleontology, 3rd edn. The University of Chicago Press, Chicago
Ross Breeders (1999) Ross 308 broiler performance objectives. Ross Breeders, Newbridge
Ruberte J, Carretero A, Navarro M, Marcucio RS, Noden D (2003) Morphogenesis of blood vessels in the head muscles of avian embryo: spatial, temporal, and VEGF expression analyses. Dev Dyn 227:470-483
Sabin FR (1917) Preliminary note on the differentiation of angioblasts and the method by which they produce blood vessels, blood plasma and red blood cells as seen in the living chick. Anat Rec 13:199-204
Sabin FR (1920) Studies on the origin of blood vessels and of red blood corpuscles as seen in the living blastoderm of chicks during the second day of incubation. Contrib Embryol 36:213-261
Sakiyama J-I, Yamagishi A, Kuroiwa A (2003) *Tbx-Fgf10* system controls lung bud formation during chicken embryonic development. Development 130:1225-1234
Salomonsen F (1967) Migratory movements of the Arctic tern (*Sterna paradisea pontoppidan*) in the southern ocean. Det Kgl Danske Vid Selsk Biol Med 24:1-37
Sato M, Kornberg TB (2002) FGF is an essential mitogen and chemoattractant for the air sacs of the *Drosophila* tracheal system. Dev Cell 3:195-207
Saxen L, Sariola H (1987) Early organogenesis of the kidney. Pediatr Nephrol 1:385-392
Scheid P (1979) Mechanisms of gas exchange in bird lungs. Rev Physiol Biochem Pharmacol 86:137-186
Scheid P (1987) The use of models in physiological studies. In: Feder ME, Bennett AF, Burggrenn WW, Huey RB (eds) New direction in ecological physiology. Cambridge University Press, Cambridge, pp 275-288
Scheid P (1990) Avian respiratory system and gas exchange. In: Sutton JR, Coates G, Remmers JE (eds) Hypoxia: the adaptations. BC Decker Inc, Burlington, Ontario, pp 4-7
Scheid P, Piiper J (1972) Cross-currrent gas exchange in the avian lungs: effects of reversed parabronchial air flow in ducks. Respir Physiol 16:304-312.
Scheid P, Piiper J (1989) Respiratory mechnics and air flow in birds. In: King AS, McLelland J (eds) Form and function in birds, vol 4. Academic Press, London, pp 364-391
Scheid P, Slama H, Piiper J (1972) Mechanisms of unidirectional flow in parabronchi of avian lungs: measurements in duck lung preparations. Respir Physiol 14:83-94
Scheid P, Worth H, Holle JP, Meyer M (1977) Effects of oscillating and intermittent ventilatory flow on efficacy of pulmonary oxygen transfer in the duck. Respir Physiol 31:251-258

Scheuermann DW, Klika E, Lasseel DG, Bazantova I, Switka A (1997) An electron microscopic study of the parabronchial epithelium in the mature lung of four bird species. Anat Rec 249:213-225

Schittny J, Burri PH (2004) Morphogenesis of the mammalian lung: aspects of structure and extracellular matrix. In: Massaro DJ, Massaro GC, Chambon P (eds) Lung development and regeneration. Marcel Dekker, New York, pp 275-316

Schittny J, Djonov V, Fine A, Burri PH (1998) Programmed cell death contributes postnatal lung development. Am J Respir Cell Mol Biol 18:786-793

Schlosberg A, Bellaiche M, Zeitlin M, Ya'acobi M, Cahaner A (1996) Hematocrit values and mortality from ascites in cold-stressed broilers from parents selected by hematocrit. Poult Sci 75:1-5

Schmalhausen II (1968) The origin of terrestrial vertebrates. Academic Press, London

Schmidt-Nielsen K (1971) How birds breathe. Sci Am 225:72-79

Schmidt-Nielsen K (1972) Locomotion: energy cost of swimming, flying, and running. Science 177:222-228

Schmidt-Nielsen K, Kanwisher J, Lasiewski RC, Cohn JE, Bretz WL (1969) Temperature regulation and respiration in the ostrich. Condor 71:341-352

Scholey K (1986) The evolution of flight in bats. In: Nachtigall W (ed). BIONA report no. 5. Gustav-Fischer, Stuttgart, pp 1-12

Schulze FE (1908). Die Lungen des Afrikanischen Strausses. S.-B Preuss, Berlin, pp 416-431

Seeherman HJ, Taylor CR, Maloiy GMO, Armstrong RB (1981) Design of the mammalian respiratory system. II. Measuring maximum aerobic capacity. Respir Physiol 44:11-23

Sekine K, Ohuchi H, Fujiwara M, Yamasa K, Yoshizawa T, Sato T et al. (1999) FGF-10 is essential for limb and lung formation. Nat Genet 21:138-141

Seko Y, Seko Y, Takahashi N, Shibuya M, Yazaka Y (1999) Pulsatile stretch stimulates vascular endothelial growth factor (VEGF) secretion by cultured rat cardiac myocytes. Biochem Biophys Res Commun 254:462-465

Seller TJ (ed) (1987) Bird respiration, vols I and II. CRC Press, Boca Raton

Serra R, Moses HL (1995) pRb is necessary for inhibition of N-myc expression by TGF-beta-1 in embryonic lung organ cultures. Development 121:3057-3066

Shalaby F, Ho J, Stanford WL et al. (1997) A requirement for flk-1 in primitive and definitive hematopoiesis and vasculogenesis. Cell 89:981-990

Shannon JM (1994) Induction of alveolar type II cell differentiation in fetal tracheal epithelium by grafted distal mesenchyme. Dev Biol 166:600-614

Shannon JM, Nielsen LD, Gebb SA, Randell SH (1998) Mesenchyme specifies epithelial differentiation in reciprocal recombinants of embryonic lung and trachea. Dev Dyn 212:482-494

Shannon JM, Gebb SA, Nielsen LD (1999) Induction of alveolar type II cell differentiation in embryonic tracheal epithelium in mesenchyme-free culture. Development 126:1675-1688

Shima DT, Kuroka M, Deustsch U, Ng YS, Adamis AP, D'Amore, PA (1996) The mouse gene for vascular endothelial growth factor: genomic structure, definition of the transcriptional unit and characterization of transcriptional and post-transcriptional regulatory sequences. J Biol Chem 271:3877-3883

Sibley CG, Alhquist JE (1990) Phylogeny and classification of birds: a study in molecular evolution. Yale University Press, New Haven

Siegwart B, Gehr P, Gil J, Weibel ER (1971) Morphometric estimation of pulmonary diffusion capacity. IV. The normal dog lung. Respir Physiol 13:141-159

Silversides FG, Lefrancois MR, Villneuve P (1997) The effect of strain of broiler on physiological parameters associated with ascites syndrome. Poult Sci 76:663-667

Smith DG, Rapson L (1977) Differences in pulmonary microvascular anatomy between *Bufo marinus* and *Xenopus laevis*. Cell Tissue Res 178:1-15

Smith JH (1985) Breeders must respond to market trends. Poult Intern 34:January
Smith RE (1956) Quantitative relations between liver mitochondria metabolism and total bodyweight in mammals. Ann NY Acad Sci 62:403–422
Snedecor GW (1956) Statistical methods applied to experiments in agriculture and biology. Iowa State University Press, Ames (Iowa)
Snyder GK (1976) Respiratory characteristics of whole blood and selected aspects of circulatory physiology in the common short-nosed fruit bat, *Cyanopterus brachyotes*. Respir Physiol 28:239–247
Soker S, Gollamudi-Payne S, Fidder H, Charmaheli H, Klagsbrun M (1997) Inhibition of vascular endothelial growth factor (VEGF) induced endothelial cell proliferation by a peptide corresponding to the exon 7-encoded domain of VEGF$_{165}$. J Biol Chem 272:31582–31588
Solomon SE, Purton M (1984) The respiratory epithelium of the lung in the green turtle (*Chelonia mydas* L). J Anat 139:353–361
Spira A (1996) Disorders of the respiratory system. In: Rosskopf W, Woerpel R (eds) Diseases of cage and aviary birds. Lea and Febiger, Baltimore, pp 415–428
Stabellini G, Locci P, Calvitti M, Evangelisti R, Marinucci L, Bodo M, Carusio A, Canaider S, Carinci P (2001) Epithelial-mesenchymal interactions and lung branching morphogenesis, role of polyamines and transforming growth factor beta1. Eur J Histochem 45:151–162
Stanislaus M (1937) Untersuchungen an der Kolibrilunge. Z Morphol Tiere 33:261–289
Stearns RC, Barnas GM, Walski M, Brain JD (1986) Phagocytosis in the gas exchange region of avian lungs. Fed Proc 45:959
Stearns RC, Barnas GM, Walski M, Brain JD (1987) Deposition and phagocytosis of inhaled particles in the gas exchange region of the duck, *Anas platyrhynchos*. Respir Physiol 67:23–36
Steen JB (1971) Comparative physiology of respiratory mechanisms. Academic Press, London
Stiles AD, Chrysis D, Jorvis HW, Brighton B, Moats-Staats BM (2001) Programmed cell death in normal fetal rat lung development. Exp Lung Res 25:569–587
Stinner JN (1982) Gas exchange and air flow in the lung of the snake, *Pituophis melanoleucus*. Am J Physiol 243:R251–257
Stinner JN, Shoemaker VH (1987) Cutaneous gas exchange and low evaporative water loss in the frogs *Phyllomedusa sauvagei* and *Chiromantis xeraphelina*. J Comp Physiol 157B:423–427
Stockard CR (1915) The origin of blood and vascular endothelium in embryos without a circulation of blood and in the normal embryo. Am J Anat 18:227–325
Stone J, Chan-Ling T, Pe'er J, Itin A, Gnesin H, Keshet E (1996) Roles of vascular endothelial growth factor and astrocyte degeneration in the genesis of retinopathy of prematurity. Invest Ophthal Vis Sci 37:290–299
Suarez RK (1992) Hummingbird flight: sustaining the highest mass-specific metabolic rates among vertebrates. Experientia 48:565–570
Suarez RK, Brown GS, Hochachka PW (1986) Mitochondrial respiration in hummingbird muscles. Am J Physiol 251:R537–R542
Suarez RK, Lighton JRB, Brown GS, Mathieu-Costello O (1991) Mitochondrial respiration in hummingbird flight muscles. Proc Natl Acad Sci USA 87:9207–9210
Suthers RA, Thomas SP, Suthers BJ (1972) Respiration, wing-beat and ultrasonic pulse emission in an echolocating bat. J Exp Biol 56:37–48
Sutherland D, Samakovlis C, Krasnow MA (1996) Branchless encodes a *Drosophila* FGF homolog that controls tracheal cell migration and the pattern of branching. Cell 13:1091–1101
Swan LW (1961) The ecology of the high Himalayas. Sci Am 205:67–78

Szebenyi G, Fallon JF (1999) Fibroblast growth factors as multifunctional signaling factors. Int Rev Cytol 185:45–106

Tanswell AK, Buch S, Liu M, Post M (1999) Factors mediating cell growth in lung injury. In: Brand RD, Coalson J (eds) Chronic lung disease of early infancy. Marcel Dekker, New York, pp 493–534

Taichman DB, Loomes KM, Schachtner SK, Guttentag S, Vu C, Williams P, Oakley RJ, Baldwin HS (2002) Notch 1 and jagged 1 expression by the developing pulmonary vasculature. Dev Dyn 225:166–175

Tebar M, Destree O, de Vree WJ, Ten Have-Opbroek AA (2001) Expression of Tcf/Lef and sFrp and localization of β-catenin in the developing mouse lung. 109:437–440

Tenney SM, Remmers JE (1963) Comparative quantitative morphology of the mammalian lung: diffusing area. Nature 197:54–56

Tenney SM, Tenney JB (1970). Quantitative morphology of cold-blooded lungs: Amphibia and Reptilia. Respir Physiol 9:197–215

Thewissen JMG, Babcock SK (1992) The origin of flight in bats: to go where no mammal has gone before. BioScience 42:340–345

Thomas SP (1975) Metabolism during flight in two species of bats, *Phyllostomus hastatus* and *Pteropus gouldii*. J Exp Biol 63:273–293

Thomas SP (1981) Ventilation and oxygen extraction in the bat, *Pteropus gouldi*, during rest and during steady flight. J Exp Biol 94:231–250

Thomas SP (1987) The physiology of bat flight. In: Fenton MB, Racey P, Rayner JMV (eds) Recent advances in the study of bats. Cambridge University Press, Cambridge, pp 75–99

Thomas SP, Suthers R (1972) The physiology and energetics of bat flight. J Exp Biol 57:317–335

Thomas SP, Lust MR, Riper HJ (1984) Ventilation and oxygen extraction in the bat *Phyllostomous hastatus* during rest and steady flight. Physiol Zool 57:237–250

Thomas SP, Thomas DP, Thomas GS (1985) Ventilation and oxygen extraction in the bat *Pteropus poliocephalus* acutely exposed to simulated altitudes from 0 to 11 km. Fed Proc Fed Am Soc Exp Biol 44:1349

Thompson KS (1971) The adaptation and evolution of early fishes. Q Rev Biol 46:139–166

Thorpe WH, Crisp DJ (1941) Studies on plastron respiration. II. The respiratory efficiency of the plastron in *Amphelocheirus*. J Exp Biol 24:270–303

Tichelaar JW, Lu W, Whitsett JA (2000) Conditional expression of fibroblast growth factor-7 in the developing and mature lung. J Biol Chem 275:111858–111864

Timpl R, Wiedemann H, van Delden V, Furthmayr H, Kuhn K (1981) A network model for the organization of type IV collagen molecules in basement laminas. Eur J Biochem 120:203–211

Tobalske BW, Hedrik TL, Dial KP, Biewener AA (2003) Comparative power curves in bird flight. Nature 421:363–366

Toews DP, MacIntyre D (1977) Blood respiratory properties of a viviparous amphibian. Nature 266:464–465

Tomanek RJ, Schatteman GC (2000) Angiogenesis: new insights and therapeutic potential. Anat Rec 261:126–135

Tomanek RJ, Holifield JS, Reiter RS, Sandra A, Lin JJC (2002) Role of VEGF family members and receptors in coronary vessel formation. Dev Dyn 225:233–240

Toth TE, Siegel PB (1986) Cellular defense for the avian respiratory tract: paucity of free-residing macrophages in the normal chicken. Avian Dis 30:67–75

Toth TE, Siegel PB, Veit H (1987) Cellular defense of the avian respiratory system – influx of phagocytes: elicitation versus activation. Avian Dis 31:67–75

Toth TE, Pyle RH, Caceci T, Siegel PB, Ochs D (1988) Cellular defense of the avian respiratory system: influx and nonopsonic phagocytosis by respiratory phagocytes activated by *Pasteurella multocida*. Infect Immunol 56:1171–1179

References

Trampel DW, Fletcher OJ (1980) Ring-stabilizing technique for collection of avian air sacs. Am J Vet Res 14:1730–1734

Tucker V (1968) Respiratory physiology of house sparrows in relation to high altitude flight. J Exp Biol 48:55–66

Tucker V (1970) Energetic cost of locomotion in mammals. Comp Biochem Physiol 34:841–846

Tucker VA (1974) Energetics of natural avian flight. In: Paynter RA (ed) Avian energetics. Nuttal Ornithological Club, Cambridge (MA), pp 298–333

Tucker VA (1998) Gliding flight: speed and acceleration of ideal falcons during diving and pull out. J Exp Biol 201:403–414

Tuyl MV, del Riccio V, Post M (2004) Lung branching morphogenesis: potential for regeneration of small conducting airways. In: Massaro DJ, Massaro GC, Chambon P (eds) Lung development and regeneration. Marcel Dekker, New York, pp 355–393

Van Furth R (1982) Cellular biology of pulmonary macrophages. Allergy Appl Immunol 76:21–27

Vidyadaran MK, King AS, Kassim H (1987) Deficient anatomical capacity for oxygen uptake of the developing lung of the female domestic fowl when compared with red-jungle fowl. Schweischr Arch Tiere 129:225–237

Visschedijk AHJ (1968) The air space and embryonic respiration. I. The pattern of gaseous exchange in the fertile egg during the closing stages of incubation. Br Poult Sci 9:173–184

Vitali SD, Richardson KC (1998) Evaluation of pulmonary volumetric morphometry at the light and electron microscopy level in several species of passerine birds. J Anat 193:573–580

Vlodavsky I, Folkman J, Sullivan R, Fridman R, Ishai MR, Sasse J, Klagsbrun M (1987) Endothelial cell derived basic fibroblast growth factor: synthesis and deposition into subendothelial matrix. Proc Natl Acad Sci USA 84:2292–2296

Vos HJ (1934) Über die Wege der Atemluft in der Entenlunge. Zeitsch Vergl Physiol 21:552–578.

Vos HJ (1937) Über das Fehlen der rekurrenten Bronchien beim Pinguin und bei den Reptilien. Zool Anz 117:176–181

Wagner GP (1989) The origin of morphological characters and the biological basis of homology. Evolution 43:1157–1171

Wallace GJ (1955) An introduction to ornithology. Macmillan: New York

Walsh C, McLelland J (1974) The ultrastructure of the avian extrapulmonary respiratory epithelium. Acta Anat 89:412–422

Wang N, Banzett RB, Butler JP, Fredberg JJ (1988) Bird lung models show that convective inertia effects inspiratory aerodynamic valving. Respir Physiol 73:111–124

Wang N, Banzett RB, Nations CS, Jenkins EA (1992) An aerodynamic valve in the avian primary bronchus. J Exp Biol 262:441–445

Warburton D, Schwarz M, Tefft D, Flores-Delgado G, Anderson KD, Cardoso WV (2000) The molecular basis of lung morphogenesis. Mech Dev 92:55–81

Warburton D, Bellusci S, Del Moral PM, Kaartinen V, Lee MD, Shi W (2003) Growth factor signaling in lung morphogenetic centers automaticity, stereotypy and symmetry. Respir Res 19:294–315

Warner RW (1972) The anatomy of the syrinx in passerine birds. J Zool Lond 168:381–393

Weaver M, Dunn NR, Hogan BL (2000) BMP-4 and FGF-10 play opposing roles during lung bud morphogenesis. Development 127:2695–2704

Weibel ER (1963) Morphometry of the human lung. Springer, Berlin Heidelberg New York

Weibel ER (1970/71) Morphometric estimation of pulmonary diffusion capacity. I. Model and method. Respir Physiol 11:54–75

Weibel ER (1973) Morphological basis of the alveolar-capillary gas exchange. Physiol Rev 53:419–495

Weibel ER (1979) Stereological methods, vol. 1. Practical methods for biological morphometry. Academic Press, London

Weibel ER (1984) The pathways for oxygen. Harvard University Press, Harvard (MA)

Weibel ER (1990) Morphometry: stereological theory and practical methods. In: Gill J (ed) Models of lung disease: microscopy and structural methods. Marcel Dekker, New York, pp 199–251

Weibel ER (1997) Design of airways and between the organism blood vessels considered as confluent tree. In: Crystal RD, West JB, Weibel ER, Barnes PJ (eds) The lung: scientific foundations. Lippincott-Raven, Philadelphia, pp. 1061–1071

Weibel ER (2000) Symmorphosis: on form and function in shaping life. Harvard University Press, Cambridge (MA)

Weibel ER, Gomez DM (1962) Architecture of the human lung. Science 137:577–585

Weibel ER, Knight BW (1964) A morphometric study on the thickness of the pulmonary airblood barrier. J Cell Biol 21:367–384

Weibel ER, Taylor CR, O'Neil JJ, Leith DE, Gehr P, Hoppeler H, Langman V, Baudinette RV (1983) Maximal oxygen consumption and pulmonary diffusing capacity: a direct comparison of physiologic and morphometric measurements in canids. Respir Physiol 54:173–188

Weidenfeld J, Shu W, Zhang L, Millar SE, Morrisey EE (2002) The Wnt7b promoter is regulated by TFT-1, GAATA6, and Foxa2 in lung epithelium. J Biol Chem 277:21061–21070

Weinstein M, Xu X, Ohyama K, Deng CX (1998) FGFR-3 and FGFR-4 function cooperatively to direct alveogenesis in the murine lung. Development 125:3615–3623

Weis-Fogh T (1964a) Diffusion in insect flight muscle, the most active tissue known. J Exp Biol 41:229–256

Weis-Fogh T (1964b) Functional design of the tracheal system of flying insects as compared with the avian lung. J Exp Biol 41:207–228

Weis-Fogh T (1967) Respiration and tracheal ventilation in locusts and other flying insects. J Exp Biol 47:561–587

Wells DJ (1993) Muscle performance in hovering hummingbirds. J Exp Biol 178:39–57

Welsch U (1981) Fine structure and enzyme histochemical observations on the respiratory epithelium of the caecilian lungs and gills. A contribution to the understanding of the evolution of the vertebrate respiratory epithelium. Arch Histol Jpn Okayama 44:117–133

Welsch U (1983) Phagocytosis in the amphibian lung. Anat Anz 154:323–327

Welsch U, Aschauer B (1986) Ultrastructural observations on the lung of the emperor penguin (*Apternodytes forsteri*). Cell Tissue Res 243:137–144

Welty JC (1979) The life of birds, 2nd edn. Saunders, Philadelphia

Wendel DP, Taylor DG, Albertine KH, Keating MT, Li DY (2000) Impaired distal airway development in mice lacking elastin. Am J Respir Cell Mol Biol 23:320–326

Wessels NK (1970) Mammalian lung development: interactions in formation and morphogenesis of tracheal buds. J Exp Zool 175:455–466

West BJ (1987) Fractals, intermittency and morphogenesis. In: Degn H, Holden AV, Olsen LF (eds) Chaos in biological systems. Plenum Press, New York, pp 305–314

West B, Zhou BW (1988) Did chickens go north? New evidence for domestication. J Archeol Sci 15:515–533

West JB (1983) Climbing Mt Everest without oxygen: an analysis of maximal exercise during extreme hypoxia. Respir Physiol 52:265–274

West JB, Mathieu-Costello O (1999) Structure, strength, failure, and remodeling of the pulmonary blood-gas barrier. Annu Rev Physiol 61:543–572

West JB, Mathieu-Costello O, Jones JH, Birks EK, Logerman RB, Pascoe JR, Tyler WS (1993) Stress failure of pulmonary capillaries in racehorses with exercise-induced pulmonary hemorrhage. J Appl Physiol 75:1097–1109

References

West NH, Bamford OS, Jones DR (1977) A scanning electron microscope study of the microvasculature of the avian lung. Cell Tissue Res 176:553-564

Wetherbee DK (1951) Air sacs in the English sparrow. Auk 68:242-244

Wigglesworth VB (1953) Surface forces in the tracheal system of insects. Q J Microsc Sci 94:507-522

Wigglesworth VB (1965) The principles of insect physiology, 6th edn. Methuen, London

Wigglesworth VB (1972) The principles of insect physiology, 7th edn. Chapman and Hall, London

Wigglesworth VB, Lee WM (1982) The supply of oxygen to the flight muscles of insects: a theory of tracheole physiology. Tissue Cell 14:501-518

Willem M, Miosque N, Halfer W, Smyth N, Jannetti I, Burghart E et al. (2002) Specific ablation of the nidogen binding site in the laminin gamma-1 chain interferes with kidney and lung development. Development 129:2711-2722

Wittenberg JB, Wittenberg BA (1989) Transport of oxygen in muscle. Annu Rev Physiol 51:857-878

Wodarz A, Nusse R (1998) Mechanisms of Wnt signaling in development. Annu Rev Cell Biol 14:59-88

Wolk E, Bogdanowicz W (1987) Hematology of the hibernating bat, *Myotis daubentoni*. Comp Biochem Physiol 88A:637-637

Woodward JD, Maina JN (2005) A 3-D digital reconstruction of the components of the gas exchange tissue of the lung of the muscovy duck, *Cairina moschata*. J Anat, 206:477-492

Xiao J, Changgong L, Zhu NL, Borok Z, Minoo P (2003) *Timeless* in lung morphogenesis. Dev Dyn 228:82-94

Xu X, Weistein M, Li C, Naski M, Cohen RI, Ornitz DM, Leder P, Deng C (1998) Fibroblast growth factor 2 (FGFR-2) mediated reciprocal regulation loop between FGF-8 and FGF-10 is essential for limb induction. Development 125:753-765

Yalden DW, Morris PA (1975) The lives of bats. The New York Times Book Co, New York

Yamaguchi TP, Dumont DJ, Conlon RA, Breitman MI, Rossant J (1993) flk-1, and flt-1-related receptor tyrosine kinase is an early marker for endothelial cell precursors. Development 111:489-498

Yancopoulous GD, Davis S, Gale NW, Rudge JS, Wiegand SJ, Holash J (2000) Vascular-specific growth factors and blood vessel formation. Nature 407:242-248

Yapp WB (1970) The life and organization of birds. Edward Arnold, London

Yuan B, Li C, Kimura S, Engelhardt RT, Smith BR, Minoo P (2000) Inhibition of distal lung morphogenesis in *Nkx2.1*(-/-) embryos. Dev Dyn 217:180-190

Yurchenko PD, Tsilibary EC, Charonis AS, Furthmayr H (1986) Models for the self assembly of basement lamina. J Histochem Cytochem 34:93-102

Zhou M, Sutliff RL, Paul RJ, Lorenz JN, Hoying JB, Haudenschild CC, Yin M, Coffin JD, Kong L, Kranias EG, Luo WL, Boivin GP, Duffy JJ, Pawlowski SA, Doetschman T (1998) Fibroblast growth factor 2 control of vascular tone. Nat Med 4:201-207

Subject Index

A

Abdominal air sac 18, 34, 35, 74, 96–99, 101, 116, 117
Abdominal cavity 15, 35, 66, 67, 96, 97, 128, 146, 164, 146
Acclimatization 10
Adaptive biology 161
Adaptive plasticity 153
Aerobic capacity 115
Aerodynamic forces 117
Aerodynamic valving 32, 66, 117
Aerosol 116
Air 3, 4, 10, 11, 14, 15, 32, 48, 66, 67, 74, 75, 94, 95, 97, 100, 104, 106, 117–119, 125, 146, 151, 152, 154, 160, 161, 164, 166, 168, 172
- capillaries 14, 21, 23, 26, 35, 51, 52, 74, 81, 92, 94–96, 104, 106, 107, 113, 145, 147, 173
- cells 163, 165, 166
- conduits 31, 74, 117, 166, 169
- duct 163, 165, 168
- flow 3, 65, 100, 101, 104, 106, 111, 116, 117, 119, 123, 173
- passages 165, 168, 172
- sac 4, 6, 7, 11, 14, 18, 32, 34, 65, 66, 96, 99, 116, 117, 123, 128, 160, 166, 172
- sacculitis 107, 112
- spaces 51, 151, 164, 165, 168
- way 13–16, 24, 31, 32, 35, 53, 55, 56, 59, 73, 78, 81, 102, 104, 106, 107, 113, 116, 160
- blood barrier 48
- breather 125, 160
- breathing 48, 116, 125, 160, 161, 168
- haemoglobin pathway 154–156
- tissue interface 146
- water interface 164, 170
- ways 14, 24, 31, 35, 53, 55, 56, 59, 102, 106

Albatross 5, 9
Altitude 10, 11
Altricial 28, 30
Alveolar macrophage 107
Alveolar stage 15, 26
Alveoli 14, 24, 59, 74, 81, 96, 107, 148
Amphibian 3, 152, 164, 165, 168
Ampibian lung 164, 165
Anaerobic 4, 6, 8, 115, 159
Analysis 65, 117, 125, 128, 154, 156
Anastomoses 18, 22, 31, 74, 92, 100, 102, 106, 160, 170
Anatomical diffusion factor 57
Anatomical valve 117
Angioblast 35, 36, 45–47, 60
Angiogenesis 45, 46, 48, 60, 61
Angiogenetic cells 36, 45, 46
Angiogenetic factor 60
Angulated 77
Anurans 164, 165
Aortic rupture 116
Apodan 164, 165
Apoptosis 21, 27, 36, 46, 51, 53
Aquatic 164, 165
Archeopteryx lithographica 5
Arctic tern 9
Argentavis magnificens 6
Arterialization 48, 104
Arteriovenous anastomoses 106
Arthropoda 5
Ascites 116
Aspiration 112, 113
Atmosphere 10, 108, 161, 170
Atria 19, 21, 28, 30, 51, 79, 81, 94, 95, 106, 107, 116
Attenuation 94, 95, 150
Avian lung 13–15, 22, 23, 31, 32, 45–47, 52–56, 59, 63, 65–67, 74, 75, 78, 81, 94–96, 100, 102, 106, 107, 111, 113, 116, 117, 128, 145, 146, 147, 149–152, 154–156, 163, 166, 173

Avian pulmonary defense 111
Avian pulmonary macrophage 107, 111–114, 116
Avian respiratory system 4, 11, 14, 15, 53, 65–67, 107, 111, 117, 128, 146, 160, 161, 163, 166

B

Back-to-front flow 32, 117
Bar-headed goose 10, 11, 118, 150
Basement lamina 19, 52
Bat 3–6, 9, 11, 128, 148, 156, 173
Bichir 161
Bifurcation 14, 31, 54, 56, 74, 106
Bimodal (transitional) breather 125, 161
Birds 3–11, 13, 15, 22, 23, 28, 30, 65, 66, 67, 79, 81, 92, 95, 97–101, 107, 111–117, 123, 128, 146–151, 156, 157, 167, 173
Blood 10, 11, 15, 21, 35, 36, 45, 46–48, 52, 53, 59, 60, 61, 67, 81, 94, 96, 102, 106, 107, 111, 119, 125, 151, 152, 155, 157, 166
Blood:
– capillaries 11, 21, 23, 26, 35, 46, 48, 51, 52, 81, 92, 94, 95, 151, 152, 164, 168
– cells 36, 46
– islands 36, 45
– monocyte 107, 111
– vessels 35, 36, 45–47, 53, 59, 60, 61, 81, 165
– volume 151
– gas barrier 21, 24, 26, 48, 51, 52, 94–96, 101, 111, 147–150, 153–157, 159, 160, 165
Body mass 3, 4, 6, 111, 112, 114, 128, 149–151, 157, 160
Body volume 128, 170
Branching 13, 14, 33, 45, 54–56, 58, 59, 63, 160
Branching morphogenesis 13, 14, 54, 56, 63, 160
Branchless 56
Breathing 10, 48, 116, 118, 119, 161, 168
Broilers 114–116
Bronchial epithelial cell 113
Bronchioalveolar tree 14, 74
Bronchopulmonary lavage 112
Bronchus 16, 18, 22, 35, 66, 73, 74, 65, 75, 76, 77, 78, 97, 166
Buccal cavity 164, 165
Budgerigar 5

Bufo marinus 164, 165
Bustard 6

C

Caecilian 164, 165
Canalicular stage 15, 24,
Capillary loading 152
Carbon dioxide 10, 15, 100, 101, 104, 106, 123, 154, 169
Carbon particles 111, 113
Cardiac output 48, 173
Carnivora 3, 13–16, 54–57, 97, 113, 168–170, 172–173
Cartilages 117, 165
Cassowary 3, 6
Caudothoracic air sac 35, 97, 99, 101, 117
Cell 13–15, 18, 19, 21, 23, 35, 36, 45–47, 51–56, 59–61, 75, 78, 94–96, 99, 107, 111, 113, 116, 149, 157, 163–165, 168, 170, 172
– division 13
– growth 54, 61
Cellular defenses 107
Centrifugal flow 106
Centripetal flow 104, 106
Cervical air sac 34, 35, 97
Chameleon 67, 166
Chemoreceptor 123
Chick lung 55
Chicken embryo 55
Chordata 5
Clavicular air sac 34, 35, 97, 99, 117
Co-current 106
Coelacanth 161
Coelomic cavity 15, 35, 66, 128, 146
Collagen 52, 54
Columnar 19, 75, 99
Comparative respiratory morphology 159
Compliance 146
Composite 128, 154
Compromises 3, 48, 172
Concentration of:
– CO_2 10, 100, 123
– haemoglobin 174
– intracellular Ca^{2+} 61
– O_2 at altitude 10
– O_2 in arterial blood 104
– O_2 in tibial trachea 173
– surfactant 95
Condor 6

Subject Index

Conductance 154, 155, 157
Connective tissue 22, 60, 78, 81, 95, 96, 99, 107, 149
Conservation 166
Constriction 66, 118, 119
Convective flow 106
Convective momentum 118, 119
Convergence 159, 173
Cost-effectiveness 160, 172
Counter-current 48, 67, 106
Craniothoracic air sac 34, 35, 97, 99, 119
Crocodilian lung 166, 167
Cross-current 48, 104, 106
Cyanocephalus volans 5
Cytokine 13

D
Deoxygenated blood 104, 106
Design 1, 2, 4, 6, 9, 11, 14, 15, 22, 23, 32, 35, 48, 66, 95, 104, 106, 116, 125, 128, 150, 153, 155, 159, 160, 172, 173
Development 3, 4, 11, 13–15, 23, 24, 27, 28, 35, 36, 46, 48, 53, 54–56, 58, 59, 60, 63, 67, 79, 95, 96, 100, 101, 111, 116, 164, 165, 170, 173
Diameter 13, 46, 60, 79–81, 92, 96, 101, 113, 119, 147, 165, 170, 172
Diaphragm 66, 156
Dichotomous branching 14, 31, 74
Differentiation 13, 14, 35, 46, 54, 59, 60
Diffuse capillary system 152
Diffusing capacity 154–156, 160
Diffusing capacity of:
- erythrocytes 155–157
- lung (total) 155–157
- membrane 155–157
- plasma 155–157
- tissue barrier 155–157
Diffusion 15, 48, 95, 106, 125, 155, 157, 159, 172, 173
Diffusional pathway 172
Dipnoi 161, 163
Direct connections 98–100
Diverticulae 97
Domestic fowl (*Gallus*) 15, 21, 22, 28, 34, 35–36, 46, 51, 52, 78, 99, 100, 101, 104, 107, 112–115, 118, 157
Double capillary system 152
Draco volans 5
Duck 15, 101, 102, 106, 112, 113, 117, 157
Dust cell 165, 168

E
Earth 4, 6, 160, 161
Ecology 3, 5
Embryogenesis 13, 15, 35, 63
Embryonic 13, 15, 24, 27, 36, 45, 46, 48, 53–56, 58, 59, 164
Embryonic lung 15, 36, 46, 56
Emu 3, 6, 98, 101, 145, 148, 149
Endoderm 13, 15, 45, 52, 53, 56, 63
Endothelial cell 35, 36, 45–47, 52, 59, 60, 61, 95, 111
Energy 3, 5, 6, 8, 146, 154
Enzymatic deficiency 111
Epithelial cell 13, 14, 18, 19, 46, 51–54, 56, 59, 61, 95, 96, 99, 113, 116, 149, 168
Epithelial-mesenchymal interaction 13
Erectile (cavenous) tissue 119
Erythroblast 36, 46
Erythrocyte 36, 45, 46, 47, 48, 151, 155, 156
Erythropoiesis 48
Evagination 116
Evolution 3, 4, 6, 11, 14, 48, 66, 96, 100, 149, 160, 161, 166, 167, 173
Exchange tissue 21, 107
Expiration 100
Extinct 5, 160, 161
Extracellular matrix 52–54, 56, 59–61, 95, 149,

F
Falcon 9
Fat cells 99
Faveoli 163
Fibroblast 51
Fibroblast growth factor 54–58
Filopodia 36, 46
Fish 3, 5, 152, 161, 168
Fitness 160
Flight 1–11, 170, 172–174
Fluid valve 117
Flying fox 6, 149
Foregut 13, 15, 53, 54, 56
Foregut endoderm 13, 53, 56
Fossorial 164
Function 3, 13, 35, 59, 61, 66, 119, 153, 159, 160, 161
Functional:
- advantage 166
- capacity 157
- concepts 66

- design   128
- differences   100
- disequilibria   115
- efficiency   168
- heterogeneity   111
- integration   1
- lungs   164
- model   154
- morphplogy   2, 11, 65
- parts   100, 154
- performance   115
- property   60, 111
- requirement   35, 65
- residual capacity   156
- significance   92, 106
- similarity   159, 160
- subunit   14, 154

G

Galliform species   22, 79, 101
Gas exchange   11, 14, 30, 48, 52, 66, 67, 79, 94, 99, 101, 104, 106, 150, 151, 154, 156, 159, 166, 173
Gas exchange:
- efficiency   172
- mantle   81
- organ   154
- pathway   173
- tissue   107
- units   96, 147
Gas exchanger   14, 15, 48, 66, 117, 125, 145, 146, 153, 157, 160, 164, 165
Genes   13
Genetic breeding   114, 115
Geodesic   22, 23
Geometry   66, 117, 123
Gills   48, 116, 146, 161, 164
Glandular stage   15
Glaucomys volans   5
Glucose   9
Goblet cell   75, 99
Golden plover   9, 101
Goose   10, 11, 118, 150
Granules   99
Granulocyte   113
Gravity   1, 6, 112, 113
Growth   13, 15, 24, 30, 34, 35, 53–56, 60, 61, 115, 116, 170
Growth factor   13, 53, 54–60
Guinea fowl   102
Guinea pig   15, 112

H

Habitat   3, 6, 128, 159, 160   164, 165
Haematocrit   157, 174
Haemoglobin concentration   174
Haemolymph   154,
Harmonic mean thickness   148, 155
Hatching   21, 27, 28, 30, 35, 52, 95, 115
Heart failure syndrome   116
Hemangioblast   36, 46
Hematogenesis   36
Hematopoietic cells   36, 45
Herring gull   5, 28, 97
Heterogenity   60, 111, 159, 163, 165, 166
Hilus   16, 74, 102, 117
Himalaya   10
Homiotherm   4
Homogenous   36, 100, 166
Horizontal septum   35, 66, 74, 96, 97
Horse   2, 115
House sparrow   79, 97, 150
Hovering   7, 8
Human   1, 2, 4, 9, 22, 26, 160
Hummingbird   6
Hypercapnia   118
Hyperpnia   118
Hypoxia   10, 11, 60, 104
Hyra arborea   164

I

Icaronycteris index   5
Immunoreactive fibers   100
In situ hybridization   63
Incubation   26, 97
Indirect connections   98–100
Infundibulae   19, 21, 51, 81, 94, 104, 106, 107, 116
Insects   4, 5, 8, 13, 15, 55, 168–170, 172, 173
Inspiration   100
Inspired air   119, 123
Internal subdivision   145, 147, 163, 164, 166, 168
Interparabronchial:
- arteries   22, 102, 106
- arterioles   104, 106
- blood vessel   46
- septa   22, 23, 78–81
- vein   106
Interstitial   51, 95, 107, 111, 170
Interstitial macrophage   107
Intrapulmonary air flow   117
Invagination   116, 168, 173

Subject Index

J
Junctional complex 99
Jurassic 5

K
Keel 7
Kiwi 3, 6, 97

L
Lamellae 146
Lamellated osmiophilic body 95, 96
Larynx 113
Laterodorsal secondary bronchus 76, 77, 81, 99, 101
Latimeria chalumnae 161
Lepidosiren paradoxa 152, 161
Leukogenetic cells 36
Lifestyle 128, 159, 160
Ligands 59, 61, 53
Liver 47, 48, 66, 97, 104, 116, 173
Lizard 5, 148, 166, 167, 168
Locomotion 1, 4, 5
Lumen 18, 47, 60, 118, 119
Lung 4, 11, 13, 14–16, 21, 23, 24, 26, 28, 31, 32, 34–36, 45–48, 51–55, 58, 59, 63, 66, 67, 74, 75, 78, 81, 95–103, 104, 106, 107, 111–113, 116, 117, 128, 145–152, 156, 157, 160, 161, 163–166, 168, 173
Lung:
– bud 13, 15, 47, 54, 55, 56, 63, 115
– cells 54, 55
– development 14, 24, 28, 53–56, 58
– endoderm 63
– growth 56
– mesenchyme 14, 54
– morphogenesis 14, 54, 57
– volume 51, 128, 145, 151, 153
– air sac system 4, 11, 65, 66, 116, 117, 123, 160
Lungfish 152, 161, 168
Lysosome 113, 116

M
Mammal 2–5, 24, 26, 27, 31, 51, 53, 54, 56, 58, 66, 74, 78, 81, 95, 96, 101, 107, 111, 112, 128, 145–149, 151, 152, 156, 157, 160, 166–168, 178
Mammalian lung 4, 14, 15, 31, 51, 54, 56, 58, 66, 74, 78, 81, 95, 96, 101, 107, 111, 145, 147, 148, 152, 156, 160, 173

Marsupial 24
Mechanical compression 96
Mechanical integrity 95, 150
Mechanical sphincter 117
Mechanical ventilation 28, 166, 172
Mediodorsal secondary bronchus 76, 77, 81, 99–101, 104, 117
Medioventral secondary bronchus 76, 77, 81, 99, 100, 102, 117, 118, 110, 123
Mesenchymal cells 18, 19, 21, 36, 45, 46, 55, 56, 60
Mesenchyme 14, 15, 21, 35, 46, 54–56, 58, 59
Mesoderm 13, 35, 45, 52, 54, 56, 61, 63
Metabolic rate 1, 5, 6, 8, 113, 149, 164
Metamorphosis 164
Micropinocytotic vesicles 95
Microvascular growth 35
Migration 4, 9, 54, 59
Mitochondria-rich cell 168
Model 14, 118, 123, 125, 153–157, 161,
Mole rat 24
Molecular factor 45, 53, 54, 60
Mononuclear phagocytic system 107, 113
Morphogenesis 13, 14, 53–57, 63, 160
Morphometric parameters 128, 157
Morphometry 101, 125, 128, 153
Mouse 5, 8, 56, 63
Mt Everest 10
Mucociliary escalator 113
Mucus 168
Mucus gland 77
Multicameral lung 166
Muscle 4, 6, 7, 9, 11, 48, 60, 99, 123, 168, 170, 173
Myotis 4, 5

N
Neoceratodus fosteri 161, 163
Neopulmo 22, 81, 99–101
Nerve plexus 99
Nonflying mammals 128, 149, 151, 156
Nonrespiratory site 95, 166
Nonsmoker 112

O
Oblique septum 97
Ostia 35, 66, 98, 99, 117
Ostrich 3, 6, 79, 92, 97, 101, 111, 123, 145, 147, 149, 150
Oxidative metabolism 111

Oxygen 7, 8, 10, 11, 15, 104, 106, 123, 125, 154–157, 159, 161, 164, 165, 167, 168, 170, 172–174
Oxygen carrying capacity 174
Oxygen consumption 4, 5, 8, 10, 115, 156, 168, 172

P
Paleolulmo 22, 81, 100
Pantodon buchhotzii 5
Parabronchial:
- gas 104, 106
- lumen 19, 22, 79, 81, 99, 104, 106, 111
- lung 14, 15, 26, 28, 31, 166
Parenchyma 21, 27, 46, 145, 147, 151, 163, 166
Passeriform species 9, 79, 80, 97
Passerine 4, 9, 101, 151
Paucicameral lung 166
Pectoral girdle 6, 8, 9, 97–99, 117, 148–150
Pectoralis muscle 6
Pelican 6
Penguin 6, 79, 92, 97–99, 149, 152, 157
Perfusion 13, 156, 159
Pericytes 60
Perikarya 52, 94
Peritoneal cavity 96, 97, 111
Permeation coefficient 154, 155
Phyllostomus hastatus 5
Phylogenetic 73, 128, 159, 160, 167, 173
Physiological diffusing capacity 156
Pigeon 5, 9, 15, 30, 81, 97, 98, 101, 107, 112
Pipistrellus 9, 149
Planum anastomotica 81, 100
Pleistocene 6
Plethodontid salamander 165
Pneumatic space 128
Pneumocyte 23, 56, 94, 164, 165, 168
Polymorphonuclear leukocyte 113
Precocial 26, 28, 30
Pressure 1, 2, 4, 10, 52, 118, 125, 149, 159, 160, 161, 172
Primate 3
Primitive species 165, 166
Primordia 34
Progenitor cell 45, 53, 60
Protopterus aethiopicus 161
Pseudostratified 75, 99
Pteropus hastatus 5
Pulmo reteformis 100

Pulmonary:
- air way 160
- artery 48, 74
- blood capillaries 95
- capillary blood 106, 151–153, 155, 157, 160
- circulation 45, 47
- circulatory system 35, 48
- defence 111, 114
- development 53
- disease 111
- endothelial cell 59
- epithelial cells 13, 14, 168
- epithelium 55, 59
- gas transfer 11
- growth 55
- hemorrhage 115
- intravascular macrophage 111
- lavage 111, 112
- macrophage 107
- mesenchyme 55
- modeling 154, 157
- morphometry 128
- phagocyte 111
- phagocytes 111
- subepithelial macrophage 111
- system 53, 60
- vascular system 53
- vasculature 35, 48, 101, 111
- vein 48
Pumping 172, 173

Q
Quantitative 60 111, 125, 128, 154, 160
Quetzalcoatlus northopi 6

R
Rana pipiens 165
Receptors 53–56, 59, 61, 63, 123
Reptile 3, 4, 128, 152, 165–167
Reptilian lung 148, 165, 166–168
Resident tissue macrophage 107, 111, 113, 114
Resistance 3, 95, 123, 155, 156
Resources 2, 161, 173
Respiratory:
- alkalosis 101
- cycle 117, 146, 174
- fluid media 125, 154, 160
- impedance 96

Subject Index

- infection   111
- macrophage   107, 111
- media   15, 125, 160
- rate   3, 118
- surface   10, 15, 23, 95, 96, 104, 106, 107, 111, 113, 114, 116, 151, 155, 159, 160, 165
- surface area   15, 22, 24, 111, 145, 146, 147, 148, 149, 151–153, 157, 160, 165
- system   3, 4, 6, 11, 14, 15, 53, 65–67, 107, 111, 112, 115, 117, 123, 128, 146, 160, 163, 166, 170, 172, 173

Rhacopholus dulitensis   5
Rhea   3, 6, 98
Ribs   16, 66, 96, 97, 128, 146
Rodentia   4

S

Saccular stage   26
Sampling   128
Sea level   10
Secretion   95, 113
Segmentum accelerans   66, 118
Selective pressure   1, 2, 4, 151, 161
Septum   35, 66, 74, 96, 97, 152
Shrew   96, 148, 149
Shunt   117
Signaling   14, 53, 54, 56, 58, 59, 60, 61, 63
Signaling molecules   59
Signals   13, 35, 53, 56
Single capillary system   152
Size   5, 8, 26, 66, 76, 77, 81
Skin   164, 165
Smooth muscle   60, 90, 123, 168
Snake   5, 113, 165, 166
Soar   1, 6
Solubility   155, 161
Songbirds   92
Speciation   5
Species   3, 4, 6, 7, 9, 15, 23, 24, 35, 53, 60, 66, 75, 78–81, 96–102, 111, 114, 117, 147, 149, 160, 161, 164
Speed   1, 4, 5, 9, 115
Sphincter   65, 117
Spiracles   168, 172
Splenic macrophage   111
Squamous:
- cell   78, 99, 149
- epithelium   99
- pneumocyte   94, 168
Stem cells   53

Strength   62, 149, 165, 170
Structural-functional correlation   159
Structural-functional inequalities   167
Structure   3, 13, 27, 35, 48, 59, 60, 65–67, 77, 81, 96, 97, 100, 101, 115, 117, 118, 125, 128, 154, 159–161, 164, 168
Subepithelial macrophage   107
Sulci   16, 102, 165
Supracoracoideus   7
Surface:
- area   9, 14, 15, 22, 24, 111, 125, 145–149, 151–157, 160, 165, 170
- density   147, 148
- resident macrophage   113, 114
- tension   96, 146–148
Surfactant   23, 95, 96, 113, 159, 164, 168
Swan   6, 10, 92, 128
Swelling   118
Swifts (Apodidae)   9
Syringeal constriction   66, 123
Syrinx   97

T

Taedia   170
Terminal gas exchange units   96, 147
Terrestrial   164, 165, 172
Tetrapod   161
Thermogenesis   4, 5, 10, 125, 155, 161, 167
Thickness   21, 52, 81, 101, 149–151, 153–155, 157, 165, 172
Thoroughbreds   115
Through-flow   32, 116
Tidal ventilation   116, 172
Tissue barrier   125, 154, 155
Trachea   2
Tracheal epithelium   169
Tracheal respiration   169
Tracheal system   13, 15, 55, 56, 168–170, 172, 173
Tracheate   169, 172
Tracheoles   168, 170, 173
Trade-off   3, 48, 172
Transcription factor   53, 54, 61, 63
Transition   48, 78, 116, 125, 166
Treadmill exercise   115
Tree-frog   113, 164, 165
Trilaminar substance   95, 96, 113
Triosseal canal   7
Tropical   161, 164, 172
Tubulogenesis   60
Turkey   97, 102, 112

Turtle 148, 152, 168
Type I cell 23, 94
Type II cell 23, 56, 59, 94, 95, 168

U
Unicameral lung 166
Urodele 164, 165

V
Vascular:
- endothelial growth factor 53, 54, 59, 60
- formation 14, 35
- growth 35, 60
- morphogenesis 35
- resident macrophage 111
- system 13, 35, 45, 53, 160
- units 46, 48, 81

Vasculoendothelial cell 45, 46
Vasculogenesis 36, 45, 46, 48, 60, 61
Ventilation 10, 11, 13, 28, 32, 97, 116, 117, 156, 159, 172, 173
Ventilatory rate 113, 172
Ventricular hypertrophy 116

Vertebrates 1, 3, 4, 8, 9, 11, 13, 14, 48, 52, 56, 95, 128, 149, 159, 165, 167, 168, 170
Viscera 35, 46, 97
Volant 1, 3, 4–6, 123, 173
Vulture 2, 10

W
Water 5, 10, 48, 116, 125, 154, 160, 161, 164, 165
Water:
- conservation 116
- vapour 10
- breather 125
- breathing 48
Whale 149
Wing 1–3, 6–9, 172
Wnt genes 61

X
Xenopus laevis 113, 165

Y
Yolk sac 45, 63